Interdisciplinary Contributions to Archaeology

Archaeology stands alone among the sciences in its attempt to enlighten us about the entire record of humankind. To cover such a broad range of time and space, archaeologists must ensure that their findings are integrated into broader spheres of scientific knowledge. The IDCA series aims to highlight the collaborative and interdisciplinary nature of contemporary archaeological research.

Topics the series has covered include:

- Paleoecology
- Archaeological Landscapes
- Statistical Approaches
- Laboratory Methods
- Human Biological and Cultural Evolution
- Human Nutrition
- Emergence of Agriculture and Pastoralism

For a copy of the proposal form, please contact Christi Lue (christi.lue@springer.com). Initial proposals can be sent to the Series Editor, Jelmer Eerkens (jweerkens@ucdavis.edu). Proposals should include:

- A short synopsis of the work or the introduction chapter
- The proposed Table of Contents
- The CV of the lead author(s)
- If available: one sample chapter

We aim to make a first decision within 1 month of submission. In case of a positive first decision the work will be provisionally contracted: the final decision about publication will depend upon the result of the anonymous peer review of the complete manuscript. We aim to have the complete work peer-reviewed within 3 months of submission.

This book series is indexed in SCOPUS.

For more information, please contact the Series Editor at (jweerkens@ucdavis.edu).

André Carlo Colonese • Rafael Guedes Milheira
Editors

Historical Ecology and Landscape Archaeology in Lowland South America

Springer

Editors
André Carlo Colonese
Department of Prehistory and Institute of Environmental Science and Technology (ICTA)
Universitat Autònoma de Barcelona
Barcelona, Spain

Rafael Guedes Milheira
Department of Anthropology and Archaeology
Federal University of Pelotas
Pelotas, Brazil

ISSN 1568-2722 ISSN 2730-6984 (electronic)
Interdisciplinary Contributions to Archaeology
ISBN 978-3-031-32283-9 ISBN 978-3-031-32284-6 (eBook)
https://doi.org/10.1007/978-3-031-32284-6

This work was supported by H2020 European Research Council (817911)

Cover illustration: Pre-Columbian fish weir, locally known as Camboa Grande, at the beach of Panaquatira (São Luis, State of Maranhão, Brazil)

This Springer imprint is published by the registered company Springer Nature Switzerland AG
The registered company address is: Gewerbestrasse 11, 6330 Cham, Switzerland

To our friend and colleague Alpina Begossi, who suddenly left us in 2023. Alpina's passionate and unparalleled work on small-scale fisheries in Brazil has inspired many of us in the field of Historical Ecology and will continue to do so for generations to come.

Foreword

Historical ecology is a research program that depending on the research problem related to the landscape of study embraces and deploys a veritable litany of tools that derive ultimately from diverse, yet ultimately connected, sciences. I hold a broad definition of science (or science in the abstract), and that is because of a theoretical attachment to the premises and perspectives that instantiate historical ecology as a distinctive method of thought and research. This fine volume of contributed chapters by specialists in anthropological archaeology, ethnography, and landscape ecology well exemplifies those premises and perspectives. Historical ecology bridges what has been called the intellectual chasm that vertically has long divided natural sciences v. social sciences, or more specifically, for a few instantiations that arise in this volume, biology v. archaeology, geoscience v. ethnography, marine science v. ethnoecology or ethnobiology. The canyon-like void between these so-called hard and soft sciences did not exist in the eighteenth-century Enlightenment because *progress* was seen to advance equally in all fields of inquiry; it only began to germinate in the minds of natural historians (Darwin, Wallace, Haeckel); in the conversations within learned societies and among their members who were students of Native America (Jefferson, Powell, Mason); and with the precursors of teleological or unilinear evolutionism in social science (Engels, Morgan, Spencer) in the nineteenth century under the all-encompassing concept of *evolution*. The embrace of evolutionism represented a positivist step beyond progress of the eighteenth century. Long-lived Herbert Spencer (1820–1903) perceived it as systematic development of simple life forms and human institutions into complex webs of material and social reality, or rather, the growth of a homogeneous concept or thing into a heterogeneous profusion of its parts. For others, it was simply 'change through time.' In mid-nineteenth century, only Darwin (and his contemporary Wallace) had a mechanism that explained it with regard to life forms, even if it did not apply neatly to cultural radiation and the accumulation of increasingly advanced technology of the Industrial Revolution. That problem of evolution—an at once logical and natural mechanism for one of its versions on the one hand v. a categorical, nominal list of stages of increasing complexity on the other—is partly what led to the separation of natural from social sciences. (It can still be seen today in anthropology, wherein certain departments—such as at Harvard University and Duke University—have seen biological anthropology split apart from sociocultural anthropology and its affiliated subfields, though the reasons are not exactly the same as they were in the nineteenth and early twentieth centuries.) Technological development is another part of the equation. The sciences that were once a positivistic whole, in the sense of August Comte[1]—ruptured apart like two sides of a geological fault line in the early to mid-twentieth century (as C.P. Snow and later Carole Crumley observed), which saw the birth of the fields known as disciplines into the late twentieth century and up to the present day.

Historical ecology contains within its primitive hard-core postulates (to borrow the phraseology of science philosopher Imre Lakatos) an implicit conceptual and methodological connection that spans the chasm between natural and social sciences, though it is not a field per se. Historical ecology is a body of theory—or a protective belt of hypotheses according to Lakatos—that radiate out from the premises that define it as a research program. I often refer to Lakatos as having had the vision to realize that the cleavage of knowledge created by the twentieth century subdivisions of the academy (and by inference learned societies, publications, and museums) is at least partly illusory. There is overlap when one delineates and investigates questions about human societies that display extensive bodies of ethnobiological knowledge (as in many tribal societies)—past and present—in their material interaction with their environments. In the study of landscape—and in this volume marine, coastal, and riverine waterscapes—even social sciences are authenticated sciences.[2] Science in the abstract only *progresses* (again, that eighteenth century notion arises) with the existence of debate, and not when confined to a monopoly of agreement among scientists themselves as to what best approaches visible truth ('seeing is believing'), or even probabilistic verisimilitude ('plausibility').

The chasm between life sciences and social sciences was essentially based on epistemological differences that derive from experimentally controlled hypothesis testing, Popperian falsificationism in a laboratory setting, and extant, standard models of how the subject matter of each field is constituted, what Thomas Kuhn labeled normal science, or day-to-day science in search of shoring up a given paradigm. The science in this volume evokes, in contrast, a considerable body of archaeological and ethnographic fieldwork that inductively drives theories which are to some extent testable outside laboratory conditions. For Lakatos, normal science was actually a fiction if it lacks debate, or a contradiction in terms. The debate herein that leads to progress of knowledge is actually the effort of all the authors to liberate the gates that block communication between disciplines on either side of the divide, in an effort to close it and unify a scientific theory of human interaction with the environment.

[1] In the American context, they were united in the study of Indigenous peoples and societies; the fusion was manifested in abstracts of meetings of learned societies, scientific publications, and in museums of natural history and ethnography in North, South, and Middle America.

[2] An ample body of literature now exists in marine historical ecology, a recent focus of research.

That effort leads to landscapes (which in its operational as opposed to vernacular sense includes seascapes). In this book, they are public architecture (Colonese et al.) and earthen buildings (Boksar et al.) formed intentionally or not (Boksar et al.) by human utilization and occupation, known nominally as *cerritos* (mounds) in Uruguay and adjacent Brazil (Gianotti et al., Milheira et al., and Mazarino et al.); *camboas* (stone weirs) in northeastern Brazil (Colonese et al.); *tekoha* (territory) in northern Brazil (Ribeiro et al.); ring enclosures with central mounds in southern Brazil (Corteletti et al.); pit houses and enclosure complexes also in southern Brazil (Carbonera and Loponte); *sambaquis* (shell mounds) in southeastern Brazil and adjacent areas to the south such as the Patos Lagoon (Milheira et al.); and charcoal forests of the past in southern Brazil (Patzlaff et al.). Where these anthropogenic structures have remained in place more or less as they were at the time people stopped building them, and if traditional agricultural societies of today still use them for food-producing practices, one can refer to the structures as *landesque capital*, a bona fide term in historical ecology today. In historical ecology, the debate, therefore, is not between different sciences, and it is not the distancing of human-centered fields from natural sciences. Rather, historical ecology spans those fields and sciences wherein the problem of the research question requires linkages in method and theory from any and all of them. This volume provides ample evidence in support of that assertion.

Tulane University
New Orleans, LA, USA
February 2023

William Balée

Preface

The tropical and subtropical lowlands of South America have some of the most diverse—yet threatened—biomes on earth and have supported human populations since the Late Pleistocene. Investigations under the broad conceptual framework of Historical Ecology and Landscape Archaeology have shown that for thousands of years humans have contributed to shaping and transforming some of these environments. Human transformation of South American landscapes and environments through time can offer comparative examples to other areas of the world; however, the nature and scale of these interactions are matters of debate and their legacy to modern ecosystems and local ecological knowledge is not fully understood. In addition to these challenges, conservation and restoration ecology in lowland South America is marked by shallow historical perspectives, while pervasive and compounded anthropogenic impacts over the last decades have accelerated biodiversity loss and population decline in a range of biomes, increasing the vulnerability of people and environments in this region. Understanding the scale of these stressors is extremely complex and exacerbated by the paucity of long-term records on plant and animal species composition, abundance and distribution pre-dating scientific observations. This sort of historical amnesia creates misconceptions that have far-reaching consequences for sustainability and conservation actions. For example, it may conceivably increase tolerance for progressive environmental degradation and contribute to setting inappropriate sustainability targets and responses by stakeholders, and hence underpin their expectations as to what are desirable and achievable states of social-ecological systems.

This volume addresses some of these issues by bringing together contributions from a range of disciplines involving archaeology and social and environmental sciences. Most of the chapters were presented in the symposium "Historical ecology and landscape archaeology in lowland South America" organized at the 20th Conference of the Society of Brazilian Archaeology in Pelotas, Brazil, in 2019. Part I (Legacies) offers studies that explore the legacy of past interactions and their contributions to informing current archaeological, conservation and development agendas and management strategies. Part II (Our Shared Past) presents case studies of human—nature interactions in the past, from plant management to aquatic exploitation. Interdisciplinary approaches are used to assess the evolution of past human practices and their implications on modern ecosystems, biological diversity and landscapes.

Here we showcase how recent collaborative research efforts between archaeologists, historians, social-environmental scientists and local stakeholders can productively advance our understanding of past and modern ecosystems, and the synergetic roles of cultural and natural drivers of change in their long-term development in lowland South America. The volume targets academics as well as the general public, local stakeholders and policy makers who may benefit from a historical perspective in attempts to balance effective solutions to anthropogenic environmental change while prioritizing human wellbeing and prosperity in the face of sustainability challenges in lowland South America.

Department of Prehistory and Institute of Environmental Science and Technology (ICTA),
Universitat Autònoma de Barcelona,
Barcelona, Spain
André Carlo Colonese

Department of Anthropology and Archaeology,
Federal University of Pelotas, Pelotas, Brazil
Rafael Guedes Milheira

Acknowledgement

We would like to thank the authors for their valuable contributions and their continuing efforts to expand Historical Ecology across the continent. We are also grateful to the Society of Brazilian Archaeology for hosting our symposium "Historical ecology and landscape archaeology in lowland South America" at the 20th Conference in Pelotas, Brazil, 2019. This work was funded by the ERC Consolidator project TRADITION, which received funding from the European Research Council (ERC) under the European Union's Horizon 2020 research and innovation programme under grant agreement N° 817911. This work was also funded by EarlyFoods (Evolution and impact of early food production systems), SGR 00527. This work contributes to the ICTA-UAB "María de Maeztu" Programme for Units of Excellence of the Spanish Ministry of Science and Innovation (CEX2019-000940-M).

About This Book

This volume showcases recent contributions to Historical Ecology and Landscape Archaeology in lowland South America and brings together contributions from a range of disciplines involving archaeology and social and environmental sciences. Part I (Legacies) offers studies that explore the legacy of past interactions and their contributions to informing current archaeological, conservation, and development agendas and management strategies. Part II (Our Shared Past) presents case studies of human—nature interactions in the past, from plant management to aquatic exploitation. Interdisciplinary approaches are used to assess the evolution of past human practices and their implications on modern ecosystems, biological diversity and landscapes.

Contents

Part I
Legacies

Chapter 1
The Legacy of Pre-Columbian Fisheries to Food Security and Poverty Alleviation in the Modern Amazon

André Carlo Colonese, Cecile Brugere, Milena Ramires, Mariana Clauzet, Rafael Brandi, Arkley Marques Bandeira, Lilia Guedes, Mario Wiedemann, Victoria Reyes-García, and Alpina Begossi

Abstract The relevance of local ecological knowledge to conservation and development agendas is gaining momentum, and the Amazon biome features as one of the most promising areas for its empirical application. Considerable attention has been given to forest composition and Indigenous land use, while coastal and marine environments have only received cursory attention. *Camboas de pedra* is the name given by coastal communities in the Brazilian eastern Amazon to permanent intertidal fish weirs/traps constructed of local stones. The structures are singular features of the present day coastal landscape of São Luís Island (State of Maranhão). They are some of the most impressive evidence of the continued use of pre-Columbian Indigenous knowledge by rural communities in this region. Our deliberative and participatory study of the *Camboas de pedra* of the beaches of Panaquatira and Boa Viagem on São Luis Island revealed that the structures continue to provide both cultural and provision services to local users, playing a pivotal role in household food security and poverty alleviation in the eastern Amazon. Our results highlight the importance of integrating local users' perceptions of heritage structures, both tangible and intangible, in order to design inclusive, equitable and sustainable conservation and fisheries management strategies.

A. C. Colonese (✉)
Department of Prehistory & Institute of Environmental Science and Technology (ICTA-UAB), Universitat Autònoma de Barcelona, Barcelona, Spain
e-mail: andrecarlo.colonese@uab.cat

C. Brugere
Soulfish Research & Consultancy, York, UK
e-mail: cecilebrugere@gmail.com

M. Ramires
Fisheries and Food Institute (FIFO), Rio de Janeiro, Brazil

Programa de Pós-Graduação, UNISANTA, Santos, São Paulo, Brazil
e-mail: milena@unisanta.br

A. C. Colonese, R. G. Milheira (eds.), *Historical Ecology and Landscape Archaeology in Lowland South America*, Interdisciplinary Contributions to Archaeology, https://doi.org/10.1007/978-3-031-32284-6_1

Keywords Amazon coast · Pre-Columbian fish traps · Local ecological knowledge · Legacy to food security and livelihood · Cultural heritage management

1.1 Introduction

Archaeological and ecological studies are transforming our understanding of the modern Amazon basin by dismissing the long held view that the region had limited forms of complex socio-economic organization in pre-Columbian times (Meggers, 1971). An increasing body of evidence is now demonstrating that forest composition, biological diversity and local ecological knowledge in large areas of

M. Clauzet
Fisheries and Food Institute (FIFO), Rio de Janeiro, Brazil

Pós-Graduação em Políticas Públicas, Estratégias e Desenvolvimento, Universidade Federal do Rio de Janeiro, Rio de Janeiro, Brazil
e-mail: mariana.clauzet@gmail.com

R. Brandi
Instituto Ambiente Humano (IAH), Batatais, São Paulo, Brazil
e-mail: rafael.brandi@ambientehumano.org.br

A. M. Bandeira
Programa de Pós-graduação Em Cultura E Sociedade, Programa de Pós-graduação Em Desenvolvimento E Meio Ambiente de Ecossistemas Costeiros e, Departamento de Oceanografia E Limnologia, Universidade Federal Do Maranhão, Bacanga, São Luís, Brazil
e-mail: arkleymbandeira@gmail.com

L. Guedes
Matis Consultoria em Arqueologia, Mirandópolis, São Paulo, Brazil
e-mail: lilia.guedes@matisconsultoria.com.br

M. Wiedemann
Programa de Pós-graduação em Arqueologia, CFCH, Universidade Federal do Pernambuco, Recife, Pernambuco, Brazil
e-mail: andomw@gmail.com

V. Reyes-García
Institute of Environmental Science and Technology (ICTA-UAB), Universitat Autònoma de Barcelona, Barcelona, Spain

Institució Catalana de Recerca i Estudis Avançats (ICREA), Barcelona, Spain
e-mail: victoria.reyes@uab.cat

A. Begossi
Fisheries and Food Institute (FIFO), Rio de Janeiro, Brazil

Programa de Pós-Graduação, UNISANTA, Santos, São Paulo, Brazil

Nepa, Capesca, UNICAMP, Campinas, São Paulo, Brazil
e-mail: albegossi@gmail.com

the basin were variably shaped by long-term socio-ecological processes initiated in the pre-Columbian era (Balée, 1993; Balée et al., 2020; De Gisi et al., 2014; Levis et al., 2017; Lins et al., 2015). Understanding the legacy of these cultural trajectories is highly relevant for a number of reasons (Balée et al., 2020; Forline, 2008), a prominent one being the growing recognition that local ecological knowledge (LEK) – broadly defined as a cumulative body of knowledge regarding socio-ecological relationships that is renewed and transmitted through generations (Berkes et al., 2000; Gadgil et al., 1993; Whyte, 2013) – plays a fundamental role in contemporary sustainable resource use (Garnett et al., 2018) and biological conservation (Braga-Pereira et al., 2021; Brondízio et al., 2021; Reyes-García et al., 2022; Walker et al., 2020). LEK is also critical to the food security of riverine and coastal fisheries in the region, notably for populations with low levels of income (Begossi, 1998; Bezerra, 2017; Furtado, 2002; Veríssimo, 1970). Benefiting from the deep historical roots of LEK and its increasing value in conservation and development agendas (Hill et al., 2020) requires a critical understanding of the origin, changing nature, and modern legacy of Indigenous and local knowledge and practices.

On São Luís Island (State of Maranhão), on the eastern Amazon, the fusion of pre-colonial and contemporary fishing practices is tangibly represented by the use of fishing weirs/traps, locally known as *Camboas de pedra,* by rural communities along the beaches of Panaquatira and Boa Viagem (Fig. 1.1a–d). *Camboas de pedra* consist of permanent intertidal dry stone walls, whose position and orientation takes advantage of the large semidiurnal tidal regime of the island (Lustosa de Alencar Neto et al., 2021). High tides submerge the structures twice a day for a few hours, permitting fishing and shellfish to enter the structures and their collection when the water recedes.

Residents of Panaquatira and Boa Viagem beaches attribute the construction of the *Camboas de pedra* to Indigenous populations that once lived in the region (Tupinambá). This is generally supported by numerous archaeological sites revealing the importance of fishing in the area since pre-Columbian times (Bandeira, 2015b; Colonese et al., 2020). Some *camboas* are positioned at different distances from the current shoreline, implying they were built over an extended period of time, with data on past sea level fluctuation (Behling et al., 2004; Cohen et al., 2005, 2009) indicating that some of the structures could have been in use at least since the Late Holocene. Significantly, European chronicles from the seventeenth century documented the use of similar structures by Tupinambá groups at Panaquatira beach, providing a *terminus ante quem* for their early implementation in the area (D'Abbeville, 2002; D'Évreux, 2002). Other structures have also been reported along the Amazon coast during the colonial and postcolonial periods (Daniel, 2004), with similar weirs documented in the states of Pará and Bahia, in the north and northeast of Brazil, respectively (Bezerra, 2017; Ott, 1944; Veríssimo, 1970). For example, on Marajó Island (state of Pará), the weirs were part of a system of colonial fisheries (Royal Fisheries, *pesqueiros reais*) that supplied fish to the emergent urban centers and to administrative services of the seventeenth century (Bezerra, 2017; Furtado, 1981).

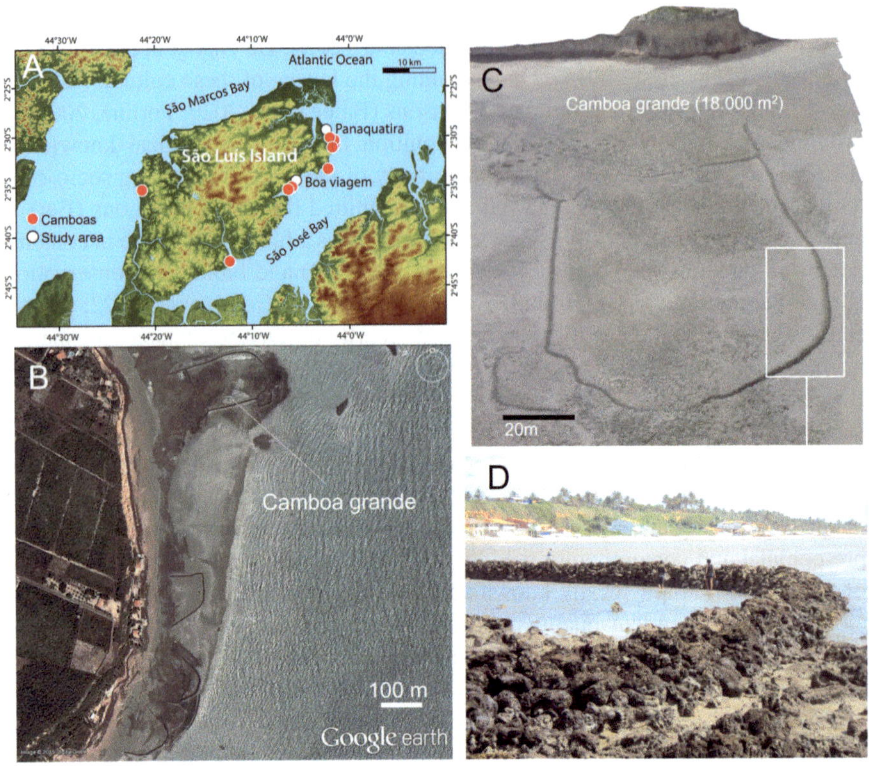

Fig. 1.1 (**a**) Location of *camboas* and the beaches of Panaquatira and Boa Viagem on São Luís Island (Maranhão), eastern Amazon; (**b**) Visible weirs during low tide at Panaquatira (©Google Earth), with (**c**) a photogrammetric view of *Camboa Grande*, also termed *Mãe de Todos* (mother of all), and its position on a petroplinthite outcrop (dark spots) at the base of the barrier cliff (back). The front wall of the adjacent structure on the bottom left is *ca.* 20 m in length. *Camboa Grande* is the largest registered weir in the study area (~18.000 m^2) with a size equivalent to two football pitches. (**d**) Detail of the deeper soft bottom area where most of the net fishing was observed

Despite historical and anecdotal evidence for the use of *Camboas de pedra* in recent times, their current function and benefits (tangible and intangible) have never been documented. Their economic and social relevance for community livelihoods remains largely unknown. This gap has potential ethical implications at the intersection of tangible heritage conservation and community livelihood. For example, the *Camboas de pedra* of Panaquatira and Boa Viagem are registered in the official Brazilian National Record of Archaeological Sites (CNSA MA00171). They are protected by Federal Law 3924 (26 July 1961) and by the Article 23 of the 1988 Constitution of the Federative Republic of Brazil. Under these regulations, any economic exploitation, destruction, or mutilation of archaeological sites is strictly prohibited.

Understanding *camboas* current use and management offers an opportunity to reflect on the modern legacy of past LEK and the significance of Brazilian archaeological heritage to the wider society (Dupeyron, 2021; Walker John, 2014). Here we contribute to filling this knowledge gap by presenting the results of a deliberative and participatory study of *Camboas de pedra* of Panaquatira and Boa Viagem. We interviewed local users and documented their perceptions of the cultural and socio-economic values of *camboas*. Our results, although preliminary in the context of a long-term research programme, reveal that *Camboas de pedra* transcend their tangible and intangible cultural heritage values, and continue to play a pivotal role in contemporary household food security and poverty alleviation in the eastern Amazon.

1.2 The Study Area

Situated in the eastern Amazon, São Luís Island (832 km^2), Maranhão, is surrounded by a large and complex estuarine system delimited by the bays of São Marcos and São José (Fig. 1.1a). The climate is tropical, with dry (January to June) and rainy (June to December) seasons. The region is part of the Upaon-Açu/Miritiba/Alto Preguiças Environmental Protection Area and lies in the transition between Amazon and Northeastern vegetation (Rêgo et al., 2018), with extensive *restinga* and mangrove ecosystems along the coast providing breeding grounds for a range of aquatic organisms (Menezes et al., 2008). The coastal environment is highly dynamic and shows relatively high biological productivity and biomass with limited seasonal variability due to the combination of semidiurnal macro-tidal regimes, mangrove vegetation, and freshwater input (Silva et al., 2018). The region presents one of the world's largest tidal fluctuations, up to 7 m, but oscillations up to 10 m have been reported (Kjerfve et al., 2002). As a consequence, exploitation of inshore coastal resources (e.g. fish, shellfish) relies on several techniques adapted to this semidiurnal tidal regime, such as permanent fish traps locally known as *curraos* (made of perishable materials such as wood, plant fibers) (Piorski et al., 2009) and *camboas* (Bandeira, 2015a). Panaquatira and Boa Viagem beaches are located in the municipality of São José do Ribamar, northeast of São Luís Island. The area has an estimated population of ca. 3000 people (IBGE 2010) and is considered rural, with household livelihoods mostly based on small-scale farming and fishing, along with an increasing share of income from market and tourism activities.

1.3 Methodology

Fieldwork was undertaken in May 2015 in Panaquatira and Boa Viagem, by a team composed of social scientists (anthropologists, ecological economists), marine ecologists, and archaeologists. Data were collected using direct observations, semi-structured interviews and a survey. Local residents and intermittent users of the coastal area were interviewed both on site and in the surrounding communities, using a snowballing approach to sampling (Biernacki & Waldorf, 1981). Direct observations (e.g. fishing, repairing gear) and semi-structured interviews were undertaken at the municipal market of São José do Ribamar. First, a survey was developed to assess the socio-economic profile of the participant and to characterize the fishing activities. This was followed by semi-structured interviews aiming to document the main fishing technique employed by the participants, and its socio-economic benefit. Both qualitative and quantitative information were collected. Participants were asked about the catch composition in the last fishing trip, their perception of catch abundance through time, and drivers of change (catch abundance and composition). Participation in the study was voluntary; the interviews and direct observations were performed only after verbal consent of the participants. Participants could decide either to have their names written down in the questionnaire or to leave it anonymous. The research and protocol were approved by the ethical committees of the Department of Archeology of the University of York (UK) and by the ethical committee of the Universidade Santa Cecília (Brazil) (protocol n. 1254012). Aerial footage (Phantom 2 with GoPro hero 2) was used to map and document participant use of the structures (IPHAN n° 01494.000593/2008-28). Photogrammetry of *Camboa Grande* was performed using Agisoft PhotoScan. Observations of geological features of the landscape were recorded on site.

1.4 Results

1.4.1 Spatial Distribution and Characterization of the Camboas

A total of 14 weirs, or camboas, were documented at Panaquatira and Boa Viagem, distributed over approximately 17 km of coastline. These were single or double structures, of variable area (up to ~18.000 m^2), height (up to ~1.8 m), and shape (trapezoidal, U- and V-shape) (Fig. 1.1b). A common feature among all the weirs was their position on top of petroplinthite outcrops, which appear to have been a constraining factor for their distribution along the shore (Fig. 1.1b, c). The weirs are typically oriented according to the tidal runoff and most of the structures have the opening oriented toward the shore (Fig. 1.1c). Some of the weirs were no longer in use and their structures appeared collapsed or partially disassembled. Others were operated and maintained by local residents (see below). Functional *camboas* usually displayed two distinct, yet continuous, internal spaces. The first consisted of a large and deeper soft bottom area, where nets can be safely operated and fishing seems to be the most frequent activity. The second were shallower rocky areas, patchily distributed, where shellfish and fish collection by hand seems to prevail.

Overall, the size of some structures implies communal labor investment for their construction and maintenance through time, and thus can be considered as examples of public architecture. This is the case of *Camboa Grande,* also termed *Mãe de Todos* (Mother of All) by local users (Fig. 1.1c, d), whose monumental size stands out in the landscape and has provided local families with marine foodstuff for hundreds of years according to historical and archaeological sources (Bandeira, 2015a;

D'Abbeville, 2002; D'Évreux, 2002), and continue to do so. Beyond providing food to human groups, *camboas* also offer discrete microenvironments, or "artificial reefs", for a range of animals, including fish, seabirds and molluscs (thus contributing to support services; see Liquete et al., 2016). We have not tested differences in species composition and diversity between *camboas* and other areas, however this could be the subject of future studies. The *Camboa de pedra* of Panaquatira and Boa Viagem can be considered examples of how pre-Columbian societies transformed their coastal environments to enhance their productivity, predictability and accessibility, and as such they align with the growing evidence of "landscape domestication" in pre-Columbian Amazon (Clement et al., 2015; Erickson, 2000; Levis et al., 2017; Watling et al., 2018).

1.4.2 Socio-economic Profiles of Camboa Users

The interviewed participants (n = 34, 8 women, 26 men) ranged from 25 to 90 years old, and most of them, although residents, were not native to the place (Fig. 1.2). All women (100%, n = 8) and a majority of the men (81%, n = 21) were involved in fishing in general. Most of the women (88%, n = 7) and a relatively smaller number of men (71%, n = 15) fished in the *camboas* (hereafter group A). Significantly, 50% (n = 11) of those who fished in the *camboas*, used them as their main fishing ground (hereafter group B), including 60% (n = 9) of the men and 29% (n = 2) of the women. Of these, 27% (n = 3) consumed only fish caught in *camboas* (Table 1.1).

Table 1.2 provides a socio-economic snapshot of the *camboa* users (groups A and B). Catches were used for household consumption by all users of the *camboas* (100%, n = 22), while fewer (50%, n = 11) caught fish to sell in the local market. A distinction can be made between participants who use the *camboas* for fishing (group A) and those who obtain the fish mainly or exclusively from the *camboas* (group B). It is worth noting the inverse relation existing between income and reliance on *camboas* as a source of food. In 2015, those who did not depend on the *camboas* (A) earned an average of $14.2 ± 74.4 (US dollars)/day, while those who relied on *camboas* as the main source of fish (B) earned an average of $5.1 ± 3.3 (US dollars)/day. This suggests that participants with lower levels of income relied more on this fishing technology for food and livelihood than participants with higher sources of income. The proportion of household children fishing in the *camboas* (not interviewed) was higher for households who depended on the structures as a source of food (B) than among those who did not depend on them (A). Other parameters remain substantially comparable between participants using *camboas* as a vital source of food or not.

Most of the participants used the *camboas* regularly (group A), from a daily to weekly basis. Among those who relied on *camboas* as their main fishing ground (group B), less than 3 h at time (2.7 ± 1.42 h) was spent in the weirs, when the tide started receding. Compared to other forms of fishing, this is not a particularly time consuming activity (Jaeschke & Saldanha, 2012; Santos, 2015), implying that the opportunity cost of *camboa* fishing is relatively low, and that *camboas* fishing does

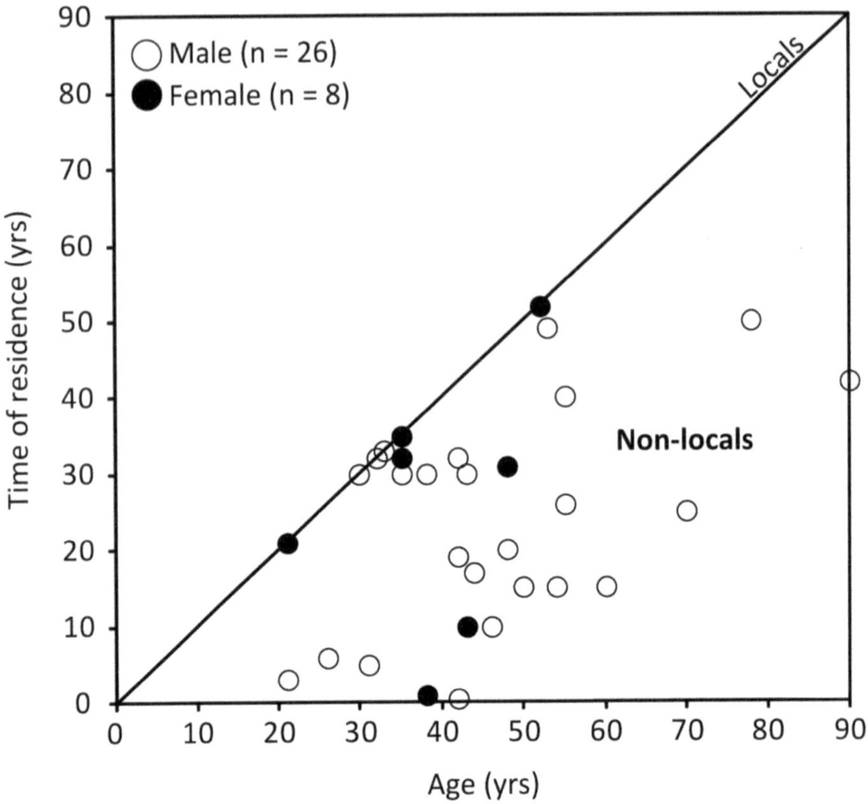

Fig. 1.2 Age distribution of participants per time of residence (years). Participants plotting on the right side of the straight line migrated to the study area at different times of their life

Table 1.1 Gender-disaggregated data on *camboas* users at Panaquatira and Boa Viagem

	Women		Men		Total	
	n	%	n	%	n	%
Participants	8	23.5	26	76.5	34	100
Involved in fishing in general	8	100	21	81	29	85
Use *camboas* for fishing (group A)[a]	7	87.5	15	71.4	22	76
Rely on *camboas* as the main fishing ground (group B)[b]	2	29	9	60	11	50
Both consume and sell *camboa*-caught fish[b]	7	100	15	100	22	100
Consume only *camboa*-caught fish[c]	1	50	2	22	3	27

[a]For those involved in fishing in general
[b]For those who use *camboas* for fishing
[c]For those who rely on fishing in *camboas* as their main activity

Table 1.2 Socio-economic profile of *camboa* users at Panaquatira and Boa Viagem

	Group A	Group B
Age (years)	41.4 ± 12.7	39.7 ± 9.4
Education (years of schooling)	6.0 ± 4.2	6.5 ± 4.3
Residence time in the community (years)	24.5 ± 13.6	21.4 ± 10.2
Experience fishing (years)	21.5 ± 13.9	15.9 ± 11.5
Income from main income generating activity ($/day[a])	14.2 ± 74.4	5.1 ± 3.3
Number of children per household	2.7 ± 2.3	2.7 ± 2.2
Number of household children who fish in general	0.5 ± 1.1	0.6 ± 1.2
Number of household children fishing in *camboas*	0.5 ± 1.2	0.9 ± 1.6
Belonging to an association (counts)	Yes: 11 No: 8 No answer: 8	Yes: 1 No: 5 No answer: 5
Total	22	11

Data are average and standard deviation. Group (A) Use *camboas* for fishing; Group (B) use *camboas* as the main fishing ground

[a]Income in US Dollar ($) assuming an exchange rate of 3.02 between Brazilian Reais (R$) and US Dollar ($) for May 2015

not substantially compete with other livelihood pursuits, as is increasingly reported for small-scale fisheries in coastal Brazil (Trimble & Johnson, 2013). The lower time commitment and low-cost entry that *camboas* fishing represents compared to offshore fishing, may explain the relatively higher numbers of women undertaking *camboas* fishing on a regular basis (Tilley et al., 2021). However, in monetary terms, most of the participants who used *camboas* as the main fishing ground (82%, n = 9) considered that fishing was not sufficient for sustaining their livelihoods, which they had to secure with other, informal regular employment (for example as carpenters, builders, housekeepers). The reverse is also possible, however, with fishing complementing household consumption in times of employment shortfall.

Fishing in *camboas* has also a considerable social dimension for those that relied on the structures as the main fishing ground (group B). Fishing is often carried out in small groups (2–3 units), generally composed of relatives (55%, n = 6) or friends (25%, n = 2), or a combination of the two (18%, n = 2). Those in the groups are generally not given fixed tasks (73%, n = 8) to perform on every fishing trip (Fig. 1.3). Thus, as demonstrated elsewhere, *camboas* fishing provides an opportunity for socializing, which, like gleaning, is particularly valued by women (Grantham et al., 2020; Whittingham et al., 2003). The social dimension of the *camboas* is also evidenced by its collective maintenance. Most of the participants who relied on *camboas* as the main fishing ground (82%, n = 9) described some form of maintenance. According to the *camboa* users, maintenance responsibilities are reported to fall equally on the *camboa* perceived 'owners' and on those who fish in them. Most *camboa* fishers who rely on the structures thus participate in the maintenance of the *camboas*, which involves mainly replacing stones when some have fallen off the structure.

Fig. 1.3 (**a**, **b**) fishing using *redes de arrasto* (beach seine) and (**c**) *tarrafa* (cast net) in deeper soft bottom areas of *camboas*. (**d**) Fish and shellfish collected by hand in shallow rocky areas. Fishing in *camboas* is often carried out in small groups, composed of relatives and/or friends

Nevertheless, access to the weirs is contentious. The majority of *camboas* fishers who relied on the structures as the main fishing ground reported issues with their use, and had diverging perceptions regarding their ownership and right of access. For example, most users (73%, n = 8) attributed the construction of *camboas* to Indigenous people, while fewer (23%, n = 3) perceived the *camboas* as a common property technology, a locally shared and inalienable heritage good that belongs to all users ("fishers are the owners"). In principle, this would mean that their access and use are legitimate. However, this was contradicted by the perception of some users that one particular individual in the community (a fisher) had an assigned right to use and decide upon the use of some *camboas*. Whilst this in itself does not necessarily prevent access and use of the *camboas* by other fishers, such lack of clarity of tenure and access rights nonetheless compromises the confidence with which fishers and their families take advantage of the *camboas* for their daily fish and livelihood needs. Subject to loose tenure and ad-hoc maintenance, *camboas* epitomise the challenges that can arise from the layering of natural (tide) and social (tenure conflict) factors in the management and conservation of coastal heritage.

1.4.3 Fishing Methods and Catch Composition

Fishing in *camboas* was mostly pursued using nets (91%, n = 10), including castnets (*tarrafa*), hand nets (*landuá, puçá, caçoeira*), beach seines (*rede de arrasto*), and to a lesser extent by hand collection (18%, n = 2) and other techniques such as machete, harpoons and hooks (9%, n = 1) (Fig. 1.3). Catches were documented from all interviewees, regardless of how they used the *camboas*. A total of 24 taxa were reported by fishers in their last "fishing trip" (Table 1.3). The catch at Panaquatira included

Table 1.3 Catch from the "last fishing trip"

Local common name	Taxa	N fishers	Price (R$/kg), H, M, L	Price ($/kg)	Decline in catches
Peixe-Pedra	Torroto grunt (*Genyatremus luteus*)	21	13–17.00; M	4.3–5.6	**
Sardinha (sardinha copeu and s. peuba)	Sardine (Clupeidae)	21	2–10.00; L	0.7–3.3	****
Pescada, pescadinha	Weakfish (*Cynoscion* spp.)	19	10–15.00; M	3.3–5.0	*
Guaravira	Cutlass fish (*Trichiurus lepturus*)	13	3–8.00; L	1.0–2.6	**
Tainha (tainha sajuba and T. pitiu)	Mullet (*Mugil* spp.)	12	8–12.00; M	2.6–4.0	***
Camarão	Shrimp	12	12–25.00; H	4.0–8.3	*
Timburu (Tibiro)	Carangid leatherjackets (*Oligoplites* spp.)	9	4–12.00; M	1.3–4.0	*
Carapitinga (Carapitanga?)	Dog snapper (*Lutjanus jocu*) and Lane snapper (*L. synagris*)	9	10.00; M	3.3	*
Corvina, corvina do sul	Whitemouth croaker? (*Micropogonias furnieri*)	7	10–15.00; M	3.3–5.0	*
Paru	Atlantic spadefish (*Chaetodipterus faber*)	7	2–3.00; L	0.7–1.0	*
Pacamão	Pacuma toadfish (*Batrachoides surinamensis*)	6	nd	nd	*
Siri	Crab	5	nd	nd	*
Arraia bicuda	Longnose stingray (*Dasyatis guttata*)	4	10–12.00; M	3.3–4.0	*
Guaja	Crab	3	nd	nd	*
Moreia	Moray (*Gymnothorax* spp.)	3	nd	nd	*
Banderado	Ariidae	2	nd	nd	*
Cuica	?	2	nd	nd	*

Only taxa mentioned by more than one fisher is reported. Price in Brazilian Reais (R$) and American Dollar ($) assuming an exchange rate of 3.02 between (R$/$) for May 2015. Brazilian prices are also ordered as *H* high (>15.00), *M* medium (10.00–15.00) and *L* low (<9.00); *nd* not declared. Decline in catches reported by participants: ****high, ***medium, **low, *not mentioned. Fish was identified based on Begossi and Ramires (2013), Frose and Pauly (2020), Martins-Juras et al. (1987), Monteles et al. (2010) and Piorski et al. (2009). Others (mentioned by one fisher): bagre (Sea catfish; *Genidens* spp.), Cangatã (Bressou sea catfish, *Aspistor quadriscutis*), xaréu (Crevalle jack, *Caranx hippos*; Horse-eye jack, *C. latus*), curiacica (not identified), favinha (not identified), papista (not identified), xixarro (not identified) *peixe-pedra* (Torroto grunt, *Genyatremus luteus*), *sardinha* (sardine, Clupeidae), *pescada* or *pescadinha* (Weakfish, *Cynoscion* spp.), *guaravira* (Cutlass fish, *Trichiurus lepturus*), *camarão* (shrimp) and *tainha* (mullet, *Mugil* spp.), among others. *Camarão* had the highest market value, followed by *peixe-pedra*, *pescada/pescadinha* and *tainha* (medium value). *Tainha* and *camarão* were the main caught taxa in *camboas*, as observed in historical times in the region (Furtado, 2002).

In terms of size, the majority of fish caught in *camboas* were reported as being small (frequency of reports: small = 7, medium = 5, large = 2, although *camarão* was reported on one occasion as medium to large in size, so there is a substantial amount of subjectivity in size reporting), and all *camboas* fishers noticed variations in the quantities of fishes caught, with 10 out of 11 (91%) indicating a decrease in quantity over the last 7–8 years (average of 7.14 ± 6.14 years). Catches of all species were said to have decreased (notably *peixe-pedra*, *sardinha* and *camarão*), although decreases in some, such as *sardinha*, were reported more frequently. According to participants who relied on *camboas* as the main fishing ground, the reasons for this are varied, but include habitat degradation (pollution, urbanization), increasing fishing pressure (large nets, illegal fishing gear), along with the increasing number of non-local and recreational fishers in general. Together, this illustrates the combined nature of the environmental and anthropogenic drivers of change impacting the fisheries resources, and the potential effects these can trigger in the *camboas* catches, and consequently the food security of those who rely on them.

1.5 Discussion

1.5.1 The Perceived Benefits of Camboas

Those who used *camboas* as the main fishing ground (group B) valued the structures primarily for their provisioning benefits (91%, n = 9), including the predictability of catch and the diversity of marine foodstuffs. By contrast, the cultural services of the structures (i.e. the cultural heritage that the *camboas* structures fulfill) were less valued among these users (27%, n = 3), as observed among other coastal fisheries (Blayac et al., 2014; Herbst et al., 2020). Most of these benefits are expressed in non-monetary terms, but are essential nonetheless for food security and for the families depending on fishing in the *camboas*. As the *camboas* users included those who fish for consumption and selling, as well as those who fish solely for consumption, *camboas* serve as a source of monetary income for some (caught species often include those considered high and medium value), and a source of high quality food protein and micronutrients for all.

However, if the contribution of *camboas* to food security is evident, their social role is not so clear-cut. On one hand, *camboas* can strengthen kinship-based relations (both familial and friendship) through the collective act of gleaning and fishing, which is often the case with other fishing technologies as well (Begossi, 2014; Blount, 2005; Donkersloot et al., 2020). On the other hand, however, and contrary to expectations for such a communal asset, they appear to play a limited role in community cohesion and identity. Evidence of this includes the tensions that exist over their tenure and the ad-hoc nature of their maintenance, which could also jeopardize their longevity and has implications for their conservation and legal protection. When a *camboa* is not maintained and falls into a state of disrepair, it tends to fall out of use and is eventually abandoned. This suggests that there is a tipping point beyond which the *camboas* will cease to perform the social, cultural, environmental and economic functions they currently have in the coastal landscape and in community livelihoods. The potential effects on community resilience and heritage, and on the loss of associated local ecological knowledge of which little is yet known, would be catastrophic given the interdependencies that exist.

Yet, the fact that these structures have endured for hundreds of years, suggests that they are in fact valued as a common property asset. The drive to maintain them, even on an ad-hoc basis, and to continue accruing the tangible and intangible benefits of having them 'there', as an integral feature of the landscape now and in the future (existence and bequest values) is stronger than the individualistic negligence or indifference that would ultimately lead to their demise. As we have demonstrated, a symbiotic relationship exists between the *camboas* and the communities of Panaquatira and Boa Viagem. The *camboas* are part of a system that contributes to community livelihood and poverty alleviation, without which the *camboas* would cease to exist due to lack of maintenance. Interventions aimed at conserving *camboas* and enhancing their biocultural heritage value without considering co-participation and inclusive, equitable co-management with current-day users would inevitably result in negative and undesirable effects through the exclusion of segments of local populations, and particularly those who are most dependent upon this heritage.

Sustainability of cultural and natural resources should be integrated with equality and social justice. Sustainability and social justice are within a "just aquatic framework" (Lopes et al., 2021) that includes recognition (supporting diversity and food security), procedural norms (rights to autonomy and governance) and distribution (equality and economic benefits). In the case of the *camboas*, the design of a conservation program that integrates its uses with the participation of the community would be environmentally and socially beneficial.

1.6 Conclusion

Local Ecological knowledge (LEK) is becoming central in cultural and biological conservation strategies in regions of rich biological diversity and particularly for people with low levels of income. In this study we documented the use of LEK by small-scale rural communities in coastal Amazon that can be traced back to pre-Columbian times. Pre-Columbian Amazonian fish weirs, or *Camboas de pedra*, are a distinctive feature of the coastal landscapes around São Luis Island, and the broader social and economic potentials of this cultural heritage are significant. They act as historical and cultural reference points for local residents and users. Our study, however, reveals that *camboas* and *camboa* users are socially and economically vulnerable. Despite a long and continuing tradition of use, the future of *camboas* appears fragile. We advocate for the integration of local users in decision making processes regarding *camboas* conservation and management strategies.

Acknowledgements This project is funded by the British Academy and Newton Fund (Newton Mobility Grants, NG140074). The work was also funded by the ERC Consolidator project TRADITION, which has received funding from the European Research Council (ERC) under the European Union's Horizon 2020 research and innovation programme under Grant Agreement No 817911, and by EarlyFoods (Evolution and impact of early food production systems), 2021 SGR 00527. This work contributes to the "María de Maeztu" Programme for Units of Excellence of the Spanish Ministry of Science and Innovation (CEX2019-000940-M) provided to ICTA-UAB. AB thanks productivity scholarships by CNPq, Brazil (grants #301592/2017-9 and 304170/2021-6). The authors thank Marcia Bezerra (Universidade Federal do Pará) for providing information on similar fish weirs in Pará. We are also grateful to Krista McGarth (Universidad Autonoma de Barcellona) for her constructive comments on the first drafts of the manuscript. The authors want to thank the people of Panaquatira and Boa Viagem for their generous participation in this research.

References

Balée, W. (1993). Indigenous transformation of Amazonian forests: An example from Maranhão, Brazil. *L'Homme, 33*(126), 231–254.

Balée, W., de Oliveira, V. H., dos Santos, R., Amaral, M., Rocha, B., Guerrero, N., Schwartzman, S., Torres, M., & Pezzuti, J. (2020). Ancient transformation, current conservation: Traditional forest management on the Iriri River, Brazilian Amazonia. *Human Ecology, 48*, 1–15.

Bandeira, A. M. (2015a). Distribuição espacial dos sítios Tupi na Ilha de São Luís – Maranhão. *Cadernos Do LEPAARQ (UFPEL), 12*(24), 59–96.

Bandeira, A. M. (2015b). Os Tupis na Ilha de São Luís-Maranhão: Fontes Históricas e a Pesquisa Arqueológica. *História Unicap, 2*(3), 79–98.

Begossi, A. (1998). Cultural and ecological resilience among caiçaras of the Atlantic Forest coast and caboclos of the Amazon. In *Linking social and ecological systems for resilience and sustainability* (pp. 129–157). The Beijer International Institute of Ecological Economics.

Begossi, A. (2014). Ecological, cultural, and economic approaches to managing artisanal fisheries. *Environment, Development and Sustainability, 16*(1), 5–34.

Begossi, A., & Ramires, M. (2013). Fish folk medicine of Caiçara (Atlantic Coastal Forest) and Caboclo (Amazon Forest) communities. In R. R. N. Alves & I. L. Rosa (Eds.), *Animals in traditional folk medicine: Implications for conservation* (pp. 91–108). Springer.

Behling, H., Cohen, M. L., & Lara, R. (2004). Late Holocene mangrove dynamics of Maraj Island in Amazonia, northern Brazil. *Vegetation History and Archaeobotany, 13*(2). https://doi.org/10.1007/s00334-004-0031-1

Berkes, F., Colding, J., & Folke, C. (2000). Rediscovery of traditional ecological knowledge as adaptive management. *Ecological Applications, 10*(5), 1251–1262.

Bezerra, M. (2017). *Teto e Afeto: sobre as pessoas, as coisas e a arqueologia na Amazônia.* GKNoronha.

Biernacki, P., & Waldorf, D. (1981). Snowball sampling: Problems and techniques of chain referral sampling. *Sociological Methods & Research, 10*(2), 141–163.

Blayac, T., Mathé, S., Rey-Valette, H., & Fontaine, P. (2014). Perceptions of the services provided by pond fish farming in Lorraine (France). *Ecological Economics, 108*, 115–123.
Blount, B. G. (2005). Small-scale fisheries in the contemporary world: Adaptations and management. *Reviews in Anthropology, 34*(1), 1–19.
Braga-Pereira, F., Morcatty, T. Q., El Bizri, H. R., Tavares, A. S., Mere-Roncal, C., González-Crespo, C., Bertsch, C., Rodriguez, C. R., Bardales-Alvites, C., von Mühlen, E. M., Bernárdez-Rodríguez, G. F., Paim, F. P., Tamayo, J. S., Valsecchi, J., Gonçalves, J., Torres-Oyarce, L., Lemos, L. P., Vieira, M. A. R., Bowler, M., et al. (2021). Congruence of local ecological knowledge (LEK)-based methods and line-transect surveys in estimating wildlife abundance in tropical forests. *Methods in Ecology and Evolution/British Ecological Society*. https://doi.org/10.1111/2041-210x.13773
Brondízio, E. S., Aumeeruddy-Thomas, Y., Bates, P., Carino, J., Fernández-Llamazares, Á., Ferrari, M. F., Galvin, K., Reyes-García, V., McElwee, P., Molnár, Z., Samakov, A., & Shrestha, U. B. (2021). Locally based, regionally manifested, and globally relevant: Indigenous and local knowledge, values, and practices for nature. *Annual Review of Environment and Resources, 46*(1), 481–509.
Clement, C. R., Denevan, W. M., Heckenberger, M. J., Junqueira, A. B., Neves, E. G., Teixeira, W. G., & Woods, W. I. (2015). The domestication of Amazonia before European conquest. *Proceedings. Biological Sciences/The Royal Society, 282*(1812), 20150813.
Cohen, M. C. L., Behling, H., & Lara, R. J. (2005). Amazonian mangrove dynamics during the last millennium: The relative sea-level and the Little Ice Age. *Review of Palaeobotany and Palynology, 136*(1), 93–108.
Cohen, M. C. L., Behling, H., Lara, R. J., Smith, C. B., Matos, H. R. S., & Vedel, V. (2009). Impact of sea-level and climatic changes on the Amazon coastal wetlands during the late Holocene. *Vegetation History and Archaeobotany, 18*(6), 425.
Colonese, A. C., Winter, R., Brandi, R., Fossile, T., Fernandes, R., Soncin, S., McGrath, K., Von Tersch, M., & Bandeira, A. M. (2020). Stable isotope evidence for dietary diversification in the pre-Columbian Amazon. *Scientific Reports, 10*(1), 1–11.
D'Abbeville, C. (2002). *História da missão dos padres capuchinhos na Ilha do Maranhão e circunvizinhanças*. Siciliano.
Daniel, J. (2004). *Tesouro descoberto no máximo rio Amazonas*. Contraponto Editora.
de D'Évreux, Y. (2002). *Viagem ao norte do Brasil feita nos anos de 1613 a 1614*. Siciliano.
De Gisi, S., Petta, L., & Wendland, C. (2014). History and technology of Terra Preta sanitation. *Sustainability: Science Practice and Policy, 6*(3), 1328–1345.
de Menezes, M. P. M., Berger, U., & Mehlig, U. (2008). Mangrove vegetation in Amazonia: A review of studies from the coast of Pará and Maranhão States, North Brazil. *Acta Amazonica, 38*(3), 403–420.
Donkersloot, R., Coleman, J., Carothers, C., Ringer, D., & Cullenberg, P. (2020). Kin, community, and diverse rural economies: Rethinking resource governance for Alaska rural fisheries. *Marine Policy, 117*, 103966.
Dupeyron, A. (2021). Capturing the impacts of archaeology for development: Opportunities and challenges in evaluating the sustainable preservation initiative in Peru. *Progress in Development Studies, 21*(2), 161–180.
Erickson, C. L. (2000). An artificial landscape-scale fishery in the Bolivian Amazon. *Nature, 408*(6809), 190–193.
Forline, L. C. (2008). Putting history back into historical ecology: Some perspectives on the recent human ecology of the Amazon Basin. *Journal of Ecological Anthropology, 12*(1), 69–74.
Frose, R., & Pauly, D. (2020). *FishBase*. World Wide Web electronic Publication. www.fishbase.org
Furtado, L. G. (1981). *Pesca artesanal: um delineamento de sua história no Pará*. http://repositorio.museu-goeldi.br:8080/handle/mgoeldi/699
Furtado, L. G. (2002). Pesqueiros reais & pontos de pesca. Traços da territorialidade haliêutica ou pesqueira amazônica. FURTADO, Lourdes Gonçalves. Boletim do Museu Paraense Emílio Goeldi. *Belém, Série Antropologia, 18*(1), 3–26.

Gadgil, M., Berkes, F., & Folke, C. (1993). Indigenous knowledge for biodiversity conservation. *Ambio, 22*(2/3), 151–156.
Garnett, S. T., Burgess, N. D., Fa, J. E., Fernández-Llamazares, Á., Molnár, Z., Robinson, C. J., Watson, J. E. M., Zander, K. K., Austin, B., Brondizio, E. S., Collier, N. F., Duncan, T., Ellis, E., Geyle, H., Jackson, M. V., Jonas, H., Malmer, P., McGowan, B., Sivongxay, A., & Leiper, I. (2018). A spatial overview of the global importance of Indigenous lands for conservation. *Nature Sustainability, 1*(7), 369–374.
Grantham, R., Lau, J., & Kleiber, D. (2020). Gleaning: Beyond the subsistence narrative. *Maritime Studies, 19*(4), 509–524.
Herbst, D. F., Gerhardinger, L. C., & Hanazaki, N. (2020). Linking user-perception diversity on ecosystems services to the inception of coastal governance regime transformation. *Frontiers in Marine Science, 7*, 83.
Hill, R., Adem, Ç., Alangui, W. V., Molnár, Z., Aumeeruddy-Thomas, Y., Bridgewater, P., Tengö, M., Thaman, R., Adou Yao, C. Y., Berkes, F., Carino, J., Carneiro da Cunha, M., Diaw, M. C., Díaz, S., Figueroa, V. E., Fisher, J., Hardison, P., Ichikawa, K., Kariuki, P., et al. (2020). Working with Indigenous, local and scientific knowledge in assessments of nature and nature's linkages with people. *Current Opinion in Environmental Sustainability, 43*, 8–20.
Jaeschke, A., & Saldanha, M. C. W. (2012). Physical demands during the hauling of fishing nets for artisan fishing using rafts in beach of Ponta Negra, Natal-Brasil. *Work, 41*(Suppl 1), 414–421.
Kjerfve, B., Perillo, G. M. E., Gardner, L. R., Rine, J. M., Dias, G. T. M., & Mochel, F. R. (2002). Chapter twenty morphodynamics of muddy environments along the Atlantic coasts of North and South America. In T. Healy, Y. Wang, & J.-A. Healy (Eds.), *Proceedings in marine science* (Vol. 4, pp. 479–532). Elsevier.
Levis, C., Costa, F. R. C., Bongers, F., Peña-Claros, M., Clement, C. R., Junqueira, A. B., Neves, E. G., Tamanaha, E. K., Figueiredo, F. O. G., Salomão, R. P., Castilho, C. V., Magnusson, W. E., Phillips, O. L., Guevara, J. E., Sabatier, D., Molino, J.-F., López, D. C., Mendoza, A. M., Pitman, N. C. A., et al. (2017). Persistent effects of pre-Columbian plant domestication on Amazonian forest composition. *Science, 355*(6328), 925–931.
Lins, J., Lima, H. P., Baccaro, F. B., Kinupp, V. F., Shepard, G. H., Jr., & Clement, C. R. (2015). Pre-Columbian floristic legacies in modern homegardens of Central Amazonia. *PLoS One, 10*(6), e0127067.
Liquete, C., Cid, N., Lanzanova, D., Grizzetti, B., & Reynaud, A. (2016). Perspectives on the link between ecosystem services and biodiversity: The assessment of the nursery function. *Ecological Indicators, 63*, 249–257.
Lopes, P. F. M., Freitas, C. T., Hallwass, G., Silvano, R. A. M., Begossi, A., & Campos-Silva, J. V. (2021). Just Aquatic Governance: The Amazon basin as fertile ground for aligning participatory conservation with social justice. *Aquatic Conservation: Marine and Freshwater Ecosystems*, aqc.3586. https://doi.org/10.1002/aqc.3586
Lustosa de Alencar Neto, W. J., Silva Soares, L., & Bandeira, A. M. (2021). Aplicação do método de análise hierárquica de processos (AHP) como modelo preditivo para mapeamento dos sítios arqueológicos na ilha de São Luís, Maranhão. *GEOgraphia, 23*(51). https://doi.org/10.22409/geographia2021.v23i51.a27944
Martins-Juras, I. da A. G, Juras, A. A., & Menezes, N. A. (1987). Relação preliminar dos peixes da ilha de São Luís, Maranhão, Brasil. *Revista Brasileira de Zoologia, 4*(2), 105–113.
Meggers, B. J. (1971). *Amazonia: Man and culture in a counterfeit paradise*. Aldine.
Monteles, J. S., de Almeida Funo, I. C., & de Castro, A. C. L. (2010). Caracterização da pesca artesanal nos municípios de Humberto de Campos e Primeira Cruz – Maranhão. *Boletim do Laboratório de Hidrobiologia, 23*(1). http://periodicoseletronicos.ufma.br/index.php/blabohidro/article/view/1879
Ott, C. (1944). Os elementos culturais da pescaria baiana. *Boletin do Museu nacional, nova série, antropologia, 4*, 1–67.
Piorski, N. M., Serpa, S. S. E., & Nunes, J. L. S. (2009). Análise comparativa da pesca de curral na ilha de São Luís, estado do Maranhão, Brasil. *Arquivos de ciencias do mar, 42*(1), 65–71.

Rêgo, J. C. L., Soares-Gomes, A., & da Silva, F. S. (2018). Loss of vegetation cover in a tropical island of the Amazon coastal zone (Maranhão Island, Brazil). *Land Use Policy, 71*, 593–601.

Reyes-García, V., Fernández-Llamazares, Á., Aumeeruddy-Thomas, Y., Benyei, P., Bussmann, R. W., Diamond, S. K., García-Del-Amo, D., Guadilla-Sáez, S., Hanazaki, N., Kosoy, N., Lavides, M., Luz, A. C., McElwee, P., Meretsky, V. J., Newberry, T., Molnár, Z., Ruiz-Mallén, I., Salpeteur, M., Wyndham, F. S., et al. (2022). Recognizing Indigenous peoples' and local communities' rights and agency in the post-2020 Biodiversity Agenda. *Ambio, 51*(1), 84–92.

Santos, A. N. (2015). Fisheries as a way of life: Gendered livelihoods, identities and perspectives of artisanal fisheries in eastern Brazil. *Marine Policy, 62*, 279–288.

Silva, M. H. L., Torres Júnior, A. R., Castro, A. C. L., Azevedo, J. W. J., Ferreira, C. F. C., Cardoso, R. L., Nunes, J. L. S., & Carvalho-Neta, R. N. F. (2018). Fish assemblage structure in a port region of the Amazonic coast. *Iheringia. Série Zoologia, 108*. https://doi.org/10.1590/1678-4766e2018018

Tilley, A., Burgos, A., Duarte, A., Dos Reis Lopes, J., Eriksson, H., & Mills, D. (2021). Contribution of women's fisheries substantial, but overlooked, in Timor-Leste. *Ambio, 50*(1), 113–124.

Trimble, M., & Johnson, D. (2013). Artisanal fishing as an undesirable way of life? The implications for governance of fishers' wellbeing aspirations in coastal Uruguay and southeastern Brazil. *Marine Policy, 37*, 37–44.

Veríssimo, J. (1970). *A pesca na Amazonia*. Universidade Federal do Pará.

Walker John, H. (2014). Reflections on archaeology, poverty and tourism in the Bolivian Amazon. *Worldwide Hospitality and Tourism Themes, 6*(3), 215–228.

Walker, W. S., Gorelik, S. R., Baccini, A., Aragon-Osejo, J. L., Josse, C., Meyer, C., Macedo, M. N., Augusto, C., Rios, S., Katan, T., de Souza, A. A., Cuellar, S., Llanos, A., Zager, I., Mirabal, G. D., Solvik, K. K., Farina, M. K., Moutinho, P., & Schwartzman, S. (2020). The role of forest conversion, degradation, and disturbance in the carbon dynamics of Amazon indigenous territories and protected areas. *Proceedings of the National Academy of Sciences of the United States of America, 117*(6), 3015–3025.

Watling, J., Mayle, F. E., & Schaan, D. (2018). Historical ecology, human niche construction and landscape in pre-Columbian Amazonia: A case study of the geoglyph builders of Acre, Brazil. *Journal of Anthropological Archaeology, 50*, 128–139.

Whittingham, E., Campbell, J., & Townsley, P. (2003). *Poverty and reefs*. Intergovernmental Oceanographic Commission of UNESCO.

Whyte, K. P. (2013). On the role of traditional ecological knowledge as a collaborative concept: A philosophical study. *Ecological Processes, 2*(1), 1–12.

Chapter 2
Creating a Collaborative Management Framework for the Conservation of an Indigenous Mounds' Landscape in the Wetlands of India Muerta (Uruguay): State of the Art and Future Perspectives

Camila Gianotti, Laura del Puerto, Lucía Courtoisie, Joaquín Aldabe, César Fagúndez, Beatriz Orrego, Cristina Cancela, Nicolás Gazzán, Joaquín Tortosa, Rodolfo Reboulaz, Mónica Quevedo, Maira Ramos, and Piero Larralde

Abstract *Cerritos de indios* (indigenous earthen mounds) represent the earliest manifestation of human constructed landscape in Uruguayan territory. Mound sites are composed of a complexity of anthropogenic structures which nowadays, after thou-

C. Gianotti (✉)
Laboratorio de Arqueología del Paisaje y Patrimonio del Uruguay, Departamento de Sistemas Agrarios y Paisajes Culturales, Centro Universitario Regional del Este, Universidad de la República, Rocha, Uruguay
e-mail: camila.gianotti@lappu.edu.uy

L. del Puerto · J. Aldabe
Departamento de Sistemas Agrarios y Paisajes Culturales, Centro Universitario Regional del Este, Universidad de la República, Rocha, Uruguay
e-mail: ldelpuerto@cure.edu.uy; joaquin.aldabe@gmail.com

L. Courtoisie · C. Cancela · N. Gazzán
Laboratorio de Arqueología del Paisaje y Patrimonio del Uruguay, Centro Universitario Regional del Este, Universidad de la República, Rocha, Uruguay
e-mail: LUCOURT@gmail.com; cristina.cancela@lappu.edu.uy; nicolas.gazzan@lappu.edu.uy

C. Fagúndez · J. Tortosa · R. Reboulaz · M. Quevedo · M. Ramos · P. Larralde
Departamento Interdisciplinario de Sistemas Costeros y Marinos, Centro Universitario Regional del Este, Universidad de la República, Rocha, Uruguay
e-mail: fagundezce@gmail.com; cfagundez@cure.edu.uy; joaquintortosavargas@gmail.com; reboulazlga@gmail.com; mquevedo0207@gmail.com; Mairaramos531998@gmail.com; quepachamamaa@gmail.com

B. Orrego
Intendencia Departamental de Rocha, Lascano, Uruguay
e-mail: beaisarojas@gmail.com

A. C. Colonese, R. G. Milheira (eds.), *Historical Ecology and Landscape Archaeology in Lowland South America*, Interdisciplinary Contributions to Archaeology, https://doi.org/10.1007/978-3-031-32284-6_2

sands of years of construction and use, represent biodiverse clusters of forest islands on a landscape dominated by wetlands and grasslands. Close to a thousand of these anthropogenic constructions are present throughout an important portion of the east and northeast of the Uruguayan territory. Nonetheless, their presence has been mainly disregarded within the official discourse on national heritage and Uruguayan history until very recently. Consequently, conservation efforts have been scarce, and the inclusion of natural-cultural heritage represented by *cerritos'* landscapes has been a consistent challenge within planning and management. Additionally, the fact that most rural areas in the country are privately owned constitutes a major obstacle for integral management of landscape heritage. In this chapter we reflect on how recent advances in land access public policies, together with transdisciplinary research and collaborative management efforts have helped shape a more promising perspective for archaeological landscape conservation in Uruguay's eastern lowlands. Considering as case-study a State-owned productive field comprising indigenous mound sites in *India Muerta*, we provide an account of a process of collaboration between the National Institute of Colonization, the University of the Republic and a group of settlers for landscape conservation and management. We argue in favor of integral planning for conservation, as well as for a place-based, collaborative approach in defining management and heritage enhancement in the region.

Keywords Landscape conservation · Collaborative archaeology · Indigenous mounds · Public land policies · Integral heritage management

2.1 Archaeological Heritage and Public Policies

Archaeological research in Uruguay has evidenced uninterrupted occupation by different human groups for at least 13,000 years (López Mazz, 2013; Nami et al., 2018; Suárez, 2019). A myriad of material evidence of human occupation attests to a prolonged history of indigenous, colonial and postcolonial life. Prehispanic earth mounds (locally known as *Cerritos de indios*), cairns, stone corrals, open-air sites, caves, hunter-gatherer camp sites, colonial houses, rural historic buildings and production facilities, rock art, petroglyphs, among other archaeological sites, pay testimony to a great diversity of socio ecological systems and livelihoods of the past (Cabrera, 2014; Curbelo, 1999; Geymonat, 2019; Lopez-Mazz et al., 2014, 2020; Milheira & Gianotti, 2020; Sotelo, 2014; Suárez et al., 2011). In spite of their profuse presence, the realities these sites attest to are not reflected in Uruguay's official history. The homogenization of an onto-epistemological discourse built upon the notion of a European-descendant identity as the foundation of the modern nation-state has perpetuated a process of invisibilization of the native people's legacy in Uruguay that has continued well beyond the nineteenth century's *charrúa* genocide and the destruction of a diversity of indigenous cultures (Rodríguez & Michelena, 2018; Rodríguez, 2017; López-Mazz, 2018). By means of the construction of a historical, geopolitical and even ecological discourse in accordance with a

European-white-male-urban national identity, the memories of indigenous cultures were mostly wiped out of Uruguayan official history, and the topography of indigenous mounds flattened down in official cartography.[1] As might be expected, not all of these traces of the past—which nowadays have become hybridized with rural, urban and industrial landscapes—are known, and only exceptionally have some been recognized and are legally protected.

More than 90% of officially recognized archaeological National Heritage values are currently condensed in urban capitals and of that number more than half is to be found in Montevideo: this shows that official heritage is unequally represented throughout Uruguayan territory, favoring urban over rural areas. Moreover, 99% of national cultural heritage is related to the country's colonial past and to European and creole legacies.[2] This situation becomes twice as paradoxical when considering that: (a) most entities susceptible of protection through being declared heritage are actually located in rural contexts dominated by agriculture under private land ownership, and (b) the vast majority of these entities attest to non-western cultures pertaining to indigenous societies and, to a lesser extent, are related to enforced settlement of enslaved groups brought from Africa. The scarce presence of these key pieces of national history and identity within national heritage policies has been pointed out by several academic researchers (*e.g.* Criado-Boado et al., 2006; Gianotti, 2005; López-Mazz, 2018; Verdesio, 2011; Vienni, 2014). This has very recently led to a regulatory shift with regards to heritage—*e.g.* through the adoption of various international regulations such as the Declaration of the Protection of Intangible Heritage 2006, the Declaration of the Protection and Promotion of the Expression of Cultural Diversity, among others. Nonetheless, the main legal norm in force regarding cultural heritage in Uruguay (Law N° 14.040 for the Protection of Heritage, approved in 1972) does not explicitly make up for these gaps, and its obsolete approach to heritage shows a certain degree of discrepancy with newly found socio-territorial needs that recent developments in the country have given rise to. ILO's Indigenous and Tribal Peoples Convention (ILO 169, 1989) has not been ratified in Uruguay either, despite the demands of local indigenous collectives (CONACHA, ADENCH, CHOÑIK,[3] among others) as discussed by several authors (López-Mazz, 2018; Rodríguez & Michelena, 2018; Rodríguez, 2017). This shows a persistent denial of indigenous identities in the country, contrary to social claims and scientific evidence (*e.g.* Sans et al., 2012).

In Uruguay as in other places, the conservation of archaeological and cultural heritage in general is often conceived as independent of environmental conservation

---

[1] In digital elevation models (30 cm resolution) recently created by Uruguay's national Spatial Data Infrastructure (Infraestructura de Datos Espaciales de Uruguay, IDEUy) earthen mounds topography is not represented, in spite of these mounds reaching between 4 and 6 m in height with 30–40 m in diameter. This is likely due to the use of algorithms specifically programmed to erase "errors" in the model, evidencing the absence of awareness of our indigenous past and its material legacy.

[2] See the official National Historical Monuments list in http://www.patrimoniouruguay.gub.uy

[3] CONACHA: Consejo de la Nación Charrúa (Charrúa Nation's Council); ADENCH: Asociación de Descendientes de la Nación Charrúa (Association of descendents of the Charrúa Nation); CHOÑIK: pampa expression meaning "we are the people".

and management (as noted by *e.g.* Fernández, 2008; Criado-Boado et al., 2006). However, policies regarding biodiversity conservation, promoted for the past two decades through the approval of several laws and decrees on the subject (such as Law N° 17.234 and Law N° 17.283 of 2000, and their decrees)[4] have begun to include cultural heritage values as a constituent part of territorial values worth of conservation and protection. As such, conservation objectives aimed at identifying, recognizing and protecting cultural values have started to be increasingly included in management plans for Protected Areas where these are present (Beovide, 2018; Brum et al., 2021; Capdepont et al., 2009; Cortes-Capano et al., 2020; Gianotti et al., 2015; Rodríguez-Gallego et al., 2012; Rivas, 2005; SNAP, 2014; Villarmarzo, 2018; among others). Similarly, the approval of a Territorial Planning and Sustainable Development Law (N°18.308, 2008) allows heritage to be considered and included as a territorial value in public policies for territorial planning and management. This law not only includes heritage as a conservation object likely to be planned for and managed, but it contemplates a comprehensive notion of heritage that integrates biological and cultural diversity as inseparable co-constituent parts, considering the different levels and dimensions in which diversity manifests—*i.e.* sites, areas, landscapes. Nevertheless, the effective implementation of these legal dispositions has not yet been fully achieved.

Recent undertakings associated with development through a 'modernization' process in Uruguay, which involve shifts in agriculture and transformations in the productive structure, have given rise to an increase in mega-entrepreneurships of several kinds—agribusiness, forestry, animal production, among others—which are significantly and negatively impacting cultural heritage and environmental configurations—human and non-human communities, habitats and regional landscapes—across the entirety of the national territory (Cortés-Capano et al., 2020; Gianotti et al., 2017; Scarlato, 2003; Rodríguez-Gallego et al., 2019) as happens worldwide (see *e.g.* Benton et al., 2003; Fernández Cacho, 2008; Modernel et al., 2016; Taylor, 2019; Escobar, 2019). The negative impact on heritage values is considered to be evident under large scale industrial undertakings—such as the installation of foreign cellulose processing plants, wind farms, and afforestation (EROT-RE, 2012; OPP, 2015). But this has not been regarded as particularly the case in other productive activities such as extensive crop-agriculture (soy, rice, improved grasslands) and the intensification of beef cattle production systems. Since 96% of the land in Uruguay is privately owned, conservation efforts towards the protection of natural and cultural heritage assets face great challenges fundamentally involving the articulation of sustainable modes of production compatible with the preservation of cultural and natural heritage values. As Criado and Barreiro (2013: 7) point out, "the reality of heritage is its management" and the existing interrelationship between archaeological heritage research, socio-economic development and the conservation of biodiversity and cultural values is undeniable (Criado-Boado et al., 2006).

[4] The Declaration of the Protection of the Environment as of General Interest, and Creation of a National System of Natural Protected Areas respectively.

This is certainly the case when it comes to indigenous mound sites: these prehistoric constructions are immersed as constituent parts of a rich and biodiverse landscape in rural productive fields of the Uruguayan lowlands (Gianotti, 2015; Gianotti & del Puerto, 2016).

2.2 Case Study: Indigenous Mounds Landscape in the Lowlands of *India Muerta*

Anthropogenic construction of earthen mounds can be found in most of the South American continent. Different manifestations of these constructions may be found along the coast of the Guianas (Rostain, 2008), in the Llanos de Moxos in Bolivia (Erickson & Balee, 2006; Prümers & Jaimes, 2014; Walker, 2008), in the delta of the Parana River and the Uruguay River (Castro, 2018; Politis & Bonomo, 2018) and in the wetlands of Pantanal (Eremites & Milheira, 2020; Eremites, 1995), also comprising mounds located along the main watercourses and deltas of the Amazon River (Roosevelt, 1991; Schaan, 2008). Altogether, these anthropic structures represent a radical change in the way of inhabiting the lowlands since the beginning of the construction of cultural landscapes that last to the present day.

Indigenous earthen mounds represent the earliest known manifestation of human constructed landscape in Uruguayan territory (Gianotti, 2015). Indigenous mounds are present throughout an important portion of the east and north east of the Uruguayan territory and southern Brazil (Milheira & Gianotti, 2020). The most ancient mound sites date back to more than 4500 years BP (before present) and are located in the Region of India Muerta (Bracco et al., 2015; Iriarte, 2006). The ancient ecological transformation of the environment of the eastern region in Uruguay's prehistoric times resulted in human-promoted biodiversity patches dominated by woody vegetation that grows on the mounds forming nowadays' peculiar forest islands that stand out on lowlands (del Puerto et al., 2023). It has been shown by del Puerto et al. (2021) that the porousness and richness of the anthropic soils that constitute the mounds and surrounding structures favour the emergence of plant and animal communities particularly rich in relative and total abundance of species, contributing to an increase in biodiversity.

India Muerta wetlands, in the eastern department of Rocha, comprise close to a thousand of these anthropogenic constructions within a 386 km^2 area: this is an average of 2.6 ancient mounds/km^2 (Gazzán et al., 2022). A total of 666 mounds have been georeferenced—with 312 mounds pending georeferencing *in situ* (*ibid.*). In this region, *cerritos* represent peculiar and biodiverse formations consisting of islands and clusters of forest islands on a lowland landscape dominated by wetlands and grasslands (*campos*). Of those hundreds of mound sites, only two are currently protected by law as National Historical Monuments[5] since 2008. Nowadays, most mounds in *India Muerta* are immersed in agricultural contexts, and multi-scale productive processes have greatly impacted these monumental landscapes with consequent losses in habitat heterogeneity (OPP, 2015).

Landscape diversity degradation in the India Muerta region has been mainly associated with rice production and cattle raising even though the region is a priority area for biodiversity conservation comprised within a Ramsar Site (Bañados del Este, Ramsar Wetlands of International Importance), one of three in the country (PROBIDES, 1999; EROT-RE, 2012). Nowadays, the majority of the remaining mound sites are located within private estates, and their conservation has been mainly dependent upon land-owners valuing the mounds as useful for providing refuge and shade for the cattle. These uses, paradoxically, contribute to the eroding of the mounds and prevent the regeneration of native forest species which have become part of them, though the impact of livestock is less drastic than the impact generated by the tilling of entire patches of land for rice, soy and eucalyptus cultivation which result in the disappearance of habitat biodiversity and the loss of archaeological heritage.

The intensification in agricultural land use—mainly hydraulic works for rice cultivation—in the north-eastern and eastern regions since the 1970s and 1980s (Bueno et al., 2021; Rios et al., 2022) resulted in an intensification in archeological research efforts carried out by the Archaeological Rescue Commission of the Laguna Merín Basin[6] (CRALM by its Spanish acronym) so as to recuperate and save as much material evidence as possible from its complete destruction (López Mazz, 2001). This led to a slow process of recognition of indigenous mounds as heritage and to great academic efforts to promote their conservation (Bracco et al., 2000; López Mazz, 2001).

2.2.1 New Paths for Archaeological Heritage and Biodiversity Conservation Through National Land Access Policies

The National Institute of Colonization (INC for its Spanish acronym) was created in Uruguay by Law (by N°11.029, 1948) as an autonomous entity with the purpose of promoting "a rational subdivision of the land and its adequate exploitation, to ensure the establishment and wellbeing of the rural worker and to promote the growth and betterment of agricultural production".[7] The creation of the INC resulted from a long history of political attempts to promote the subdivision and fairer distribution

[5] Ministerial Resolution N° 444/008: Declaration of Historical Monument. Indigenous Mound Sites of the area of India Muerta. Rocha. Retrieved from: https://www.impo.com.uy/bases/resoluciones/444-2008/1

[6] CRALM operations were conducted by archaeologists of the Faculty of Humanities of the University of the Republic (UdelaR).

[7] https://www.colonizacion.com.uy/es_ES/-que-es-el-inc-

of land in Uruguay since 1815, and the enactment of the aforementioned law was not without lengthy debate (Vassallo & Ferrerira-Chávez, 2014). Its creation was simultaneous with an increase in the still ongoing rural exodus phenomenon (*ibid.*) linked to multifactorial causes which resulted in a great decrease of family-based small-scale livestock production systems throughout the country (Camors, 2015). Since then, efforts in promoting and implementing land access policies have oscillated with changes in government and national interests, and it was not until recently since 2005) that a sound shift in classic land access policies started to take place, organizing rural settlement efforts around alternative production models (Vassallo & Ferreira Chávez, 2014) and favoring, in the best of cases, the settlement of families of rural workers and small producers in state-owned fields, while also incorporating gender perspectives into policy making (Camors, 2015). Nowadays, the INC has 615.534 hectares (ha) under diverse modes of ownership. This represents 4% of total productive land and 27% of family-based production in Uruguay (DIEA-MGAP, 2019), and 88% of the land owned by INC is assigned through lease to rural families and rural workers.

The INC's presence in the Uruguayan eastern region, especially in Rocha Department, has been scarce. The demand for land by rural workers and small producers gave way, in 2015, for the acquisition by the INC of estate N° 735, 1000 ha in *Rincón de la Paja*, *India Muerta* (INC, 2016a, b), see Fig. 2.1. INC officials[8] in charge of surveying the newly acquired field, mainly composed of wetlands, lowlands and grasslands fit for grazing, detected there the presence of several indigenous mound sites and required an archaeological assessment of the mounds by the University of the Republic (UdelaR by its Spanish acronym). A diagnosis of the archaeological heritage within the field was carried out in 2016 (CURE-UdelaR).[9] Thirty-two indigenous earthen mounds and other anthropic structures were identified and their conservation was recommended (Gianotti & del Puerto, 2016).

The land was assigned in 2017 by the INC to an association of ten families of local rural workers—called "*Cerros Lindos* group"[10]—to develop a production project of collective sustainable cattle raising. As recommended by Gianotti & del Puerto (2016), 21.5 ha comprising three archaeological mound sites (see Fig. 2.1) within the field were assigned to CURE-UdelaR for research, conservation and monitoring purposes. An agreement was signed between INC and the University in 2018, through which co-management of the Research, Monitoring and Conservation Units (RMCUs) corresponding to the three main archaeological mound sites was

[8] Agric. Eng. Ana Lía Larrosa and MSc. Antrop. Verónica Camors (both INC officials in Uruguay's eastern region) played key roles throughout this entire process and in conservation and management efforts since the very beginning, not only by establishing contact with CURE as of the identification of indigenous mounds within the field, but by actively taking part in research and outreach activities.

[9] Eastern Region University Centre (CURE by its Spanish acronym) is part of the UdelaR's decentralization process and promoting the presence of higher education throughout the Uruguayan territory, outside the capital.

[10] *Cerros Lindos* may be translated as "beautiful earthen mounds".

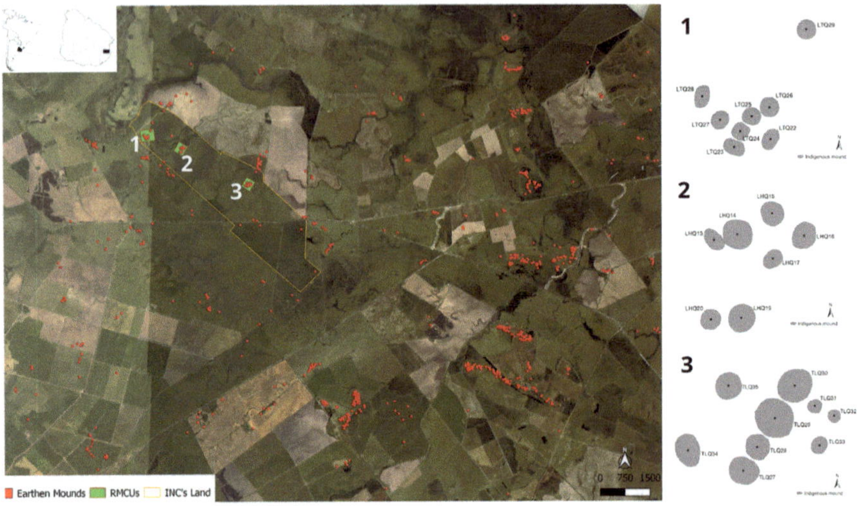

Mound	Date 14C	Sample	Reference
LTQ26	1807 ± 24 (AA112120)	sediment	ANII Project
LTQ26	4133 ± 28 (AA112119)	sediment	ANII Project
LTQ25	2101 ± 31 (AA112115)	human bone	ANII Project
LTQ25	2775 ± 52 (AA113490)	sediment	ANII Project
LTQ25	1180 ± 31 (AA113491)	sediment	ANII Project
LTQ25	2220 ± 33 (AA113494)	human molar	ANII Project
LTQ28	1681 ± 34 (AA113492)	charcoal	ANII Project
LTQ28	1801 ± 39 (AA113493)	charcoal	ANII Project

Mound	Date 14C	Sample	Reference
LHQ14	3758 ± 59 (AA113487)	charcoal	ANII Project
LHQ14	3540 ± 39 (AA113488)	charcoal	ANII Project
LHQ14	1692 ± 65 (AA113489)	sediment	ANII Project

Mound	Date 14C	Sample	Reference
TLQ29	2490 ± 70 (URU201)	sediment	Bracco et al 2015
TLQ29	2940 ± 80 (URU208)	sediment	Bracco et al 2015
TLQ27	3020 ± 80 (URU203)	sediment	Bracco et al 2015
TLQ29	3060 ± 110 (URU200)	sediment	Bracco et al 2015
TLQ30	3180 ± 80 (URU196)	sediment	Bracco et al 2015
TLQ27	3320 ± 80 (URU209)	sediment	Bracco et al 2015
TLQ27	3350 ± 80 (URU204)	sediment	Bracco et al 2015
TLQ29	3560 ± 70 (URU202)	sediment	Bracco et al 2015

Fig. 2.1 Study area location in *India Muerta* (Rocha Dpt., Uruguay), delimitation of the RMCUs within INC's field, and distribution ^{14}C dates of the indigenous mounds in each archaeological site

arranged. The Agreement created a Follow-up Committee—integrated by representatives of INC, CURE-Udelar and *Cerros Lindos* group—with the purpose of developing and evaluating a collaborative and inter institutional framework for the articulation of sustainable production and the conservation of heritage in the RMCUs. This gave way for the launching of an interinstitutional collaborative project to work towards heritage conservation and management and sustainable livestock production as a basis for local social development. The spatial definition of the RMCUs involved interinstitutional and community discussions to agree on the physical delimitation and implementation of a wire fence on their perimeters. Figure 2.1 shows the three resulting areas, with the names of the archaeological sites they encompass. RMCU I (8.9 ha) corresponds to the mound site known as *La Tapera*, RMCU II (5.75 ha) corresponds to the mound site *Los Talitas* and RMCU III (6 ha) corresponds to the mound site known as *Los Huesos*. A team of archaeologists, biologists, agronomists, social assistants, rural workers and their families, veterinary doctors, accountants, among other community and professional agents together with local institutions (INC, MEC,[11] IDR,[12] Cultural Centre of Lascano[13]) participated in a 2-year process of research, integral planning and monitoring towards the creation of a landscape management model for the conservation of heritage.

The process involved actions on various fields of research, efforts of promotion of heritage assets within the *cerritos* landscape, and the design of a long-term integral management plan. The experience arose a series of questions that we try to address in this chapter: how is it possible to integrate different disciplinary frameworks into one inter transdisciplinary framework for heritage management? Are standard management manuals useful and their application feasible in the complex reality of a *cerritos* landscape? What does it actually mean to integrate the sense of place and a relational-affective dimension of landscape into a management plan? What is the best way of fixating diverse relational processes within strategic planning for the conservation and management of *cerritos* landscape? How can these questions be answered considering the plural onto-epistemologies that make the thick fabric of heritage?

Cerritos' landscape is here considered through the notion of archaeological landscape as constituted of three interrelated aspects: the natural environment, the human society, which transforms and builds the environment through its action, and culture, through which humans think and give meaning to the place (Criado-Boado, 1993). This definition is broad and focuses on the interrelationality between all three aspects, but also allows for the contemplation of a wide variety of ways of describing landscape as defined by other authors (*e.g.* Ingold, 2000; Taylor, 2019), and of the affective, topophilic dimensions (Tuan, 2007) which are related to the experience of inhabiting the landscape (Ingold, 2000) and to the sense of place (Escobar, 2001). Based upon this background, in this chapter we present the advances in a collaborative, interinstitutional and transdisciplinary experience

---

[11] MEC: spanish acronym for the National Ministry of Education.

[12] IDR: spanish acronym for the Municipal Government of Rocha.

[13] Lascano is the most important town in the region, and its cultural center is a pillar in the regions' community life.

carried out with the aim of developing research, conservation and management of an archaeo-ecological landscape in the lowlands of *India Muerta* (Rocha, Uruguay). This experience represents a first step forward towards the consideration and integration of landscape heritage conservation and management within land access public policies promoted by the State in recent years.

2.3 Methodological Approach

2.3.1 The ANII-FMV Project and the Challenges of Describing, Characterizing, Planning for the RMCUs Through Multiple onto Epistemologies

Our methodological approach was based upon the Value Chain of Heritage (VCH) model created by Criado (1996) as a broad framework for the progressive integration and reintegration of different instances of heritage value and significance which emerge throughout the activation of a patrimonialization process (Barreiro et al., 2022). The VCH model was originally formulated as an ideal sequence of instances that every patrimonialization process should go through, and it includes seven phases which go from identification, diagnosis and localization, to a descriptive research phase of (historical, cultural, scientific, communitarian) sense-making, to the promotion, enhancement, and ingress of heritage elements into communication and social use circuits (Criado, 1996). Further research on the VCH has drawn attention to the importance of incorporating a participatory dimension to the model to allow for a multivocal emergence of heritage as *process* and of heritage as *product* (Cuesta et al., 2009; Criado & Barreiro, 2013). Recent works posit the need for generating a co-theorized and co-produced heritage activation from the very beginning of the patrimonialization processes (Sanchez-Carretero et al., 2019). Within this conception, collaborative undertakings (Roura-Expósito et al., 2018) emerge as a new horizon of possibilities to think and to make heritage together; the experience we account for in this chapter shows some results of such an undertaking. Collaboration is here taken to be the possibility for and the necessity of producing collectively through intercultural, intergenerational, interdisciplinary and multi ontological dialogues (Escobar, 2010, 2019). Methodologically, this approach has involved a radical implementation of the ethnographic method—*i.e.* ethnographic interviews, participant observation, social cartographies, workshops and other collective instances—throughout most active phases that the project involved (Guber, 2001; Vaughan, 2018; Rappaport, 2007; Risler & Ares, 2013). Thus, and considering Escobar's (2001) understanding by which "culture sits in places", each step in the VCH involved a reflective and collective process for defining landscape in-place, from the perspective of inhabiting and being part of its life (Barreiro et al., 2022).

Hamilakis' (2015) novel approach to multi-temporality in archaeology and his proposal of a sense-based way of researching and experiencing archaeological heritage opens up a whole new path for heritage studies and conservation. He poses questions on time which have guided our work throughout the multiple processes of research and characterization of the area. What time(s) does this landscape represent is one of these questions. *Cerritos* are pluritemporal structures in themselves. The evidence of varied human uses have been unearthed by burrowing animals such as foxes (*Pseudalopex gymnocercus* and *Cerdocyon thous*), lizards (*Tupinambis merianae*) and nine-banded armadillos (*Dasypus novemcinctus*) and, in some cases, constructive strata have been moved about by trees deeply rooted in the heart of the mounds. The anthropology of time as understood by Fabian (2019) and Hamilakis (2015) brings a complex understanding of temporality that integrates sensorial experiences, affect and memory, and this framework allows for a new understanding of *cerritos* landscape. The coexistence of past, present and future as constitutive elements of current landscapes has also been the subject of reflection in Abram's (1996) work on landscape ecology. These sometimes invisible histories of landscapes should be and are the target of conservation and evaluation (Yang & Han, 2020). The embodiment of time within landscape has been often compared to the nature of palimpsests and the instrumental integration of this fourth dimension and of intangible heritage values for the creation of planning and management tools for specific landscapes has only recently began to be explored (*e.g.* Criado Boado & Barreiro, 2013; Taylor, 2019; Yang & Han, 2020).

This theoretical-methodological framework involves an ontological reflection on materiality, immateriality, human and non-human entities and, specially, the relations that intertwine in the field of experience that a landscape is composed of. In practical terms, our endeavor contemplated a variety of dimensions of the landscape (ecological, archaeological, social, economic, symbolic, etc.) from an interdisciplinary perspective so as to establish a complex baseline from which to discuss a common ground for conservation and actual and future uses of the RMCUs. Each dimension involved research actions which are briefly described below, and which were reflected upon and integrated with ethnographic information so as to define heritage values and strategies for their management and conservation.

2.3.2 Research Actions

Research for the identification and diagnosis of ecological and archaeological values within de RMCUs involved working with the group of settlers "*Cerros Lindos*" and other local actors so as to co-activate a patrimonialization process for the mounds' landscape. The work implied participatory workshops and social cartography (methods proposed by Risler & Ares, 2013 and CIMAS et al., 2009), the co-creation of a visual identity for "*Cerros Lindos*" group and a process of co-construction of governance through the implementation of the aforementioned Follow-up committee as a meeting ground. The participatory dimension of the process was essential for a place-based collective identification, definition and mapping of heritage values and the design of strategies for their evaluation, monitoring, promotion and conservation, especially regarding "invisible" landscape heritage (Yang & Han, 2020). This was possible through the theoretical-methodological incorporation of key aspects such as the local involvement and appropriation of values regarded as constitutive of identity and a sense of rootedness (Escobar, 2001; Masterson et al., 2017; Taylor, 2019).

The temporal thickness of landscape is seldom considered in management plans but it has been shown to be essential for assessing the sustainability of landscape and its values (Yang & Han, 2020). For an archaeological characterization of the RMCUs, planimetric surveys and digital elevation models, based on procedures applied on other mound sites (*e.g.* Bonomo et al., 2011; Cancela et al., 2016; Gianotti, 2005), were done for RMCUs I and II using total station and submeter GPS. Data obtained allowed for a morphological characterization of each anthropic structure, a quantitative estimation of dimension, volume and surface area, as well as for designing monitoring indicators to assess the impact of cattle and other degradation factors (*i.e.* ground compacting, erosion). A stratigraphic and geochronological characterization of the mounds was done through sediment sampling on mounds and other anthropic structures of all three RMCUs with AMS Soil Sampler kit. Samples were ^{14}C dated, and physical, textural, mineralogical and geochemical analyses were carried out (del Puerto et al., 2021). Samples were also analyzed by FTIR and pyrolysis. Micromorphological analyses for the characterization of the mound constructive layers were carried out in national and international labs following standard protocols applied in other mound sites (del Puerto et al., 2016, 2021; Kaal et al., 2017; Suárez & Gianotti, 2013). Bio siliceous particles—phytoliths and diatoms recovered *in situ*—were subjected to paleoethnobotanical analyses, following the same procedures applied for other mound sites (del Puerto, 2015), with the aim of generating information on paleoenvironmental conditions, building material input areas and existing plant resources potentially used by mound-builders.

Three excavation campaigns of RMCU II (*Los Talitas*) were carried out during 2019–2020. Interventions included surveys and stratigraphic excavations in micro-reliefs, mounds and adjacent plains within the sites. Materials recovered are being analyzed in the Laboratory of Archaeology and Territorial Analysis of CURE-UdelaR (Rocha). Lithic analysis follows the criteria and categories proposed by Orquera & Piana (1986), which have been implemented in other mounds sites (Gazzán, 2018). Bone materials were studied through morphotechnological analysis, reassembly studies and morphological reconstruction according to the general criteria established by Moreno (2014) and Bica (2020) with the aim of identifying animal species used and processed for consumption and tool manufacture in mound-builder communities.

A survey of environmental heterogeneity was done both for the entire field and for all three RMCUs through field work and satellite photography analysis. Within the RMCUs, a survey of arboreal-shrub vegetation of the RMCUs I and II (*La Tapera* and *Los Talitas*) was done, with abundance quantified as the number of tree and shrub individuals; herbaceous species were registered as the frequency of occurrence (Kent, 1992). α, β and ɣ biodiversity was characterized and biodiversity values were compared between both RMCUs. α biodiversity was estimated as an exclusive species % among semi-detached plots. Resulting lists of species for the RMCUs were compared to available species composition and ethnobotanical information for the region so as to identify differential distribution of species of agricultural, medicinal or other economic value (Campbell et al., 2006; Erickson & Balée, 2006; Martin, 2001). Other actions involved research on grazing intensities and their impact on herbaceous coverage of the ground and fodder supply. This was done by measuring seasonal dry biomass production within and outside the RMCUs on grassland patches, together with measurements of grass height, so as to define and evaluate good production practices within the RMCUs in collaboration with the settlers (based on methods proposed by Haydock & Shaw, 1975 and adjustments presented by Do Carmo et al., 2015).

A bird and mammal species survey was also carried out with the objective of determining if mound sites constitute habitat patches for endangered species, and whether these patches significantly increase the region's biodiversity. To this end, forest and grassland bird species under different grazing intensities were identified in all RMCUs and in control areas located elsewhere in the field. Mammals were surveyed by means of search and registration of footprints and by the application of trap cameras within the *cerritos'* forests. Species ID was done with the aid of field guides for identifying birds (e.g. Narosky & Izurieta, 2015) and mammals (e.g. González & Martínez-Lanfranco, 2010).

2.3.3 Planning and Management for Conservation

During 2020, a group of environmental management students worked together with biology, archaeology and agricultural science researchers, the local community and the institutions with the purpose of elaborating a participatory and comprehensive management plan for the conservation of the archaeological and ecological landscape in articulation with livestock production. Considering that heritage constitutes a controversial scenario where a multiplicity of diverse, and sometimes conflicting, narratives, legitimacies, worldviews, interests, and sense-making converge (Criado & Barreiro, 2013; Jiménez-Esquinas, 2016; Smith, 2006, 2011), creating a management plan for conservation and use of the RMCUs involved the need for finding a common ground upon which mutual trust could be established and an integral management structure accorded.

Approaches to biodiversity conservation conventionally used in protected areas (Granizo et al., 2006; CMP, 2013) provide useful tools for the systematic identification of conservation targets, drivers of degradation and management measures aimed at conservation. The incorporation of cultural values has been less explored in such manuals, and participatory approaches have not been developed in depth, largely due to the division between nature and culture as the normative ontological basis for conceiving design, management and conservation in general, which have characterized the instrumentalization of management procedures in Uruguay's protected areas (Cortes-Capano et al., 2020; Gianotti et al., 2015; Villarmarzo, 2018). Thus, as far as has been possible, a participatory process combining several planning methods and approaches (e.g. Granizo et al., 2006; CIMAS et al., 2009; Criado Boado & Barreiro, 2013; CMP, 2013; Risler & Ares, 2013) was developed so as to allow for the emergence of an open, inclusive and empathic framework for decision-making and for establishing action-orienting strategies. The process was carried out from a perspective of equality that includes and celebrates diversity in multiple manifestations: gender diversity, age diversity, neuro physical and social diversity and the diversity in human-environment modes of relation. Through this, the strengthening and widening of interinstitutional and social place-based networking was made possible and this worked as the basis upon which the Integral Management Plan (IMP) was conceived: as a roadmap in order to achieve both ecological and productive sustainability and heritage conservation and enhancement.

2.4 Results

2.4.1 *Diagnosis of* Nature-Cultures*: Landscape as Palimpsest*

Results showed that indigenous occupation associated with the building of mounds within the three RMCUs occurred between 4200 and 1600 ^{14}C year BP (see Table on Fig. 2.1). *Cerritos* are mainly the result of the accumulation of anthropogenic sediments rich in organic remains with strong evidence of thermal-alteration. Fire played a central role in the anthropogenic sediment formation processes for mound building, possibly evidencing a combination of cleaning and thinning activities in domestic and productive spaces, cooking food, technological production and waste management (del Puerto et al., 2021). In addition to this, the presence of human remains confirms the funerary character of some of the mounds. Dates obtained for these human remains (2300 ^{14}C year BP) push back in time the chronologies for the funerary use of indigenous mounds in Uruguay.

Preliminary analysis of materials recovered on-site has led to the identification of a wide range of resources that were used and consumed by indigenous mound builders. Indigenous use, processing and consumption of resources include of several cervid species (*Blastocerus dichotomus* and *Ozotoceros bezoarticus*), armadillo (*Dasypus novemcinctus*), rodents *(Myocastor coypus, Cavia aperea) and* birds *(Rhea americana).* Bone remains of different freshwater fish species such as catfish (Siluriformes) and eel (*Synbranchus marmoratus*), and estuarine fish species (e.g. *Micropogonias furnieri*) have been found in the excavation of *Los Talitas* site (RMCU II), evidencing interaction with coastal areas as has been found in other mound sites within Uruguay's north eastern and eastern regions (Bica, 2020; Moreno, 2014; Pintos, 2000).

Within the RMCUs, present species composition and richness evidence shows that mound sites concentrate a higher alfa and beta biodiversity, and that the landscape's habitat heterogeneity is entirely represented within the sites. Different environmental units were identified within the RMCUs (see Fig. 2.2). High dense grasslands dominated by *Panicum prionitis*—locally known as *pajonales*—*campos* grasslands and *varges,*[14] together with the mound forests, are represented within all RMCUs, thus providing refuge and resources for complex animal and plant communities. Results show how ecological management in prehistoric times resulted in the promotion of patches of biodiversity dominated by the presence of woody vegetation on the earthen mounds which modulates and enriches the current biodiverse landscape (Barbero et al., 2019; Triunfo et al., 2018). Among wood species, 19 were identified *on* the mounds (13 tree spp. and 6 shrub spp.), of which 32 are native spp. and 2 exotic, while only 2 woody spp. were found unconnected to these structures (Triunfo et al., 2018). Mound forests present a unique plant species composition which combine tree and shrub species characteristic of riparian and coastal forest

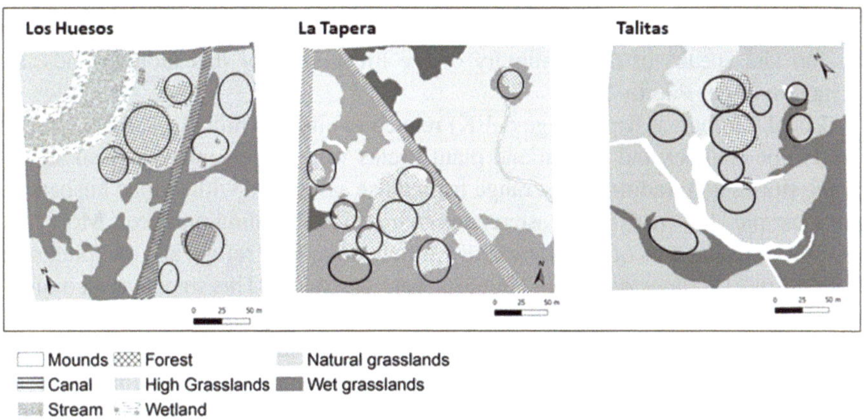

Area (ha)	Los Huesos	La Tapera	Talitas
Total area	6.11	8.96	5.75
Wet grassland	0,16	0.48	1.17
High grassland	0.49	4.8	0.39
Natural grassland	3.94	2.32	3.54
Forest	0.63	1.11	0.25
Wetland + Stream	0.82	-	-

Fig. 2.2 Surface area of different environmental units identified within the RMCUs

[14] *Varges* is the local name given to a type of wetland, characteristic of these lowlands, dominated by native *Luziola peruviana* and *Paspalum hydrophylum* which are highly appreciated by cattle producers for their nutritive quality for cattle raising.

and ombú (*Phytolacca dioica*) groves. Herbaceous composition in the surrounding lowlands is dominated by uruguayan *campos* (grasslands, e.g. *Paspalum notatum, Axonopus affinis, Paspalum dilatatum, Stenotaphrum secundatum*) and varge native species (e.g. *Paspalum hydrophylum, Luziola Peruviana*).

Birdwatching within all four environmental units for RMCU I resulted in the identification of 80 bird spp., of which 27 were observed exclusively within mound forests, 12 exclusively associated to *varges*, 6 were registered only on *campos* grasslands and 6 only in high dense grasslands (Triunfo et al., 2018). Camera traps placed within the forests registered the presence of 9 mammals (Barbero et al., 2019; Triunfo et al., 2018), 4 of which (*Pseudalopex gymnocercus, Dasypus novemcinctus, Leopardus geoffroyi* and *Cerdocyon thous*) are currently classified as endangered/priority for conservation (Soutullo et al., 2013: see Fig. 2.3).

Ethnographic research has shown that there is a vast corpus of oral history embodied in the local community and alive through memories and local forms of relating to place (see Fig. 2.4). The name of mounds sites within and outside of the study area refers to their local history. *La Tapera*, for instance, refers to the presence of an anthropic brick construction and exotic vegetation which attests to the historical occupation of the mounds in the last century, a use that remains common in the region and speaks of the continuity of the life history of this mound landscape (Gianotti & del Puerto, 2016).

Local Ecological Knowledge (LEK) related to climatic processes, the history of landscape changes and animal and plant cycles were documented, as well as legends, stories and traditions that range from those associated with animal husbandry, for example "la yerra",[15] to those related to traditional healing practices. More than 50-year-old accounts of historic use of earthen mounds as refuge and camping sites for "nutria" (*Myocastor coypus*) hunters were registered. This trade, very common in *India Muerta* and which ended simultaneous to the draining of the wetlands in the 1980s, was practiced during the rainy season and depended on the use of hand-made boats to navigate the marshes and swamps. Management of native forest species growing upon the mounds, specially of native fruit trees, such as *Butia odorata*, evidence ways of inhabiting and managing the mounds' biodiversity. These intangible assets, understood as part of *India Muerta*'s landscape's invisible heritage (*sensu* Yang & Han, 2020), need a proper way of being registered and promoted for achieving the aim of holistic conservation and management of the archaeological-ecological landscape as a whole.

[15] "Yerra" is the name given locally to a traditional annual celebration in which cattle branding and neutering is done. Traditionally, the event involves the entire community of rural families and neighbors, with ritual use of the *lasso* for capturing the cattle and a feast where food and drink is shared between all involved.

Fig. 2.3 Some plant and animal species identified within the RMCUs using trap cameras and surveys *in situ*. (1) Martín pescador chico (*Chloroceryle americana*). (2) Roseate spoonbill (*Platalea ajaja*). (3) Birdwatching. (4) Zorro perro (*Cerdocyon thous*) (5) Lagarto overo (*Salvator merianae*) (6) Gato montés melánico (*Leopardus geoffroyi*). (7) Tatú (*Dasypus novemcinctus*). (8) Pasto Tubiano (*Paspalum hydrophylum*). (9) Gramillón (*Stenotaphrum secundatum*). (10) *Abutilon grandifolium*

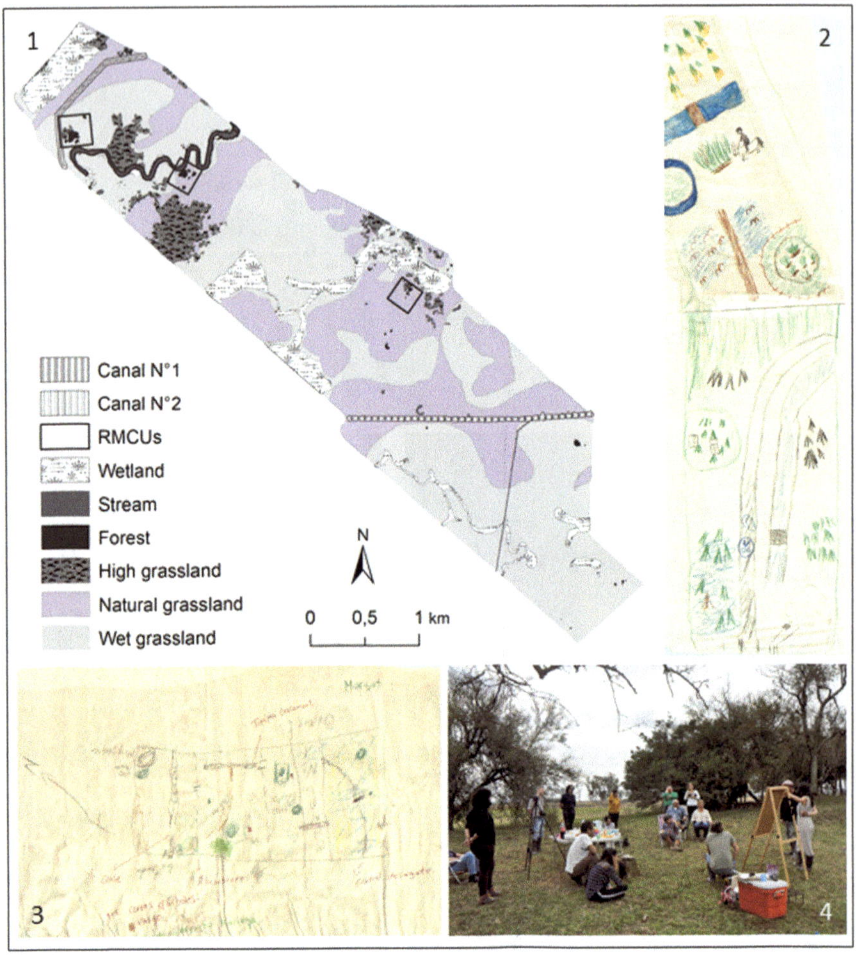

Fig. 2.4 (1) Map of environmental and cultural landmarks within the field (INC estate n° 735) created by settlers through instances of collective mapping (social cartography) and ethnographic interviews; (2 & 3) maps with representations of the landscape made by two settlers and (4) participatory social mapping workshop—held in July 2020

2.4.2 *Participatory Planning Process for Heritage Conservation: Co-creation of an Integral Management Plan (IMP)*

A conservation vision (description of the desired condition that is expected to be achieved for the area through management *sensu* CMP, 2013) was collectively elaborated for the RMCUs and expressed as:

> an area of reference with regards to integral management, which is representative of the consolidation of a shift in land access and heritage management and conservation public policies; the area, under family-based production, visibilizes native peoples' legacy through the preservation of the indigenous mounds landscape, *campos* and wetlands, and is managed through the collaboration between institutional and local social agents, generating opportunities for socioeconomic, productive and cultural sustainable development that articulates with the conservation of heritage in all its manifestations.

For the community of settlers, the construction of a vision implied the involvement of local memories and their interrelation with other places, and objectifying the *sense of living* in the lowlands. At the same time, and paradoxically, the creation of a vision opened up as a *possibility* for the community, through the intervention of the institutions (the INC and the UdelaR), for the policies of conservation and access to land. The cultivation of a sense of place has increasingly been shown to play an important role in people's motivations for environmental stewardship and adaptation to environmental changes (Chapin & Knapp, 2015; Cortes-Capano et al., 2020; Masterson et al., 2017; Raymond et al., 2016). Thus, the vision involved a place-based co-construction of the meaning of the RMCUs in context as a rooted experience of this specific location and a way of relating to it (Escobar, 2001; Tuan, 2007).

Conservation targets are conventionally defined as specific elements that are chosen because of their relevance relative to an area's identity and their ability to operate as umbrella objects for the protection of the entire system of which they are a part of (Granizo et al., 2006; CMP, 2013). Conservation targets for the RCMUs were determined through collective and participatory instances between academics, institutions and the Cerros Lindos group. Both scientific, technical and communitarian information was integrated, defining five conservation targets for the RCMUs: (1) the *indigenous mounds landscape*, (2) *ecosystem diversity*, (3) the *sensorial landscape*, (4) *local knowledge and practices* and (5) *the relations upon which local familiar-associative production is sustained.* The process of defining conservation targets emphasized the relevance of the sensorial experience of landscape as crucial to overall conservation (as proposed by Hamilakis, 2015), as well as the importance of identity in defining a sense of place (Escobar, 2001; Taylor, 2019; Tuan, 2007). These key elements are part of the invisible, relational and symbolic dimension of landscape which has seldom been considered in management plans for areas with similar values in Uruguay.

Potential threats to the conservation of the aforementioned targets were defined and analyzed with the aim of preserving and enhancing them through five management strategies which were established for a 10-year period (Fig. 2.5), and a grazing protocol for the RMCUs was outlined. Conservation actions were established in accordance with the objectives, degradation factors and conservation strategies, through the evaluation of available capabilities for their realization—see Fig. 2.5.

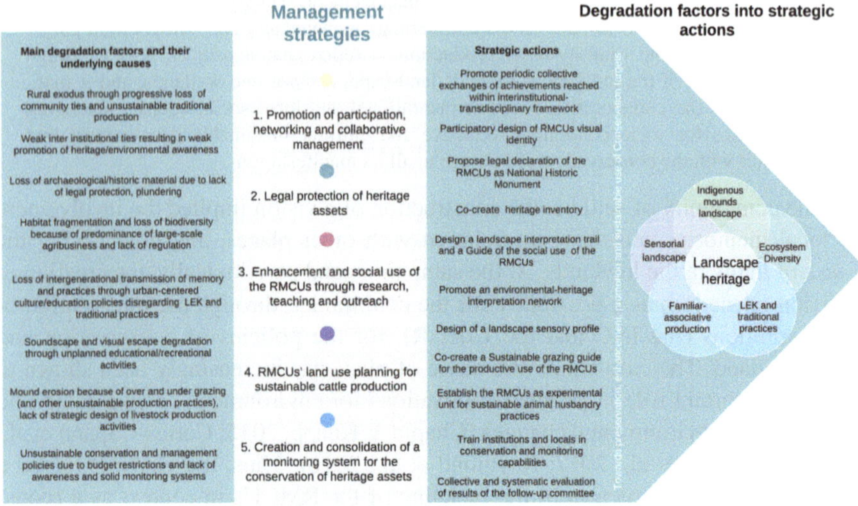

Fig. 2.5 The main degradation factors identified are addressed through five Management Strategies, transforming each factor into an opportunity for the implementation of a strategic action. All strategic actions have positive impacts on the conservation targets identified as assets of this Landscape Heritage through integral management

2.5 Discussion and Final Reflections

2.5.1 On Transdiscipline, Integrality and Landscape Conservation: The Need for a Relational, Place-Based Approach to Heritage and Environmental Management

The indigenous mound landscape is a palimpsest, formed by a plurality of material and immaterial strata, by a complexity of past and present uses and valuations. Based on this principle, this project involved a diversity of actors and their different ways of understanding the environment, its history, the character of the landscape and the parts that compose it. The participatory process showed that various conservation targets in the area, which were first characterized from a scientific perspective, are interwoven with the affective, sensorial and memory-based assessments of local actors and their identity. Their perspectives bring forth elements that, in themselves, become conservation assets constituent of the landscape's invisible heritage. Conventionally defined as local ecological knowledge, this invisible heritage includes a vastness of traditional knowledge and practices which account for the complex processes that occur in the region of *India Muerta*. These non-institutionalized, place-bound forms of identity and worldviews constitute the deep roots of a sense-of-place that is embodied in the rural and *rururban* (Gorenstein et al., 2007) local communities and subsist in a diversity of interactions between them and the environment of the lowlands of Rocha.

Ontological and epistemological dualism in science, and therefore in the field of conservation, has been contested from several theoretical frameworks (Abram, 1996; Descola, 2012; Escobar, 2001, 2010; Haraway, 2003; Ingold, 2000 among many others). It has been shown that social struggles are often strongly related, if not originated, by environmental conflict, which in turn has its root in dualistic modern thinking (Escobar, 2001, 2010, 2018, 2019; Gazzano & Achkar, 2013). In this case study, the issues related to conservation and degradation problems were found to be inextricably linked to resource-exploitation oriented land tenure, to the erasure of local identities and to losses in archaeological heritage, cultural diversity and environmental heterogeneity. Decolonizing processes in Uruguay are necessary and urgent, and they involve recovering identities and communities' memories from a historical and systemic erasure of the rural as diverse. This requires the conservation and protection of material and immaterial heritage values, as well as recognition of the indigenous and afro origins of the Uruguayan rural proletariat. In this respect, ancestry has been shown to be a central theme to identity (Taylor, 2019) both from the standpoint of the indigenous and afro cultures' legacy as well as from the conservation of nature and native biodiversity. Ancestry is latent in the region of *India Muerta*; it has been unlinked and uprooted but it persists (López Mazz, 2018; Rodriguez & Michelena, 2018) as an invisible heritage that is alive and remembered through diverse ways of relating to the indigenous mounds landscape and the reconstruction of family genealogies. The landscape arises as materialization of all those past and contemporary livelihoods, knowledge and management practices which become the actual conservation targets identified.

Environmental and heritage management emerged as interdisciplinary fields for planning of spaces as places and for the conservation and enhancement of the natural-cultural elements within them though. In Uruguay, this aim is still far from being entirely achieved. Our work has shown that its achievement is only feasible through an horizontal co-design with local communities and through the incorporation of sensorial, affective and memory-based understandings of and engagements with landscape. In this sense, we reaffirm the need to implement collaborative designs within the political technologies of planning and management of heritage and territories. It is one of the essential alternatives to build a sustainable future that seeks and is based upon the "radical interdependence" between the human and non-human worlds (Escobar, 2018).

This experience has shown that through collaboration and a place-based approach it will be possible to face the conservation challenges identified, and those to come. Furthermore, considering that the INC owns a large surface area of rural land (615.534 ha) distributed in fields throughout the country, this experience opens up new possibilities for comprehensive heritage conservation and management in other INC or privately owned fields in Uruguay. One of the historical uses given to mound sites in colonial and post-colonial times in the India Muerta region has been tightly linked to traditional cattle rising. Working towards the reconciliation of production and conservation in this case implied taking into account the important role of mounds in providing refuge for the cattle in flood seasons and shade in the summer. In the context of planning for conservation and management of the RMCUs, this issue has been of central importance, especially considering the importance of preserving traditional, family-based activities in rural contexts. Precisely one of the results recently achieved was the participatory design and implementation of a grazing protocol as a tool for co-management and evaluation of the responsible production practices within the indigenous mound sites. This has been applied and perfected through monthly monitoring cattle input and impact on grass height and forage supply. The instrumentalization of this protocol worked as a common ground for discussing deeper issues related to heritage, identity and threats to the human-environment relations that sustain the conservation of mounds' landscape and their natural and cultural components.

The archaeological viewpoint applied in this case study allowed for a multi-temporal, ethnographic approach to the multiple meanings of these indigenous mounds and the livelihoods and identities inextricably associated with them. It also has shown that materiality and immateriality intermingle in such a way in this landscape that management should be approached as a situated practice, considering the indigenous mounds and their characteristic components as multi temporal entities (Hamilakis, 2015), so as to visualize and contextualize their intangible assets for the development of management strategies and decision making. Within this framework, planning and management involved opening up the scope of meaning for the endangered entities—archaeological, ecological, sensorial, relational and symbolic—that became our conservation targets, bringing into presence multiple perspectives for landscape heritage sustainability.

2.5.2 Challenges and Achievements in the Current Scenario (2020s): Research and Planning for Landscape and Heritage Management in an Uncertain World

The IMP for the RMCUs was designed in the beginning of the 2020s, in a context of great social, ecological and economic transformations occurring globally and locally. In Uruguay, changes in national policies affecting every area of life nationwide—economy, education, culture, health, security, production, conservation—are becoming effective through the installment of a newly elected national government. This, together with several measures installed through a controversial Urgent Consideration Law[16] and the uncertainty that the global pandemic has put in place for worldwide markets as well as for governments in South American countries, add

[16] Urgent Consideration Law (National Law N° 18.889) (a.k.a. LUC) is a legislation package which introduces over 476 articles affecting health, education, production, security, environmental conservation, among other areas of life. Even though the Uruguayan Constitution allows for the proposal of urgent laws—approved within a 90 day period in parliament—LUC has been broadly criticized nationally and internationally for having been introduced with scarce democratic debate in a scenario of pandemic crisis, and for potentially limiting human rights and broadening the power and discretionality of private corporate interests, among other issues.

up to a current scenario of budget restrictions, changes in culture, education and rural production policies, development and intensification of luxury tourism, and economic development policies which are exerting pressure on rural communities and small producers. In Uruguay this is threatening the continuation of decentralization efforts in culture and education, as well as imposing restrictions on the INC's capacity of developing fairer policies for land occupation. In this context of uncertainty, elements of change may pose grave threats to landscape conservation and management. Nevertheless, they may also constitute an opportunity for propelling a shift to a more sustainable way of relating to landscape, its history and a place-based approach to learning, educating and promoting heritage values within local communities. As Taylor (2019: 61) pointed out, culture inevitably changes regardless of the implementation of strategic actions, but may also be changed because of them. Thus, a collaborative, reflective, and critical approach oriented to landscape and heritage conservation and management is of the utmost importance in an uncertain and changing local and global context.

The need to find ways out from the global monoculture-based development unsustainable model installed in *India Muerta* to alternative place- and identity-based management strategies posed the challenge of working within established power dynamics of government and the academia and involved a radical openness towards the incorporation of local perspectives, affects and worldviews. To speak about and towards a more sustainable and inclusive way of relating to the landscape means to explore multivocal discourse and experience and to integrate relational realities to avoid the disintegration processes associated with disciplinary analytical thinking and the global extractivist development model (Escobar, 2001). This is possibly the greatest challenge that both archaeology and the biological sciences are facing today with regards to conservation in the lowlands of South America. In the experience we presented in this chapter, a sort of 'disassembling-reassembling' (Escobar, 2018) of academic models for research and management as transitions out of the asymmetric power relations these usually stand upon (Escobar, 2001, 2019; Criado & Barreiro, 2013) have been necessary so as to move towards a co-construction of horizontal relations capable of making present, transforming and sustaining a self-organized collective that is rooted in place (Escobar, 2001, 2019). Our work constitutes the first experience of transdisciplinary research-planning-management for archaeological-ecological heritage in a rural context in Uruguay. It is a contribution to thinking about the conservation and sustenance of relations within *India Muerta*'s landscape itself. The collaborative methodology for research ans management implied a reformulation of the role of institutions—in this case, of the INC and the UdelaR and the future of these institutions and their policies within a context of political change.

The construction of place and the process of re-rootedness of culture (Escobar, 2001) were found to be underexplored within the globalized paradigms of environmental and heritage conservation commonly implemented. In the experience we presented, both the settlers' and the academics' understanding of the landscape and their engagement with it have been transfixed by the challenges of implementing a democratic appropriation of the past to think the present and future of the places the RMCUs have come to be. An awareness of the human-environment dialectic has emerged, empowering communities and institutional actors and allowing for the unfolding of a potential for sustainable change both from the perspective of the settlers and INC and UdelaR academics. In this sense, the globalized importance of conservation of nature has become more of a defense of identity, justice and culture in place through the lens of integral, place-based research, planning and management. How are we to live together in these local-global emerging worlds? We consider this to be the most relevant question regarding landscape heritage conservation and management. The question remains open, and it must be answered plurally and in place.

Acknowledgements This work has been carried out within the framework of ANII – FMV_1_2017_1_136446 project, with the support of other research and outreach projects of the University of the Republic: CSIC I + D Ecological management and environmental domestication, and CSIC I + D Group (494) – PIARPA. We are deeply grateful to Verónica Camors, Ana Lía Larrosa and Zoia Peirano from INC for their support and constant collaboration, to the group of settlers "Cerros Lindos" for transiting this process with us, to Martin Abreu who collaborated in the identification of animal species in camera traps, and to Mercedes Rivas, Lorena Rodríguez and Martín do Carmo for their accompaniment and support in the elaboration of the integral management plan.

References

Abram, D. (1996). *The spell of the sensuous: Perception and language in a more than human world*. First Vintage Books edition.

Barbero, S., Pereyra, A., & Méndez, M. (2019). In Informe de Planificación Ambiental (Ed.), *Plan piloto ambiental para la conservación del Patrimonio Cultural bio-arqueológico en predios con producción ganaderas en las planicies bajas del este del Uruguay* (46 pp). Centro Universitario Regional del Este.

Barreiro, D, Gianotti, C., & del Puerto, L. (2022). Cerros Lindos. De la Cadena de Valor del Patrimonio al Patrimonio como medio de Valorización. En *Anales de Arqueología y Etnología, 77*(2), 131–161. Dossier Puesta en Valor Sostenible. Paisaje y Patrimonio.

Benton, T. G., Vickery, J. A., & Wilson, J. D. (2003). Farmland biodiversity: Is habitat heterogeneity the key? *Trends in Ecology and Evolution, 18*(4), 182–188.

Beovide, L. (2018). La zonificación arqueológica en la cuenca del río Santa Lucía, Uruguay, como herramienta de gestión territorial: sinergias y conflictos en el desarrollo sustentable. *Revista Memorare, 5*(2), 112–113.

Bica, C., (2020). *Peces y pesca en las tierras bajas de la Laguna Merín. Análisis de la ictiofauna recuperada en el sitio arqueológico CH2D01 (Rocha, Uruguay)*. Master's thesis (123 p), Instituto de Ciências Humanas da Universidade Federal de Pelotas.

Bonomo, M., Politis, G., & Gianotti, C. (2011). Montículos, jerarquía social y horticultura en las sociedades indígenas del Delta del Río Paraná (Argentina). *Latin American Antiquity, 22*(3), 297–333.

Bracco, R., Cabrera, L., & López-Mazz, J. M. (2000). La Prehistoria de las Tierras Bajas de la Cuenca de la Laguna Merín. In A. Durán & R. Bracco (Eds.), *Arqueología de las Tierras Bajas* (pp. 13–38). Imprenta Americana.

Bracco, R., Inda, H., & del Puerto, L. (2015). Complejidad en montículos de la cuenca de la laguna Merín y análisis de redes sociales. *Intersecciones en Antropología, 16*, 271–286.

Brum Bulanti, L., Florines, A., & del Puerto, L. (2021). Procesos de Patrimonialización y Conflictos Ambientales en una Zona Costera del Uruguay Natural. *Cadernos do Lepaarq, XVIII*(36), 11–32.

Bueno, C., Alves, F., Pinheiro, L., Perez, L., Agostini, V., Fernandes, E., Möller, O., Weschenfelder, I., Pinho, G., Wallner-Kersanach, M., Moura, R., Durán, J. M., Etchevers, I., Costa, L., Werlang, C., Bortolin, E., Machado, E., Figueira, R., Ferreira, P., Andrade, C., Fornaro, L., & García-Rodríguez, F. (2021). The effect of agricultural intensification and water-locking on the world's largest coastal lagoonal system. *Science of the Total Environment, 801*, 149664.

Cabrera, L. (2014). Arte rupestre de la región norte del Uruguay. In C. J. Bitencourt (Ed.), *Arqueología Ibero-Americana e Transatlántica. Arqueologia, Sociedade e Território*. Habilis Press.

Camors, V. (2015). *Política de acceso a la tierra y desarrollo rural del Instituto Nacional de Colonización en Uruguay. La situación de las productoras familiares y asalariadas rural y las alternativas a las desigualdades existentes (2014–2015)*. Master's thesis (99 p), FLACSO.

Campbell, D., Ford, A., Lowell, K., Walker, J., Lake, J., Ocampo-Raeder, C., Townesmith, A., & Balick, M. (2006). The feral forests of the eastern Petén. In W. Balée & C. Erickson (Eds.), *Time and complexity in historical ecology: Studies in the Neotropical lowlands* (pp. 21–55). Columbia University Press.

Cancela, C., Machado, A., Gianotti, C., & Sotelo, M. (2016). Tecnologías geoespaciales para el registro y estudio de estructuras monticulares en piedra. In *Actas del XIX Congreso Nacional de Arqueología Argentina* (Vol. 54, pp. 1155–1161). Facultad de Ciencias Naturales e Instituto Miguel Lillo de la Universidad Nacional de Tucumán.

Capdepont, I., Marozzi, O., Villarmarzo, E., Gianotti, C., Sotelo, M., & Carve, F. (2009). *Proyecto catalogación del Patrimonio Cultural del Área Laureles-Cañas (Departamentos de Tacuarembó-Rivera)* (Memoria final). LAPPU – SNAP. FHCE.

Castro, J. C. (2018). La entidad arqueológica Goya-Malabrigo en el río Uruguay. In G. Politis & M. Bonomo (Eds.), *Nuevas perspectivas sobre la entidad arqueológica Goya-Malabrigo* (pp. 217–245). Editorial UNICEN.

Chapin, F. S., & Knapp, C. N. (2015). Sense of place: A process for identifying and negotiating potentially contested visions of sustainability. *Environmental Science & Policy, 53*, 38–46.

CIMAS, Alberich, T., Arnanz, L., Basagoiti, M., Belmonte, R., Bru, P., Espinar, C., García, N., Habegger, S., Heras, P., Hernández, D., Lorenzana, C., Martín, P., Montañés, M., Villasante, T. R., & Tenze, A. (2009). *Manual de Metodologías Participativas*. Observatorio Internacional de Ciudadanía y Medio Ambiente Sostenible.

CMP. (2013). *Open standards for the practice of conservation*. Version 3.0. 47 pp. https://cmp--openstandards.org/wp-content/uploads/2014/03/CMP-OS-V3-0-Final.pdf

Cortés-Capano, G., Toivonen, T., Soutullo, A., Fernández, A., Dimitriadis, C., Garibotto-Carton, G., & Di Minin, E. (2020). Exploring landowners' perceptions, motivations and needs for voluntary conservation in a cultural landscape. *People and Nature, 2*, 840–855.

Criado, F. (1996). Hacia un modelo integrado de investigación y gestión del Patrimonio Histórico: la cadena interpretativa como propuesta. *PH Boletín, 16*, 73–78.

Criado, F., & Barreiro, D. (2013). El Patrimonio era otra cosa. *Estudios Atacameños Arqueología y Antropología Surandinas, 45*, 5–18.

Criado, F., & González Méndez, M. (1994). La Puesta en Valor del Patrimonio arqueológico desde la perspectiva de la arqueología del paisaje. In *Cuadernos de conservación arqueológica: reflexión y debate sobre teoría y práctica* (Vol. III, pp. 55–75). Consejería de Cultura y Medio Ambiente. Instituto del Patrimonio Histórico.

Criado-Boado, F. (1993). Límites y posibilidades de la Arqueología del Paisaje. *Spal, 2*, 9–55.

Criado-Boado, F., Gianotti, C., & López Mazz, J. M. (2006). Arqueología aplicada al Patrimonio Cultural: la cooperación científica entre Galicia y Uruguay. In G. Muñoz & C. Vidal (Eds.), *Actas del II Congreso Internacional de Patrimonio Cultural y Cooperación al desarrollo* (pp. 165–186). Universidad Politécnica de Valencia.

Cuesta, A., Dimuro, J., Gianotti, C., & y M. Muttoni. (2009). De la investigación a la construcción participativa del patrimonio. Un programa de educación patrimonial y divulgación de la cultura científica en Uruguay. *ARKEOS, 4*(11), 1–15.

Curbelo, C. (1999). Análisis del uso del espacio en San Francisco de Borja del Yí (Depto. De Florida, Uruguay). In A. Zarankin & F. Acuto (Eds.), *Sed Non Satiata. Teoría Social en la Arqueología Latinoamericana Contemporánea* (pp. 97–116). Del Tridente.

del Puerto, L. (2015). *Interrelaciones humano-ambientales durante el Holoceno tardío en el este del Uruguay: cambio climático y dinámica cultural*. PhD thesis, PEDECIBA, Facultad de Ciencias, Universidad de la República.

del Puerto, L., Gianotti, C., & Inda, H. (2016). Gestión del medio y producción de recursos en las tierras bajas del Noreste de Uruguay: análisis paleoetnobotánico del sitio Pago Lindo. *Cadernos do LEPAARQ, XIII*(25), 198–222.

del Puerto, L., Gianotti, C., Bortolotto N., Gazzán, N., Cancela, C., Orrego, B., & Inda, H. (2021). Geoarchaeological signatures of anthropogenic soils in Southeastern Uruguay: Approaches to formation processes and spatial-temporal variability. Geoarchaeology 37:180–197.

del Puerto, L., Inda, H., Gianotti, C. Fagúndez, C., Suárez, D., Rivas, M., Bortolotto, N., & Leal, A. (2023). Pre-Columbian mounds harbor distinctive forest communities in the Southern Campos of American Pampas. *Human Ecology*. https://doi.org/10.1007/s10745-023-00389-x

Descola, P. (2012). *Más allá de la naturaleza y de la cultura*. Ed. Amorrortu.

DIEA-MGAP. (2019). *Anuario Estadístico Agropecuario*. 255 p. Retrieved from: https://descargas.mgap.gub.uy/DIEA/Anuarios/Anuario2019/Anuario2019.pdf

Do Carmo, M., Soca, P., Cardozo, G., & Balzarini, M. (2015). Ajuste de Carga animal en base a oferta de Forraje. In L. Allende & M. Piedracueva (Eds.), *Cartilla técnica, aportes al diseño*. Proyecto GFCC-DGDR-MGAP.

Eremites de Oliveira, J. (1995). *Os argonautas Guató: aportes para o conhecimento dos assentamentos e da subsistência dos grupos que se estabeleceram nas áreas inundáveis do Pantanal Matogrossense*. Master's thesis, PUCRS, Instituto de Filosofia e Ciências Humanas.

Eremites, J., & Milheira, R. (2020). Etnoarqueología de dois aterros Guató no Pantanal: Dinâmica construtiva e história de lugares persistentes. *MANA, 26*(3), 1–39.

Erickson, C., & Balee, W. (2006). *Time and complexity in historical ecology. The historical ecology series. Studies in the Neotropical Lowlands*. Columbia University Press.

EROT-RE. (2012). *Estrategias Regionales de Ordenamiento Territorial y Desarrollo Sostenible*. MVOTMA-DINOT.

Escobar, A. (2001). Culture sits in places. *Political Geography, 20*, 139–174.

Escobar, A. (2010). Ecologías Políticas Postconstructivistas. In M. Redclift & G. Woodgate (Eds.), *International handbook of environmental sociology* (2nd ed.). Elgar/Universidad de Carolina del Norte.

Escobar, A. (2018). Habitabilidad y diseño: la interdependencia radical y la terraformatividad de las ciudades. *Astrágalo: Cultura de la Arquitectura y la Ciudad, 25*, 19–44.

Escobar, A. (2019). Transiciones civilizatorias. In A. Kothari, A. Salleh, A. Escobar, F. Demaria, & A. Acosta (Coords.). *PLURIVERSO. Un diccionario para el posdesarrollo* (pp. 458–462). Icaria Editorial S.A.

Fabian, J. (2019). *El tiempo y el otro. Cómo construye su objeto la antropología* (p. 280). Universidad de los Andes, Bogotá: Ediciones Uniandes.

Fernández Cacho, S. (2008). El Patrimonio Arqueológico en el planeamiento medioambiental. In *Patrimonio Arqueológico y Planificación Territorial. Estrategias de gestión para Andalucía (N°143)* (pp. 89–117). Junta de Andalucía y Universidad de Sevilla.

Gazzán, N., (2018). *Análisis de la distribución espacial del material lítico del sitio Pago Lindo (Caraguatá, departamento de Tacuarembó)*. Master's thesis, Facultad de Humanidades y Ciencias de la Educación. Universidad de la República.

Gazzán, N., Cancela-Cereijo, C., Gianotti, C., Fábrega-Álvarez, P., del Puerto, L., & Criado-Boado, F. (2022). From mounds to villages: The social construction of the landscape during the middle and late Holocene in the India Muerta lowlands, Uruguay. *Land, 11*(3), 441.

Gazzano, I., & Achkar, M. (2013). La necesidad de redefinir ambiente en el debate científico actual. *Gestión y Ambiente, 16*(3), 7–15.

Geymonat, J. (2019). La Calera de las Huérfanas. *Revista Raíces, 221*, 26–28.

Gianotti C. (2005 *Cooperación científica, desarrollo metodológico y nuevas tecnologías para la gestión integral del Patrimonio arqueológico en Uruguay*. TAPA 36 (160 p). Instituto de Estudios Galegos Padre Sarmiento (CSIC).
Gianotti, C. (2015). *Paisajes arqueológicos de las tierras bajas del este y norte del Uruguay*. PhD thesis, Facultade de de Xeografía e Historia. Universidad de Santiago de Compostela.
Gianotti, C., & del Puerto, L. (2016). *Diagnóstico patrimonial para el manejo de sitios monticulares en el Padrón 3406, Rincón de la Paja, Lascano* (Technical report) (78 p). LAPPU - CURE, Universidad de la República.
Gianotti, C., Villamarzo, E., Piazza, N., Nin, M., Rodríguez, L., & Lembo, V. (2015). El Paisaje Cultural Laguna de Rocha como objeto focal de conservación: propuesta para su integración dentro del plan de manejo de un área protegida. In M. Medina (Comp.). *Paisaje, patrimonio, proyecto, desarrollo local. Paisajes culturales en Uruguay* (pp. 15–31). CSIC.
Gianotti, C., Gascue, A., del Puerto, L., Inda, H., & Villamarzo, E. (2017). Archaeological heritage on the Atlantic Coast of Uruguay: Heritage policies and challenges for its management in coastal protected areas. In T. Dawson, C. Nimura, E. López-Romero, & M. Y. Daire (Eds.), *Public archaeology and climate change* (pp. 149–161). Oxbow Books.
González, E. M., & Martínez-Lanfranco, J. A. (2010). *Mamíferos de Uruguay. Guía de campo e introducción a su estudio y conservación*. Ediciones de la Banda Oriental. Vida Silvestre & MNHN.
Gorenstein, S., Napal, M., & Olea, M. (2007). Territorios agrarios y realidades rururbanas. Reflexiones sobre el desarrollo rural a partir del caso pampeano bonaerense. *Revista eure, XXXIII*(100), 91–113.
Granizo, T., Molina, M. E., Secaira, E., Herrera, B., Benítez, S., Maldonado, O., Libby, M., Arroyo, P., Ísola, S., & Castro, M. (2006). *Manual de Planificación para la conservación de las Áreas, PCA*. The Nature Conservancy & USAID.
Guber, R. (2001). *La Etnografía. Método, Campo y Reflexividad*. Grupo Editorial Norma.
Hamilakis, Y. (2015). *Arqueología y los sentidos. Experiencia, memoria y afecto*. JAS Arqueología Editorial.
Haraway, D. (2003). *The companion species manifesto* (112 pp). Chicago University Press.
Haydock, K. P., & Shaw, N. H. (1975). The comparative yield method for estimating dry matter yield of pasture. *Australian Journal of Experimental Agriculture and Animal Husbandry, 15*, 663–670.
INC [Instituto Nacional de Colonización]. (2016a). In A. L. Larrosa & V. Camors (Eds.), *Informe sobre la potencial demanda y destino del inmueble n° 735, Lascano, Departamento de Rocha* (Regional Lavalleja. Technical Report). Instituto Nacional de Colonización.
INC [Instituto Nacional de Colonización]. (2016b). *Estudio socio productivo del inmueble N°735 (Lascano, Rocha) Regional Lavalleja* (Technical Report. Expediente N° 2016-70-1-00740). Instituto Nacional de Colonización.
Ingold, T. (2000). *The perception of the environment: Essays on livelihood, dwelling and skill*. Routledge.
Iriarte, J. (2006). Landscape transformation, mounded villages and adopted cultigens: The rise of early formativa communities in South-Eastern Uruguay. *World Archaeology, 38*(4), 644–663.
Jiménez-Esquinas, G. (2016). De "añadir mujeres y agitar" a la despatriarcalización del patrimonio: la crítica patrimonial feminista. *Perspectivas Revista ph, 89*, 137–140.
Kaal, Knop, E., Zoller, L., Ryser, R., Gerpe, C., Hörler, M., Fontaine, C., et al. (2017). Artificial light at night as a new threat to pollination. *Nature Biology Letters, 548*, 206–209.
Kent, M. (1992). *Vegetation description and analysis: A practical approach*. Wiley.
López Mazz, J. (2001). Las estructuras tumulares (cerritos) del litoral atlántico uruguayo. *Latin American Antiquity, 12*(3), 231–255.
López Mazz, J. M. (2013). Early human occupation of Uruguay: Radiocarbon database and archaeological implications. *Quaternary International, 301*, 94–103.
López Mazz, J. M. (2018). Sangre indígena en Uruguay. Memoria y ciudadanías post nacionales. *Athenea Digital. Revista de Pensamiento e Investigación Social, 18*(1), 181–201.

López Mazz, J., Buffa, V., de León, V., & y Cancela, C. (2014). La localidad histórico arqueológica del Río San Salvador (Soriano, Uruguay). *Revista del Museo de Antropología, 7*(2), 285–292.
López Mazz, J., Marín, C., Dabezies, M., & y Tejerizo, C. (2020). Arqueología y memoria de la esclavitud africana en la frontera uruguayo-brasileña: el caso de la Estancia de los Correa (Rocha, Uruguay). *Arqueología, 25*(3), 1–21.
Martin, G. (2001). *Etnobotánica. Manual de Métodos*. Editorial Nordan.
Masterson, V. A., Stedman, R. C., Enqvist, J., Tengö, M., Giusti, M., Wahl, D., & Svedin, U. (2017). The contribution of sense of place to social–ecological systems research: A review and research agenda. *Ecology and Society, 22*(1), 49. https://doi.org/10.5751/ES-08872-220149
Milheira, R., & Gianotti, C. (2020). The earthen mounds (*Cerritos*) of southern Brazil and Uruguay. In C. Smith (Ed.), *Encyclopedia of global archaeology*. Springer Nature. https://doi.org/10.1007/978-3-319-51726-1_3025-1
Modernel, P., Rossing, W. A. H., Corbeels, M., Dogliotti, S., Picasso, V., & Tittonell, P. (2016). Land use change and ecosystem service provision in Pampas and Campos grasslands of southern South America. *Environmental Research Letters, 11*, 11. https://onlinelibrary.wiley.com/doi/abs/10.1111/gfs.12445
Moreno, F. (2014). *La gestión de los recursos animales en la Prehistoria del Este de Uruguay (4000 años AP – Siglo XVI)*. PhD thesis, Departamento de Prehistoria, Universidad Autónoma de Barcelona. Barcelona.
Nami, H., Florines, F., & Toscano, A. (2018). New Paleoindian finds, further fell points data, and technological observations from Uruguay: Implications for the human peopling in Southeastern South America. *Archaeological Discovery, 6*, 21–37.
Narosky, T., & Yzurieta, D. (2015). *Guía para la Identificación de las Aves de Argentina y Uruguay* (15th ed.). Vazquez Mazzini Editores.
OPP. (2015). *Reporte Uruguay 2015* (Technical report). Oficina de Planeamiento y Presupuesto. Retrieved from: https://www.opp.gub.uy/sites/default/files/documentos/2018-10/Reporte%20Uruguay%202015.pdf
Orquera, L., & Piana, E. (1986). *Normas para la descripción de objetos arqueológicos de piedra tallada* (Contribución científica, publicación especial 1). CADIC.
Pintos, S. (2000). Economía Húmeda del Este de Uruguay: el Manejo de Recursos Faunísticos. In A. Durán & R. Bracco (Eds.), *Arqueología de las Tierras Bajas* (pp. 249–270). Imprenta Americana.
Politis, G., & Bonomo, M. (Eds.). (2018). *Goya-Malabrigo: Arqueología de una sociedad indígena del noreste argentino* (Colección Saberes) (368 pp). Editorial UNICEN.
PROBIDES. (1999). *Plan Director. Reserva de Biosfera Bañados del Este/Uruguay* (Technical report) (159 p). PNUD. Retrieved from: https://www.probides.org.uy/reserva-biosfera.php
Prümers, H., & Jaimes, C. (2014). 100 años de investigación arqueológica en los Llanos de Mojos. *Arqueoantropológicas, 4*, 11–54.
Rappaport, J. (2007). Más allá de la escritura: la epistemología de la etnografía en colaboración. *Revista Colombiana de Antropología, 43*, 197–229.
Raymond, C. M., Raymond, C. M., Bieling, C., Fagerholm, N., Martinlopez, B., & Plieninger, T. (2016). The farmer as a landscape steward: Comparing local understandings of landscape stewardship, landscape values, and land management actions and land management actions. *Ambio, 45*, 173–184.
Ríos, C., Lezama, F., Rama, G., Baldi, G., & Baeza, S. (2022). Natural grassland remnants in dynamic agricultural landscapes: Identifying drivers of fragmentation. *Perspectives in Ecology and Conservation, 20*(2022), 205–215.
Risler, J., & Ares, P. (2013). *Manual de mapeo colectivo. Recursos cartográficos críticos para procesos territoriales de creación colaborativa* (80 p). ICONOCLASISTAS. Tinta Limón.
Rivas, M. (2005). Desafíos y alternativas para la conservación in situ de los palmares de Butia capitata (MART.) BECC. *Agrociencia, IX*(1), 161–168.
Rodriguez, M. (2017). Reemergencia indígena en los países del Plata: Los casos de Uruguay y de Argentina. *Conversaciones del Cono Sur, 3*(1), 1–7.

Rodríguez, M., & Michelena, M. (2018). Memorias Charrúas en Uruguay: reflexiones sobre la reemergencia indígena desde una investigación colaborativa. *Revista Abya Yala, 2*(2), 180–210.

Rodríguez-Gallego, L., Nin, M., Suárez, C., & Conde, D. (2012). *Paisaje Protegido Laguna de Rocha, propuesta de plan de manejo* (Management plan. Technical report). Futuro Sustentable Srl.

Rodríguez-Gallego, L., Alfaro, M., Carranza, A., Clavijo, C., De León, G., Franco-Fraguas, P., Lozoya, J. P., Meerhoff, M., & Rivas, M. (2019). *Síntesis de la revisión de antecedentes sobre el estado de los ecosistemas y la biodiversidad en Uruguay y las causas de su degradación* (Technical report for National Environmental Plan for sustainable development, Serie 1) (Vol. 1). CURE-UdelaR.

Roosevelt, A. C. (1991). *Moundbuilders of the Amazon: Geophysical archaeology on Marajó Island Brazil*. Academic.

Rostain, S. (2008). Agricultural earth works on the French Guiana Coast. In H. Silverman & W. Isbell (Eds.), *Handbook of South American archaeology* (pp. 217–233). Springer.

Roura-Expósito, J., Díaz, A., Ruíz-Blanch, A., Sánchez-Carretero, C., & Cortés-Vázquez, J. (2018). Repensando las prácticas académicas: el diseño colaborativo de un proyecto de investigación a partir de la metodología IAP. *Revista de Dialectología y Tradiciones Populares, LXXIII*(2), 407–424.

Sánchez-Carretero, C.; Albadelejo, J.M; Ruíz-Blanch, A.; Roura-Expósito, J. eds. (2019). El imperativo de la participación en la gestión patrimonial. CSIC.

Sans, M., Figueiro, G., & Hidalgo, P. C. (2012). A new mitochondrial C1 lineage from the prehistory of Uruguay: Population genocide, ethnocide and continuity. *Human Biology, 84*(3), 287–305.

Scarlato, G. (2003). *Gestión ambiental de los humedales de la Cuenca de la Laguna Merín 28. Ordenamiento Territorial y desarrollo sostenible. Los problemas en el caso del arroz en Uruguay* (34 pp). CIEDUR. Retrieved from: http://bibliotecavirtual.clacso.org.ar/ar/libros/uruguay/ciedur/scarlato.rtf

Schaan, D. (2008). The nonagricultural chiefdoms of Marajó Island. In H. Silverman & W. Isbell (Eds.), *Handbook of South American archaeology* (pp. 339–357). Springer.

Smith, L. (2006). *Uses of heritage* (351 pp). Routledge/Taylor & Francis Group.

Smith, L. (2011). El "Espejo Patrimonial". ¿Ilusión narcisista o reflexiones múltiples? *Antípoda Revista de Antropología y Arqueología, 12*, 39–36.

SNAP. (2014). *Documento técnico N° 2; Fortalecimiento de la efectividad del Sistema Nacional de Áreas Protegidas incluyendo el enfoque de paisaje en la gestión* (Technical report URU/13/G35). DINAMA-MVOTMA.

Sotelo, M. (2014). Cairnes y vichaderos en las tierras altas de Uruguay. *Revista del Museo de Antropología, 7*(2), 309–316.

Soutullo, A., Clavijo, C., & Martínez-Lanfranco, J. A. (Eds.). (2013). *Especies prioritarias para la conservación en Uruguay. Vertebrados, moluscos continentales y plantas vasculares* (222 pp). SNAP/DINAMA/MVOTMA y DICYT/MEC.

Suárez, R. (2019). High resolution AMS 14C dates for late Pleistocene Fishtail technology from the Tigre site, Uruguay river basin, South America. *Quaternary Science Reviews, 213*(1), 155–161.

Suárez, R., Leigh, D. S., & Trindade, M. (2011). First early human occupation in caves and Rockshelters in Uruguay and diverse landscapes utilized by early South Americans. *Current Research in the Pleistocene, 28*, 170–172.

Suárez-Villagrán, X., & Gianotti, C. (2013). Earthen mound formation in the Uruguayan lowlands (South America): Micromorphological analyses of the Pago Lindo archaeological complex Mound-builders. *Journal of Archaeological Science, 40*(2), 1093–1107.

Taylor, K. (2019). New lives, new landscapes. Landscape, heritage and rural revitalization: Whose cultural values? *Built Heritage, 3*, 50–63.

Triunfo, M., Barbero, S., & Pereira, J. (2018). *Descubriendo nuestro legado indígena: Una mirada hacia la diversidad biocultural de las tierras bajas de India Muerta* (Scientific Report. Taller Interdisciplinario de Tópicos Regionales II, Licenciatura en Gestión Ambiental). CURE-Udelar.

Tuan, Y. F. (2007 [1977]). *Topofilia. Un estudio de las percepciones, actitudes y valores sobre el entorno*. Melusina.

Vassallo, M., & Ferreira Chávez, E. (2014). Colonización y nuevas formas de acceso a la tierra de productores familiares: enseñanzas de la Colonia Maestro Soler en Uruguay. *Revista NERA, 17*(24), 146–166.

Vaughan, L. (2018). Mapping the spatial logic of society. In L. Vaughan (Ed.), *Mapping society: The spatial dimensions of social cartography* (pp. 1–23). UCL Press.

Verdesio, G. (2011). Entre las visiones patrimonialistas y los derechos humanos: Reflexiones sobre restitución y repatriación en Argentina y Uruguay. *Corpus. Archivos virtuales de la alteridad americana, 1*(1), 1–9.

Vienni, B. (2014). *La socialización del conocimiento científico como problema interdisciplinario: el caso del patrimonio arqueológico de Uruguay*. PhD thesis, Universidad de Granada.

Villarmarzo, E. (2018). *Gestión Integral del Patrimonio Arqueológico Costero. Investigación y Extensión en dos casos de estudio: Parque Nacional Cabo Polonio y Paisaje Protegido Laguna de Rocha*. PhD thesis, Facultad de Ciencias Sociales, UNICEN.

Walker, J. (2008). The Llanos de Mojos. In H. Silverman & W. Isbell (Eds.), *Handbook of South American archaeology* (pp. 927–940). Springer.

Yang, C., & Han, F. (2020). A digital information system for cultural landscapes: The case of Slender West Lake scenic area in Yangzhou, China. *Built Heritage, 4*(1), 3.

Chapter 3
Through Zo'é Paths and Capoeiras: Networks of Relations in the Guianas and the "Sparse Population" Question

Fabio Augusto Nogueira Ribeiro, Claide de Paula Moraes, and Raoni Bernardo Maranhão Valle

Abstract This chapter presents initial results of anthropological and archaeological research about the history of indigenous occupation in the region of the Brazilian Amazon currently inhabited by the Zo'é. Through information collected in two journeys to the Erepecuru-Cuminapanema interfluve, we explore what we understand to be points of convergence between the ethnology of the Guianas, the archaeological perspective of deep history, and some assumptions of Historical Ecology. By means of this approximation, our objective is to continue deconstructing some of the equivocate images around Zo'é sociality, as well as engage in a critical dialogue with interpretations surrounding human occupation of Amazonia. In particular, we argue that accounts about events and trajectories of important people, material evidence found during these journeys, and the Zo'é mode of occupation of the interfluves, point to the persistence of extensive networks of relations in space and time.

Keywords Brazilian Amazon · Zo'é · Oral history · Community identity · Material culture

An initial version of this work was presented in the IV International Meeting of Amazonian Archaeology (EIAA), hosted in the city of Trinidad, Bolivia, in October 2016.

F. A. N. Ribeiro (✉)
Center for Amerindian Studies (Cesta), University of São Paulo/USP, São Paulo, Brazil
e-mail: fabio@povosisolados.org

C. d. P. Moraes · R. B. M. Valle
Anthropology and Archaeology Program/Federal University of Western Para (PAA/UFOPA), Santarém, Brazil
e-mail: claide.moraes@ufopa.edu.b; raoni.valle@ufopa.edu.br

A. C. Colonese, R. G. Milheira (eds.), *Historical Ecology and Landscape Archaeology in Lowland South America*, Interdisciplinary Contributions to Archaeology, https://doi.org/10.1007/978-3-031-32284-6_3

3.1 Introduction: The Zo'é and the Erepecuru-Cuminapanema Interfluve

The Zo'é[1] are speakers of a Tupi-Guarani language and live in northern Pará state in an area of forest situated between the Erepecuru (or Paru do Oeste) river and the Cuminapanema river, both sub-affluents of the left margin of the Amazon river. Together with the Wajãpi and Emerillon, they are the only speakers of Tupi-Guarani languages currently living in the western portion of the region known as the "Guianas"[2] in Americanist literature, an area that is principally occupied by Carib-speaking peoples. Today, they live in the Zo'é Indigenous Land (*Terra Indígena Zo'é*, henceforth TIZ), an area officially recognized by the Brazilian state in 2009 that measures 6865 km^2 in area and houses some 320 people organized in different groups (*wan*) that engage in high residential mobility.

While references to the presence of the Zo'é in this area are scarce in the documental sources, information from the oral history of this group (Gallois & Havt, 1998), as well as those about the Carib occupation in the Erepecuru basin (Nimuendaju, 1944; Frikel, 1958) indicate that the Zo'é came to occupy the Erepecuru-Cuminapanema interfluve relatively recently, probably during the colonial period. In this sense, although the available information is insufficient to be able to define their migration routes and the circumstances that led them to their current territory, Gallois and Havt (1998) raised the hypothesis that the Zo'é, like the Wayãpi, are the descendants of Tupi-Guarani-speaking groups of Western Amazonia. Settled in religious missions on the left bank of the Amazon river during the seventeenth and eighteenth centuries, at some point in colonial history, these peoples (or perhaps a fraction of them, or new collectives) would have taken refuge in the large region of forest intersected by the Cuminapanema and Erepecuru rivers and other affluents of the Amazon.

In this movement, it is certain that people that identified (or came to identify themselves) as Zo'é entered into relations with non-Zo'é peoples, classified as *apam, tapy'yj* and *kirahi,* the last term being used mainly for non-indigenous people. Bearers of diverse characters, these different kinds of people ended up modulating both the territorial movements and the very constitution of the Zo'é collective. In the oral literature there are abundant stories of ancient alliances, parties, wars, exchanges, kidnapping and revenge. These same accounts indicate that the Zo'é became progressively deprived of amicable relations and, in the not-too-distant past

[1] The Brazilian state considers the Zo'é to be an indigenous of "recent contact", in agreement with the classification of the Fundação Nacional do Índio (Funai). Since 2011, a series of research activities and indigenist actions have been developed with the aim to promote greater autonomy and control for the Zo'é over their own territory and their relationships with non-Zo'é peoples (both the State and neighboring indigenous groups).

[2] The Guianas correspond roughly to the northeast region of South America, delimited to the south by the Amazon river, to the west by the Negro-Orinoco rivers, and to the northeast by the Atlantic. The focus of the present work is in the eastern portion of the "Guianas", close to the border between Brazil and Suriname.

(not traceable genealogically), opted to confine their movements to the headwaters of rivers, far away from large river courses. The absence of references to the Zo'é in the regional ethnographic and historiographic literature written before their "official contact" in the 1980's, can be taken as indicative of their relative (rather than absolute) isolation, given that these stories run throughout Zo'é accounts of past events, and that these relations are at the base of what make the Zo'é a differentiated collective.

This scenario of relative "isolation" lasted until 1987, when fundamentalist missionaries from the New Tribes Mission (NTM) established permanent "contact" with the Zo'é. Due to the many subsequent deaths of Zo'é people as a result of infectious diseases introduced by the missionaries, in 1991 Funai removed the missions from the indigenous territory (Gallois & Havt, 1998). Since then, the "Tupi of Cuminapanema" began to be depicted as one of the last "isolated" indigenous peoples on the planet to maintain a "pristine" and culturally "untouched" way of life. Although criticized through ethnography (see Gallois, 2015), these equivocate images about the Zo'é continue to be reiterated in the present.[3]

By presenting a discussion based upon shared approximations between ethnology, history and archaeology in the South American lowlands (Viveiros de Castro, 2002; Neves, 2012; Gallois, 2016), we seek to undo these images of the Zo'é as a "Stone Age", "semi-nomadic" people that live in a "paradise" surrounded by a "see of Christianized Karibs". The ethnography of Havt (2001) is an excellent description of how Zo'é construction of territory (*tekoha*), built through movements of concentration and dispersion, is far from a system of wandering "semi-nomadism", and is modulated by factors that are ecological (seasonal availability of resources), socio-political (maintaining good space between local groups; trajectories of important chiefs), and cosmological (relations of alterity, deaths, etc.), while also presenting clear evidence for a pendular model of "walking in one's own footsteps". Furthermore, Ribeiro (2020) shows how contemporary relations between the Zo'é and neighboring indigenous peoples make little sense if read in the evolutionist (or acculturative) sense of "Isolated Indians x Christianized Indians", and that the context of ancient and extensive gift exchange networks must be considered.

Rather than think of the Zo'é as the "last isolated groups" or as victims of (American) nature and/or (colonial European) history, we address the manner in which the Zo'é inhabit the Erepecuru-Cuminapanema interfluve from a long-term historical perspective (Neves, 2012; Heckenberger, 2005). The current chapter seeks to enrich the dialogue between these disciplines and Zo'é knowledge using

[3] In 2008, a journalist writing for The Sunday Telegraph affirmed that "[...] the Zo'é live a Stone Age existence". In the same way, an article published in a widely-circulate book entitled *Povos Indígenas no Brasil – 2006/2010* ("Indigenous People of Brazil – 2006/2010"), Cartagenes (2011: 318) characterized the Zo'é as "the instigating, autonomous Tupi encrusted on the borders of a sea of Christianized Karibs" (author's translation from Portuguese). In 2013. In an article published by *Trip* magazine, the photographer Rogério Assis characterized the Zo'é as a "semi-nomadic" people. Finally, in 2020, renowned photographer Sebastião Salgado, who travelled to the TIZ in 2009, stated, "I took a journey to paradise" ("*Fiz uma viagem ao paraíso*") in a report published in the Folha de São Paulo newspaper.

material collected during two fieldwork seasons carried out with these people. We aim to explore what we understand to be some points of convergence between anthropological perspectives of networks of relations in the Guianas (Gallois, 2005; Havt, 2001; Perrone-Moisés, 2006; Ribeiro, 2020), archaeological perspectives of deep history, and some assumptions of Historical Ecology (Balée & Meredith, 2020; Balée & Erickson, 2006; Balée, 1998). Independently of their (recent) occupation of the interfluves, the Zo'é mode of existence is tied to the continuity of millennial-scale ecological and political processes, of which they are both heirs and constructors/innovators.

We believe this via permits us to think about the chronological depth of such networks and refute the idea that Amazonian interfluves are hostile refuges containing scant resources that pre-Columbian populations merely "passed through" before European invasion. In particular, we offer a criticism of relatively recent viewpoints, based on paleoecological indicators, that reiterate this "sparse population" model (Piperno et al., 2015, 2017, 2021; McMichael et al., 2012, 2014, 2017; Bush et al., 2015). We demonstrate that historic, ethnographic and archaeological data (both regional and collected in the TIZ), as well as indigenous accounts, go against the notion that the *terra firme* was disconnected from the *várzea*, as well as the idea that denser *terra firme* occupations only occurred as a result of colonialism.

3.2 Two Journeys with the Zo'é

In 2013 and 2015, we made two journeys with the Zo'é through the Erepecuru-Cuminapanema interfluve. Our objective was to reach some Zo'é *capoeiras*[4] (*taperet*) and register what the Zo'é had to say about them. On the first journey, we visited the region of Iwi'ara, an old Zo'é village situated in the Kare river basin (affluent of the Cuminapanema). On the second, aside from visiting diverse *capoeiras* in the Pupuruni stream (affluent of the Erepecuru) region, we also traveled to the *Cachoeira da Paciência* (Patience Waterfall), located in the extreme north of the TIZ (Fig. 3.1).

We had the privilege to travel with remarkable people. On the first two journeys we were accompanied by Kwai, a man with expert knowledge who lived most of his life in villages in the Pupuruni basin, and his wife Terõ, daughter of Kisiriguhu, one of the great chiefs of Iwi'ara. Apart from this couple, we were accompanied on both journeys by other experts (like Namihu and Sarará), important chiefs (like Simirã, Toke and Tereren) and their respective families. These journeys were especially rich because, apart from the men and women, we were also accompanied by elders and children. Our camps in these journeys were, in fact, small, transitory villages.

[4] Capoeiras (em Zo'é: *taperet*) are places of old villages. Composed largely by of shrubs, grasses and lianas as well as domesticated plants.

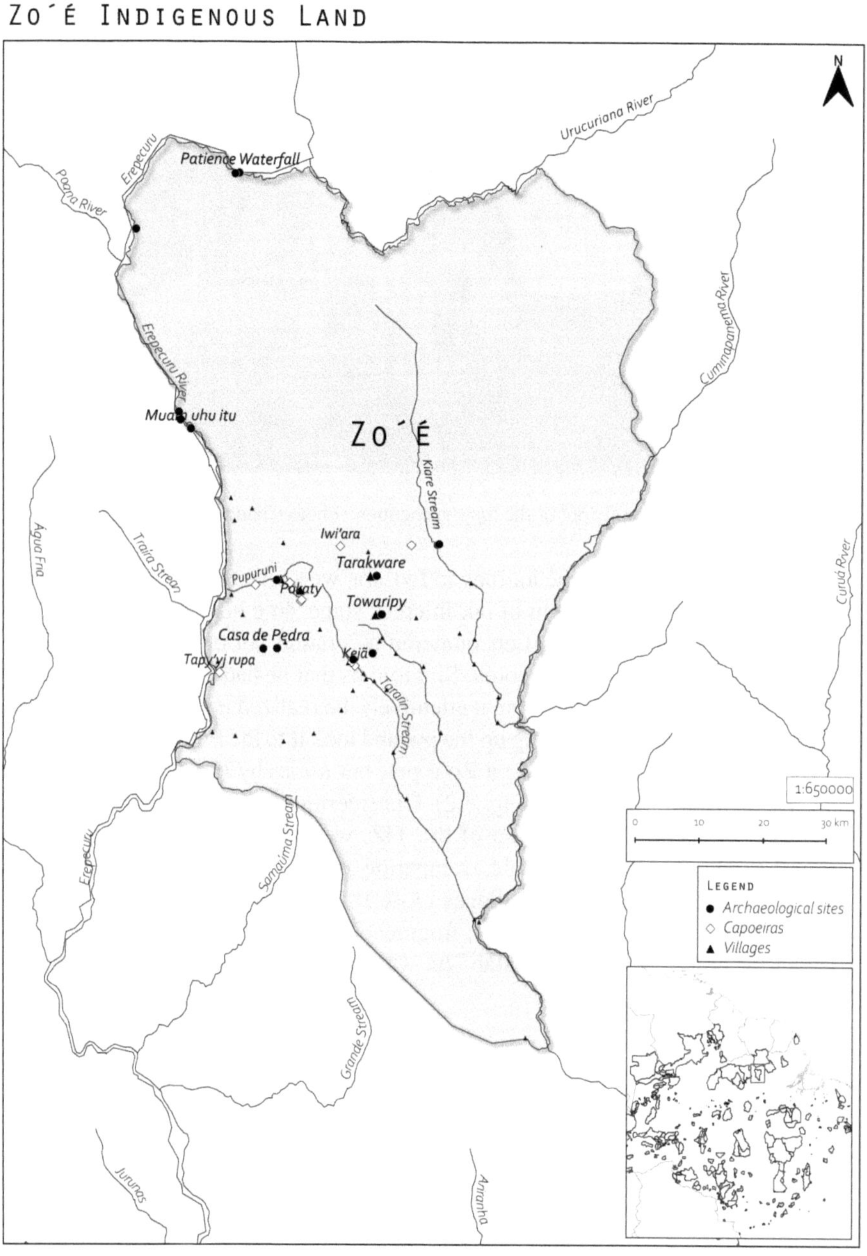

Fig. 3.1 Zo'é Indigenous Land highlighting the current villages and some of the evidence of ancient occupations. (Map by Guilherme Siviero)

Fig. 3.2 Siju and the ceramic pot of the *tapy'yj* enemies. (Photo Claide Moraes/2013)

In Kejã,[5] the day before the journey to Iwi'ara, we spoke to various people about the objective of the trip. In light of our interest, some Zo'é informed us that several years ago, the shaman Siju had encountered a ceramic pot on the margins of the Sarapejuk stream. In his own account, Siju told us that he had tripped on something resembling a stone. After observing it attentively, he realized it was an upside-down, semi-buried ceramic pot. Siju dug up the pot and took it to the Funai post. According to him, its form did not resemble a Zo'é pot, but a *tapy'yj tapimã* – that is, a pot made by their *tapy'yi* enemies (Fig. 3.2). Considering that the Sarapejuk stream is located in the central-south region of the TIZ, we realized that this was the first material evidence for recent Zo'é occupation of the Erepecuru-Cuminapanema interfluve. During visits to the villages close to the Funai base, we were able to confirm the presence of *terras pretas*, fragments of old ceramics, as well as a large complex of polishers fixed on a slab (*Itu* waterfall), on the way to the villages of Poda'y and Tahiripa.

3.2.1 Journey to Iwi'ara

Located on the margins of the Bureaka stream, affluent of the Kare river, Iwi'ara was one of the Zo'é villages occupied during the period before permanent contact with the *kirahi* (Gallois & Havt, 1998). The Kare river constitutes an important frontier in the geography of Zo'é relations, since *Kare rowaj*, or "the other side of the Kare", is considered the territory of their *tapy'yj* enemies. Located in a transition

[5]An old Zo'é village situated in the Tararin river basin (affluent of the right margin of the Cuminapanema), Kejã is the place in the TIZ where the FUNAI post is situated.

zone between tall forest (*ka'a uate*) and savanna (*jun*), Iwi'ara was abandoned at the end of the 1990's in function of the death of chief Sarakut and is now covered in *capoeira* vegetation. Iwi'ara seemed to us an ideal place to begin researching the history of indigenous occupation, since it represents a frontier zone for the Zo'é and was occupied for a long period of time. Among the Zo'é that willingly participated in the journey, Kwai and Terõ were the only ones that knew how to get to Iwi'ara. Most did, however, remember the region of the Bureaka headwaters as being abundant in animals – principally those that thrive in forest/transition areas – as well as the large houses and *sepy*[6] made by Sarakut, Kisiriguhu and other chiefs of Iwi'ara. Thus, from the point of view of the Zo'é, the journey was an opportunity to hunt and to revisit (or, in the case of the young people, get to know) an old, big village.

After 2 days of walking from Kejã, observing signals, remains and histories (and staying in the villages of Tarakware and Barawa), we arrived in Iwi'ara around midday. While part of the Zo'é stayed to set up camp at the margins of the Bureaka, we went to visit the old village in the company of the others. In Iwi'ara, the quantity of wooden posts (*kita aret*), rubbish heaps (*ity*) and domesticated plants (principally açaí, cashew, jenipapo and urucum) we encountered fit the information provided by the Zo'é that the village had been occupied for a long period of time and by a considerable number of people. Kwai and Terõ pointed to the locations of houses and recalled the many villagers of Iwi'ara, as well as the names of the owners of the cashew trees, urucum bushes and communal houses. They quickly found several other vestiges of human occupation: ceramic fragments, stones used to break Brazil nut seed cases, animal bones, etc.

While we wandered about the place, the young woman Byby announced she had found a stone axe blade (*jy uhu*) in one of the rubbish heaps (Fig. 3.3). Kwai then proceeded to make a long account on the subject. According to him, the Zo'é do not know how to manufacture stone blades, but they find them buried in diverse places within their territory. The Zo'é attribute the technique of making stone axe blades to *Tupã*, a demiurge associated with thunder.[7] To make them, Tupã manipulated the stones (*te*) in the same way that women manipulate clay (*jyk*) to fabricate ceramic vessels. During a celestial walk, Tupã let the blades fall to the earth because his basket had holes in it (for another version of this episode, see Gallois, 2016). It was during the time of the late Sihet[8] that a woman called Sijã found the first stone blade while digging up yams in her garden in the village of Koporuhu. The object was

[6] Party in which a fermented drink of the same name is consumed.

[7] Alfred Métraux, mentioning the work of the priest Fernão Cardim, notes that the Tupinambá of the sixteenth century associated Tupã with "tools to open fields", similar to the association made by the Zo'é between Tupã and stone axe blades: "Cardim assures that the Indians intended to have received from Toupan the wood used to dig and the plants they cultivate" (Métraux, [1928] 1979: 11).

[8] According to Kwai, Sihet was a Zo'é man that lived in the region of the Pupuruni stream, particularly in the locale called Jawaraty (lit. "place of the jaguars"), where Sihet is buried. The Zo'é attribute practically all of their current specific knowledge to Sihet, including knowledge about historic events, songs and parties, about foods, different beings, curing practices and other technical knowledge (for instance, the use of the lip adornment *embe'pot*).

Fig. 3.3 Kwai and Byby examine a stone axe blade found in Iwi'ara. (Photo Fabio Ribeiro/2013)

called *Tupã ruwai*, "the penis of Tupã", due to its phallic form. After much examination, the blade was washed and hafted, and with it the Zo'é began to cut wood and large trees.

These and other blades found afterwards were sharpened on stone slabs on the margins of river courses, on the many stools of the "grandfather-vulture" (*uruwu ramuj*) made by the *Bahit*, small subaquatic beings with human form that are considered by the Zo'é as the owners of the stones. It was only with this sudden technological discovery that the Zo'é began to open larger fields in the forest. Before this event, small fields of manioc, banana, potato and yam were opened using a slower process by which small- and medium-sized trees were felled with the use of wooden lances and fire. These days, even with metal axes available, the Zo'é continue to use small stone axes for some activities like, for example, the extraction of fibers from the bark of the Brazil nut tree, which they make into hammocks.

While we walked around the area of the old village and the adjacent savanna, the other part of the group had stayed to make camp (*tapaby*) on the bank of the Bureaka stream. When we returned, we were surprised by the speed in which the Zo'é had transformed that small area of forest.[9]

[9]An area of approximately 10 × 10 m was prepared for the camp. Its preparation consisted in the removal of the leaf litter layer, understory vegetation, shrubs and small trees, which were accumulated in heaps that encircled the clean area. Fire was not used in any of the stages of camp preparation. Some posts were erected by hitting the pointed stakes into the soil. After putting up the hammocks and a partial cover, an intense activity of processing of the collected vegetables and hunted animals took place. Three large fires were maintained to smoke the game. Other smaller fires were lit to cook or to prepare or fix guns and tools. At night, braziers from the fires were moved close to the hammocks to combat the cold that sets in in the morning. These activities were repeated during the 3 days we stayed there.

During these days, after mealtimes, Kwai and Namihu gave accounts about Iwi'ara and, especially, Kunamiju. As noted by Gallois and Havt (1998) and Havt (2001), the Zo'é associate the Iwi'ara region to the story of Kunamiju – a man who escaped with his family to the other side of the Kare river (*Kare rowaj*) after a conflict with the other Zo'é. For a certain period, Kunamiju's people (*Kunamiju iwan*) maintained contact with those who stayed in Iwi'ara; however, they eventually ended up far from the village and contact with them was lost.

During the other 2 days in which we stayed in Iwi'ara, we made detailed records of the site and the activities carried out by the Zo'é. On one of those days, while some men were busy hunting two peccaries that were later roasted in *moquens* (smokers), the women made carrying baskets with açaí leaves in the same area where the Iwi'ara houses once stood (Fig. 3.4). Kwai and Kuru located the cemetery (*takuje*) and Kwai gave two long accounts: one about Zo'é funerary practices, and the other about the history of the split of the Iwi'ara group. With the help of some of the young people, we also proceeded with the localization of several other rubbish heaps and with the georeferencing of the site.

Fig. 3.4 Jipiho, your son and Byby putting açaí in a carrying basket fabricated *in situ*. (Photo Claide Moraes/2013)

On the third day, once the camp had been closed and we were ready to continue the journey, we drew a plan (Fig. 3.5) of the dismantled camp, trying to plot evidence of the activities that had taken place there and of the principal transformations of the landscape. The rubbish heaps now contained some charcoal from the fires that had been cleaned, as well as seeds of açaí, peppers, urucum, cotton and Brazil nut, and processed animal bones, feathers and fish scales. No artefact, entire or fragmented, remained at the locale, except for small pieces of arrows and vegetable fibres.

While the visual aspect of the Iwi'ara is currently that of a forest in regeneration, in an aerial photo of the locale (Fig. 3.6), one can see that this regeneration was influenced by the activities that had occurred there over a longer timescale. The concentrations of açaí palms, for instance, practically define the limits of the *capoeira*.

After 3 days in Iwi'ara, we returned to our journey, visiting Mytunem, Bosua dypyke, Tarakware and Towaripy along the way. At the time, Mytunem was a small fishing camp made by the Barawa local group, situated close to the mouth of the Bureaka stream. Bosua dypyke, located on the confluence of the Bureaka and Kare river was a small field in 2013, and is today (2021) a village occupied by Namihu and his children.[10] From there, we continued to the village of Tarakware, also Namihu's. In this village we found areas *of terra preta* and more white quartz flakes,

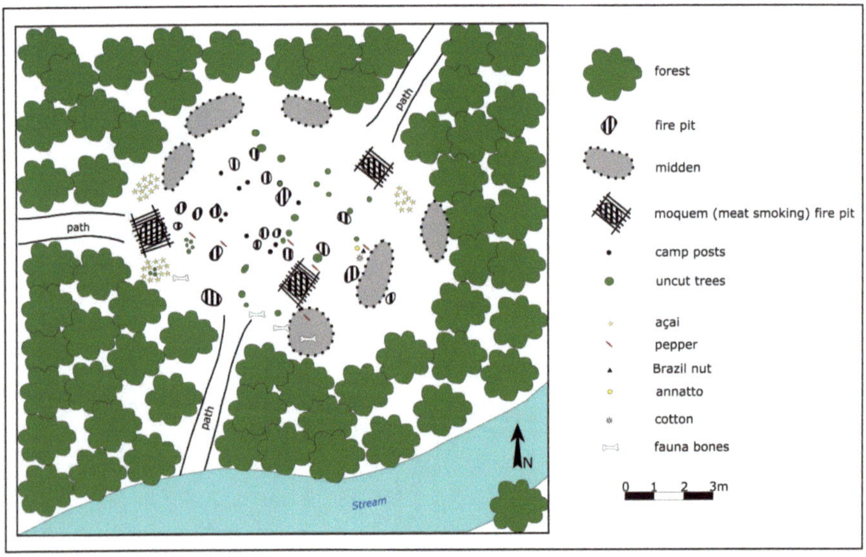

Fig. 3.5 Plan of the Iwi'ara camp. (Claide Moraes/2021)

[10] The locale is also situated in an ecotonal zone since upstream of the Bureaka mouth the Kare runs through the savanna, as can be observed in satellite images of the TIZ. By the quantity of lithic workshops (we found several grinders and flakes made from milky quartz) and Zo'é oral accounts, we can infer that the area has a long history of Zo'é and non-Zo'é indigenous occupation.

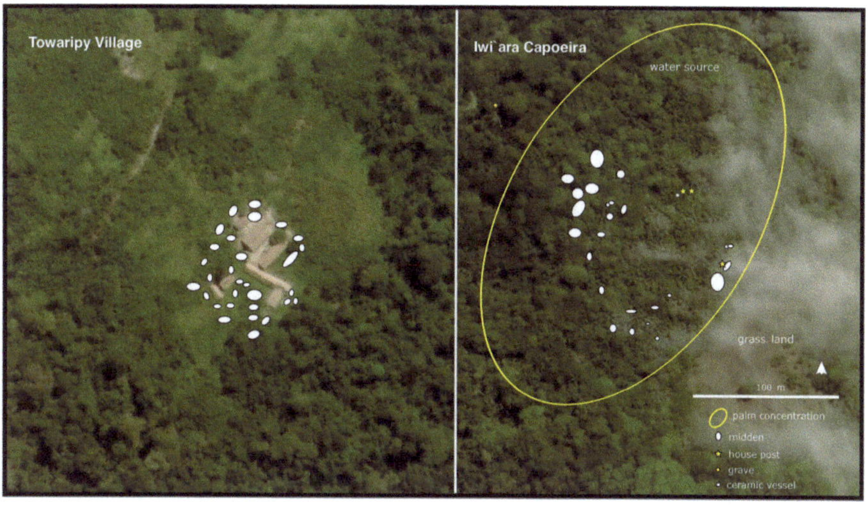

Fig. 3.6 Aerial view of the Towaripy e Iwi'ara locations showing concentrations of palm trees in the area of the old village. (Image of MapsApple and drawing by Claide Moraes/2021)

the latter being present on the terrace of the water course present in the locale. Kwai and Namihu did not associate the flakes or the *terra preta* with the activities of the Zo'é. We asked Namihu (chief of the Tarakware local group who, in the past, decided to open a field and build a house there) what factors led him to choose that specific place to live. He presented three reasons: the presence of water (*y*); the presence of *terra preta* (*ywy byk*) and its proximity to the Barawa local group (*Barawa iwan*), with whom Namihu maintained an alliance. This answer offered us an important clue to understanding Zo'é settlement patterns: the choice of living place is conditioned by factors that are geographic and environmental (water, *terra preta*, resource availability), as well as political (marriages, alliances).

From Tarakware we went to Towaripy in the company of Kuru. After arriving there in the morning, and Kuru receiving us in his house (Fig. 3.7), we conducted a survey of the village with him: quantity and quality of rubbish heaps, sizes of the houses, field system, location of *terras pretas*, distribution of artefacts, variety of domesticated plants, etc. Since Towaripy was one of the biggest Zo'é settlements at that time, we decided to adopt it as a reference for ethnographic and archaeological comparisons with Iwi'ara.

3.2.2 *Journey to Erepecuru*

In 2015, we traveled to the region of the Erepecuru river in the company of several Zo'é. We left from Kejã in the direction of the village of Pokaty (see Fig. 3.1), situated in the Pupuruni river basin (affluent of the left margin of the Erepecuru river).

Fig. 3.7 Communal house of Towaripy. (Photo Claide Moraes/2013)

On the following day, on the trail between Pokaty and Towari abura rupa, Kwai showed us the following places: Pokaty taperet; Kirahi taperet; Wajkyt jukeha rupa and Dybusi taperet. As Kwai explained, in the past the first three places had been stage to a conflict involving the Zo'é, the *kirahi* – then allied with the Zo'é – and the *apam* enemies.

It is important to mention that during this walk through *capoeiras*, specifically when we were in Pokaty taperet (which judging by its characteristics was an area occupied for a long time), Kwai and Terõ recognized the location of one of the houses based upon the presence of an enormous soursop tree (*rasiku*) that, according to them, was planted by a man named Tihekururum. They also recognized the "place where Wajkyt was murdered" by the presence of a hog plum tree (*tapery*). Terõ said, "*Apam tapery aret*" (lit: "the hog plum that belonged to the Apam"). It was an interesting experience to observe how the Zo'é establish temporal markers from elements in the landscape. When we were close to Towari abyra rupa, Kwai showed us Dibusi taperet. Soon afterwards, at the end of the afternoon we arrived in Towari abura rupa, the village of San.

The following day, when on the trail that connects Towari abyra rupa to Pupuruni hemba, Kwai took us to a place called *Wajkyt re poroke*. He explained to us that *Wajkyt jukeha rupa,* the *capoeira* we had visited the day before, was where the *apam* had killed Wajkyt, while *Wajkyt re poroke* was the place where they disemboweled him. This locale is an immense slab situated on the riverbed of the Pupuruni stream. While walking by the slab we found a series of marks on the stones which, to an archaeologist, would be considered a "lithic workshop", a strong indicator that the past inhabitants of this place used the slab to fabricate or sharpen stone axe blades. Kwai, however, in analyzing the marks (Fig. 3.8), said that they were made

Fig. 3.8 Kwai analyzes marks of the Bahit on a stone slab in the Pupuruni steam. (Photo Fabio Ribeiro/2015)

by *Bahit*, small beings of human form (*awa kõ rapijet;* "identical to humans"), the size of children, that live in the depths of the water. The *Bahit* made marks on the stones in the past and continue doing so. For example, at the same moment we were talking about *Bahit* in *Wajkyt re poroke*, San commented that one of the marks on the slab had been made by the *Bahit* during last year's rainy season.

We arrived in Pupuruni hemba at the beginning of the afternoon. We realized that, in the period between 2012 and 2015, what was a temporary camp had become an enormous field with a large house in the center, owned by Toke.

We went up the Erepecuru to the Cachoeira da Paciência waterfall. During the journey we identified the site of Cajuaçú, registered by Frikel and classified by Hilbert (1982) as a Koriabo site; as well as several rock engravings (Fig. 3.9) upon various outcrops, principally in the area of the waterfall where Coudreau (1901) and Rondon (1953) also identified engravings. Upon seeing them, Kwai commented "*Bahit ate te kusiwet*" ("it was *Bahit* that drew on the stones"). However, in this same conversation, Kwai told us that the Zo'é did not have knowledge of these rock engravings. "The late Sihet did not speak of engravings written on the stones. The

Fig. 3.9 Rock engraving in Patience Waterfall. (Photo Raoni Valle/2015)

Zo'é know only about the marks of the *Bahit*". During the journey, we understood that the Zo'é consider both marks caused by severe weather and rain action, as well as lithic workshops and rock engravings, to be marks of the *Bahit* (e.g., Valle et al., 2018). On one of the slabs where we found an anthropomorphic figure, the Zo'é mentioned that it was the image of the body of the *Bahit* (*Bahit tepot rangue*).

In one of our nightly conversations, Sarará commented that there was an *apam* capoeira (*apam taperet*) in the proximity of Wikihe, the place where the late *Duby uhu* used to live.

> In Wikihe, there is a waterfall on the way, it's there that there is an old *apam* path. I saw the stone, just like a house, and it's there that the apam used to stay. In that time Towari abura rupa already existed. The *apam* dug a lot, and made a house. *Apam* sleep underneath the stone. There are pot sherds there. A lot. Only Wikihe contains capoeira of the *apam*. I think it was during that time that they killed Duby uhu.

On the return trip from Cachoeira da Paciência, Sarará took us to this place. After descending the Erepecuru, we arrived in the village *Tapy'yj rupa* ("living place of tapy'yj") where the trail to Wikihe begins. This place, situated on the margin of the Erepecuru, is one of the villages of Simirã's people and was once a *quilombo*. In other words, Simirã opened a village on top of an old Afro-American village (*quilombo*). Going by the map presented in Cruls ([1930] 1973), it is probably the "old *mocambo* of Santana" mentioned by Padre Nicolino (1946). On this day, the Zo'é even told us that on the other side of the Erepecuru river, exactly in front of this place, there are lime trees that once belonged to the quilombolas (*kirahi byk limão aret;* "the lime of the black *kirahi*"). On the day we passed through this village, we found ceramic fragments with *caraipé* temper on the bank of the Erepecuru. The Zo'é do not use *caraipé* temper, reason for which they doubted whether it was *apam*, *tapy'yj* or quilombola pottery.

The next day, we started our walk towards Wikihe. About halfway there, when we arrived at the waterfall mentioned by Sarará, we took another trail which, an hour later, led us to the rockshelter where Sarará said the *apam* that killed *Duby uhu* had lived. Many ceramic fragments and flaked quartz were visible on the ground surface. On that day, we slept in Wikihe and on the next we returned to Kejã, marking the end of our expedition.

Up to this moment, we have registered 38 archaeological sites in the TIZ (Fig. 3.1 and Table 3.1), and the Zo'é know of a lot more places that we have not yet visited. There is a large diversity of sites: rockshelters (1), lithic workshops on rocky outcrops (9), rock engravings (11), ceramic sites with *terra preta* (13) and *capoeiras* of old Zo'é villages (4). Other points with permanent marks are also recognized as locales frequented by other peoples with which the Zo'é maintained relations (*quilombolas*, *apam* and *tapy'yj*). These connections show that what we classify as archaeological sites are persistent and important places for the mode in which the Zo'é construct their territory (*tekoha*) in diverse areas of the Erepecuru-Cuminapanema interfluve.

3.3 Following Connections and Trajectories

In this section, we present a deeper discussion of some aspects of Kunamiju's trajectory and the story of Wajkyt's death in light of the ethnology of the Guianas. Other interesting connections, such as those concerning the stone axe blades and the *Bahit* will be addressed in future works.

3.3.1 Networks in Movement in the Guianas

Despite the evidence of the continuity of exchange networks through time and space in a vast region of the Guianas (see Farage, 1991; Gallois, 1986; Porro, 1985), the first modern, regional ethnological syntheses (Rivière, 1984; Overing, 1983/1984) portrayed these indigenous peoples as formed by dispersed groups characterized by

Table 3.1 Archaeological sites mapped in Zo'é Indigenous Land

Site	Village	Terra preta	Midden	Pottery	Lithic	Rock art	Polisher	Capoeira	Rock shelter	Lat.	Long.
Japukei	X	X	X	X				X		0°14′34.44″S	55°55′6.41″W
Kirahi byk taperet		X	X	X				X		0°14′8.40″S	55°54′50.95″W
Wajkyty taperet		X	X	X				X		0°13′15.18″S	55°55′34.81″W
Pokwatv Taperet		X	X	X				X		0°14′12.41″S	55°54′55.76″W
Dibusi taperet		X		X						0°12′45.82″S	55°55′50.02″W
Towariabyrupa	X	X	X	X				X		0°12′27.00″S	55°56′35.60″W
Apam taperet		X	X	X						0°12′55.58″S	55°58′43.89″W
Waikyt Reproke							X			0°12′31.51″S	55°56′55.14″W
Cachoeira Cajuaçú-Muam uhu itu		X		X		X	X			0°0′7.18″S	56°4′21.75″W
Bahira Kusiwet – antropomorfo grande						X	X			0°01′15.5″N	56°5′15.9″W
Cachoeira Paciência – complexo Bahira Kusiwet					X	X				0°20′56.77″N	56°0′10.16″W
Paciência -Polidor						X				0°20′56.45″N	56°0′28.47″W
Bahira Kusiwet meio do rio						X				0°16′17.09″N	56°8′47.17″W
Bahira Kusiwet 3 figuras						X				0°1′15.70″N	56°5′26.3″W
Bahira Kusiwet geométrico						X				0°0′33.59″N	56°5′5.42″W
Pedra Gravada						X				0°30′57.81″S	56°6′49.03″W
Casa de Pedra				X	X				X	0°18′6.50″S	55°58′7.01″W
Itu polidor							X			0°18′7.81″S	55°56′54.20″W
Tapy'yj rupa				X						0°20′6.15″S	56°1′49.65″W
Bahira Kusiwet Máscara						X				0°16′37.57″N	56°8′56.38″W
Polidores							X			Erepecuru River	
Bahira Kusiwet2 antropomorfos						X				0°0′39.60″N	56°5′1.13″O

Bahira Kusiwet pedra do acampamento						X				Erepecuru River	
Kejä	X	X		X						0°19′39.24″S	55°50′11.69″W
Terra Preta Itu	X	X	X	X						0°19′14.62″S	55°50′35.60″W
Complexo de Polidores Itu							X			0°19′2.61″S	55°50′32.59″W
Polidor boca do jacaré							X			0°18′26.63″S	55°48′56.88″W
Terra Preta Tawaripv	X	X	X	X						0°15′13.47″S	55°48′32.79″W
Terra Preta Tarakware	X	X	X	X	X					0°12′11.34″S	55°49′2.74″W
Iwi'ara			X	X	X			X		0°9′39.14″S	55°51′37.68″W
Bahira polidores 5							X			0°9′37.45″S	55°43′26.03″W
Bahira polidores 6							X			0°9′37.13″S	55°43′25.67″W
Bahira polidores 4										0°9′36.59″S	55°43′25.40″W
Praia Kare					X					0°9′34.35″S	55°43′18.04″W
Bahira polidores 1					X					0°9′33.90″S	55°43′16.86″W
Bahira polidores 2					X					0°9′33.90″S	55°43′17.16″W
Bahira polidores 3					X					0°9′34.51″S	55°43′17.63″W
Brake Taperet		X	X					X		0°9′39.77″S	55°45′37.84″W
Bahira Kusiwet Kare						X				0°27′44.97″S	55°36′22.79″W

socio-political autonomism, prescriptive endogamy and inside/out binarism. Using a comparative ethnographic approach focused on the analysis of kinship systems, Rivière (1984) outlined "invariants" in the settlement patterns, village composition, categories of social classification, and relations of autonomy and dependence of local groups. On this last point, he sketched a model of "political economy of people" in which the ideally endogamic relationships between groups are a function of pressure generated by the scarcity of human resources, in particular, women.[11]

Viveiros de Castro (1987) made important reflections on the "minimalist" model that provided avenues for the research proposed in this chapter. In an essay about the theory of alliance in Amazonia, the same author states that:

> [...] both within and outside Guiana, finally, the atomic and autarchic character of Amazonian social structures was submitted to close historical-ethnographic scrutiny [...] It could be time to attempt a comparative analysis of supralocal morphologies and processes in Amazonia [...] My impression is that these morphologies are not generally segmentary, but instead inductive and not totalizable, being of the network type. The local groups and regional agglomerations are more-or-less transitory condensations of these polycentric networks, commanded by a contractile regime of alliance, and not by any finalized structure of descendance and territory (Viveiros de Castro, 2002: 104–105, authors' translation from Portuguese).

The "close historical-ethnographic scrutiny" mentioned above, elaborated by authors that were already studying exchange circuits in the Guianas since the 1970's, is replete with evidence concerning the dynamic of these "polycentric networks".[12] In the same way, the thematic research which resulted in the collection "Networks of Relations in the Guianas" (*Redes de Relações nas Guianas*, Gallois, 2005) brought a fundamental contribution to regional ethnology. In particular, its perspective about Amerindian modes of relations produced an alternative image in which the articulation between different scales and dominions of trade networks between different people became evident. The volume innovatively proposed a comparative "transversal" exercise centered not only on kinship, but also on the articulation of war, trade and shamanism, and spatio-temporal relationships in the constitution of Guianese socialities. This sparked an important debate with Peter Rivière that was published in the journal *Mana* (Rivière et al., 2007), in which the differences between the two comparative projects became evident.

---

[11] Viveiros de Castro (2002: 326–327) compares this model of a "political economy of control" and the adaptationist perspective of ecological anthropology. While the model proposed by Rivière stems from the assumption that scarcity of people can explain Amazonian social morphologies, "adaptationist theories part from the marginalist assumptions of resource scarcity and cost/benefit optimization (yield-to-effort ratio) that assume an infused rationality of evolutionary character, ruled by thermodynamic parameters" (authors' translation from Portuguese).

[12] As well as the studies already mentioned, other notable works include those of Butt-Colson (1973), Thomas (1982), Arvello-Jimenez and Biord (1994), Dreyfus (1993) and Howard (1993). The latter two texts, particularly, make considerations about Guianense history and morphology that – in our opinion – contribute to a more credible construction of sociality when compared to the models of Rivière e Overing.

Afterwards, Perrone-Moisés and Sztutman (2010) made relevant contributions to the ethnographic and historical perspective presented here. In their article on the "Tamoio confederation", the authors address the complex question of Amerindian ethnonyms and of the "social units" adopted in colonial and modern descriptions: "what unit could this be, whose dynamic alternatively includes and excludes certain groups that, from the modern point of view, should all be united?" (p. 417, authors' translations from Portuguese).[13] In sequence, they make a powerful argument about the forms of Amerindian politics:

> "The Tamoio confederation did not constitute a movement in the direction of a chiefdom, a proto-state, or a state-like formation. Rather, it signaled the possibility of reversible transformations that cannot be understood as mere reactions to the Conquest, and which concern the pendular mechanisms that are proper to indigenous political action [...] Re-reading *Society Against the State* through the bias of "dualism in perpetual disequilibrium", such as unmasked by Levi-Strauss, we would be inclined to conclude that the coastal Tupi, who are Amerindians that avoid fixation –on the poles of both dispersion and unification – maintain themselves in a pendular movement between one and the other. Unification was a concrete possibility, repeatedly realized and always doomed to be dismantled, representing the opposite of dispersion, which was the solution to such refusal. Always almost united, interrupted by dispersion which does not come to full fruition. Re-reading Clastres through the metaphor of the "war machine" of Deleuze & Guattari, we could say that in every line of escape, in every dissolution, resides the possibility of a new composition of collectives, of a new segmentation, of a re-territorialization (Perron-Moisés & Sztutman, 2010: 425).

As noted by Perrone-Moisés (2006), from a comparative perspective, there is notable similarity between the morphology of the Tamoio confederations and a "type of Guianese confederation", since both are marked by the possibility of "reversible transformations" – in other words, by the pendular movement between unification and dispersion. In this text, the author had the merit to detect that both of the models that configure the poles of debate around the "Guianese paradox" (minimalism vs. networks of relations) are, in fact, constitutive poles of Guianese socialities whose dynamic is exactly that of the formation and dissolution of collectives, of distance and approximation, of alternation between moments in small settlements and moments of concentration in larger villages.

Finally, Perrone-Moisés (2015), in defending the proposal to apply the axis *Festa e Guerra* (Party and War) to our understanding of Amerindian social philosophy, ends up making a comparative exercise around modes of oscillation between these two poles, marked by the pairs host/guest (Party) and friend/enemy (War). In doing so, the author shows the specific ways in which this oscillation is related to the movement of distance/approximation and dissolution/formation of people and collectives and highlights that, in the long term, these contrasting forms end up defining a movement of continuing variation. It is through this path of Party and War that structures become history: "like the 'opening to the event', each party produces a

[13] By touching on this point, the authors were entering into the contemporary ethnographic critique of the concept of "society" in anthropology, associated with the works of Roy Wagner, Marilyn Strathern, Bruno Latour and Eduardo Viveiros de Castro, and to the emergence of analytical categories such as "socialities", "networks" and "fractals" to describe these relations.

collective, always temporarily" (Perrone-Moisés, 2015: 81, authors' translation from Portuguese).

3.3.2 Kunamiju's Trajectory

The accounts tell of a series of comings and goings made by Kunamiju along a long path that started from the mouth of a stream called *Buhu kwara hemba* (a tributary of the right bank of the Kare). After crossing the Kare over a bridge (*ruwowo*), crossing the path of the *tapy'yj* and two other rivers, Kunamiju arrived at the locality of Tetam. According to the Zo'e, Tetam was the abode of other indigenous groups and appeared "just like a city". In Tetam, Kunamiju became friends with the local people, who gave him a dog called *jawara wata* (lit: "jaguar that walks"). On several occasions when Kunamiju and his people attended *sepy* festivals in Iwi'ara, Kunamiju brought his dog, together with machetes and axes which were then exchanged with those from Iwi'ara. The Zo'é who stayed in Iwi'ara, however, never went to Tetam. Kunamiju did not want them to!

According to Sarará, the journey between Iwi'ara and Tetam used to take approximately 10 days. When talking about the path of Kunamiju, the Zo'é always point in the direction northeast of Iwi'ara as being the route to the locality of Tetam, which makes us think that Tetam is nothing more than a Zo'é memory of "a kind of Guianese confederation" addressed by Perrone-Moisés (2006). Looking at the map, the region to the northeast of Zo'é territory is the location of Apalai and Wayana territory (Paru de Leste Indigenous Land and the region east of the Parque do Tumucumaque Indigenous Land).

Accounts that mention that Kunamiju obtained machetes, axes and a dog in Tetam, are clear evidence of the insertion of Kunamiju into regional exchange networks, as registered by diverse authors (see Barbosa, 2005) and mentioned in accounts presented by Ribeiro (2020). Kunamiju, however, wanted to maintain a monopoly over trade relations with the inhabitants of Tetam, which was the reason for which he did not want the Zo'é from Iwi'ara to go to this locality. This monopolization of trade relations in the Guianas has already been observed by Gallois (1986) in relation to the Wayana and the Wajãpi. The Wayana, in this case, impeded the Wajãpi from establishing direct exchange with the *maroons* coming from the Guianas.

Kunamiju's trajectory points to a fundamental aspect of the morphology of Guianese networks: small villages like Iwi'ara were contemporary with large settlements like Tetam. It is interesting to note the similarity between Sarará's comment (that Tetam is "like a city") and the description made by Perrone-Moisés (2006: 3) from the chronicle of Tony, in which: "the Wayana village seems to be exceptional in its architecture and urbanism: surrounded by vast clean lands with four triple paths [...] that meet in a straight line in the middle of the village, in a 'kind of public plaza' [...] The houses make way for the paths, forming 'types of streets'" (authors' translation from Portuguese).

By way of his long route, Kunamiju oscillated between these two poles: he obtained things in Tetam and participated in *sepy* parties in Iwi'ara, even though Iwi'ara had been the stage of a war from which Kunamiju escaped. Kunamiju's oscillation confirms the argument made by Perrone-Moisés (2015) that in the Guianas, "both the formation of large agglomerates and their dispersion into small 'local groups' are constant, in the way that one can consider both to be 'typical' in the region" (p. 8, authors' translation from Portuguese). In other words, large population concentrations, far from indicating the formation of "chiefdoms" or "proto-states" (an evolutionist argument), would only be one pole in the model of "dualism in perpetual disequilibrium" (Lévi-Strauss, [1991] 1993). Networks of relations in the Guianas are characterized by exactly these movements of concentration and dispersion, approximation and distance, and oscillation between party and war. Movements that, in the long term, generate transformations: Kunamiju and his people, with the passing of time, did not return to Iwi'ara, and ended up becoming *others*.

3.3.3 The Death of Wajkyt

Broadly speaking, Kwai told us that the *apam*, whose villages were in the vicinity of the Erepecuru, killed and ate a Zo'é man called Wajkyt, which is why one of the localities we visited was called *Wajkyt jukeha rupa*, which literally means "place where Wajkyt was murdered". In retaliation, the *kirahi* allied to the Zo'é who at that time lived in the Pokaty region – which is why another of the locations we visited is known as *kirahi taperet*, "kirahi capoeira" – killed the *apam* and thus avenged Wajkyt's death. This episode is also mentioned by Gallois and Havt (1998).

While throughout this research we have registered Kwa'i's version of Wajkyt's death, in the following paragraphs we discuss the version narrated in 2017 by Namihit in the village of *Towari abyra rupa* and presented in Ribeiro (2020). The account of Namihit presents the episode of Wajkyt's death in four movements.

First movement: After killing Wajkyt close to the place that came to be the *capoeira* of Pokaty, the *apam* take him to the place known as *Wajkyt re poroke*, disembowel him, and continue to the village situated before the mouth of the Pupuruni stream; in this local, the *apam* cook Wajkyt in a large vessel and eat him;

Second movement: The Zo'é follow the trail of the *apam* to the village, remain hidden and, after the *apam* eat Wajkyt, blow arumã powder (*ruwy*) so that the *apam* fall asleep; the Zo'é then tie down the house of the *apam* with string made from carauá (*kurawa*) and burn it, killing the *apam*; the only young *apam* that was able to escape is clubbed to death;

Third movement: The *apam* from another village situated at the mouth of the Pupuruni followed the Zo'é to Pokaty and upon arrival asked them, "who was the killer?". The Zo'é do not tell them however (saying: "I wasn't there"). At the end of the afternoon, the *sepy* party begins, but the *apam* do not mix with the Zo'é.

The following day, at the end of the party, when the Zo'é women are serving the *sepy* drink, the Zo'é owner (*jet*) of the party sings a song to an *apam* woman saying that he would kill her husband; the *apam* once again ask who had killed the other *apam*, and the Zo'é continue to say, "I don't know, it wasn't me". One Zo'é, however, does not think well (*damuruhaj*) and ends up telling them. The party then becomes a war. A Zo'é man called Namikwet is killed with a club to the head (*ne kan nupã*). In revenge, *kirahi* friends of the Zo'é kill some of the *apam* with guns. The other *apam* flee in the direction of the mouth of the Pupuruni.

Fourth movement: The *kirahi* friends of the Zo'é then follow the *apam* until the village at the mouth of the Pupuruni, kill them all and throw them in the Erepecuru river.

The accounts of the story of Wajkyt point to several interesting aspects of this nexus of Guianese networks: cannibalism, revenge, party, war and alliances with the *kirahi*. The *apam*, as far as we know, frequented Zo'é villages. So much so that, after the Zo'é avenged the death of Wajkyt, the *apam* of the village situated at the mouth of the Pupuruni river followed them until Pokaty and participated in a party there (although they did not mix with the Zo'é). This party, in turn, was probably a celebration of the revenge of Wajkyt's death. The song chanted by the chief, however, made the party degenerate to war, and in doing so reiterated the oscillatory movement: parties generate wars that generate parties that generate wars.

The mention of both *apam* and *kirahi* people in the Pupuruni river basin (the entirety of which falls into the TIZ today) confirms two observations made by Gallois and Havt (1998): firstly, that some – if not all – of the areas occupied today by the Zo'é were also enemy territory in the past; secondly, that the Zo'é were already allied with the *kirahi* at this time. Some versions of this episode inform that these *kirahi* were *kirahi byk* (black people). In fact, the quilombola (*maroon*) presence in the Erepecuru is indicated by authors such as Nicolino (1946), Coudreau (1901), Rondon (1953), Cruls [1930] 1973, Andrade (1995) and Funes (1995).

3.4 Rethinking the "Sparse Population" Question

Despite the Zo'é's recent occupation of the region, the preliminary archaeological data show significant evidence of an interfluvial area that has been occupied by indigenous groups in a considerable way since very ancient times. The exploratory data we present are the first to have been collected for the Erepecuru-Cuminapanema interfluvial region. Previous studies were briefly mentioned only in reference to rock engravings and some *terra preta* areas on the margins of the Erepecuru river (Pereira, 2004).

While we did not conduct systematic surveys, the initial results obtained without subsurface interventions and through the recording of signals of human presence, guided by paths that are maintained and frequented by the Zo'é, show that the area has high archaeological potential, and it is an attractive – not inhibitive – place for human occupation, 170 km north of the main Amazon river channel. Some of this evidence was discovered during the journeys, but the vast majority of the sites were already known to the Zo'é. For example, many of their current villages are archaeological sites with *terra preta* (*ywy byk*). The criteria used by the Zo'é to define village location open various possibilities for dialogue with studies that have analyzed data, projected models and offered interpretations about the history of human occupation in Amazonia.

Recent discussions have relit the debate about the scale of human impact in Amazonia and the legacy that these activities had on modern environmental composition (Piperno et al., 2015, 2017, 2021; McMichael et al., 2012, 2014, 2017a, b; Levis et al., 2017; Junqueira et al., 2017; Watling et al., 2017a, b; Bush et al., 2015; Clement et al., 2015). These discussions have essentially brought back the debate around the possibility of human occupation on the margins of large rivers – principally floodplains – and in large swathes of interfluvial zones classified as *terra firme*. The proponents of the "sparse population model" use the floodplain/*terra firme* dichotomy as a principal explanation of Amerindian social morphologies, while proponents of the "landscape domestication model" focus on the role that indigenous action had in transforming the Amazon forest and producing biodiversity.

Piperno et al. (2015), for instance, mention Meggers (1971) affirming that, "the older idea that prehistoric Amazonia supported mainly small and highly mobile human populations who exerted little impact on their environments is being abandoned". Citing several studies that attest to large settlements and areas of transformation, the authors support what would be the "new vision" about an Amazonia that was densely occupied in the past. However, at the end, they conclude that, "prehistoric humans preferred to settle in habitats where resources were more abundant and easily captured, fertile soils were available, and transportation routes were closer at hand, rather than those posing adverse conditions for those factors" and that "the *varzea* versus *terra firme* distinction may not have worn out its relevance after all".

In opposition to this new uptake of old ideas, Clement et al. (2015), using the conceptual baseline of landscape domestication, affirm that "the immense diversity of social, cultural and historical processes that shaped Amazonia during the Holocene, situates this vast area in the company of other world anthromes" and "that the potential of human influence requires recognition of cultural and historical continuity with many indigenous peoples today". In the same direction, Watling et al. (2017a), investigating landscape transformations in the geoglyph region of Acre highlight paleoecological evidence for significant forest management in the interfluve that did not involve large-scale deforestation. In this way, adherents to the "sparse population model", still framed by "environmental determinism", suggest that the floodplain and *terra firme* are two distinct environments that resulted in distinct modes of human occupation in Amazonia, while the proponents of the "landscape domestication model", aware of recent advances in ethnology, history and archaeology, call for the dissolution of this analytical dichotomy.

Contrary to what authors of the "sparse population model" defend, Meggers (1971) did not defend the idea of a *terra firme* that was empty and unaltered by humans. Meggers (1971: 146) suggests two principal points that guarantee the balance between a human population and the environment it inhabits: "(1) development of measures to maximize food yields in a determined area and (2) not allowing that the population grows or becomes concentrated to the point that the local environment is put in danger". In a way, Meggers' point 1 fits neatly with what Clement and Cassino (2018) define as "landscape domestication". The main divergence between the contested archaeology in Meggers' work and Historical Ecology is that several other forces are at play, principally in relation to point 2. Meggers used demographic data of populations from the twentieth century to justify her environmental limitation model without considering the historic processes of population decline these populations underwent because of colonization, incurring that which Anna Roosevelt (1989) characterized as the danger of "ethnographic projection". And, above all, these data was used to infer a barrier between riverine and *terra firme* indigenous populations that perhaps never existed in pre-colonial times.

Among the archaeological evidence that best demonstrates the non-existence of rigid frontiers between riverine and *terra firme* populations, we highlight here stone axe blades. It is well-known that the availability of lithic raw materials suitable for blade production is scarce in large portions of floodplain regions. The relationship that the Zo'é have with these blades, the accounts of technological change that followed their adoption, and the abundance of polishing workshops in the TIZ leaves little doubt over the importance of the interfluve for the production and circulation of this highly significant tool. Also of relevance is the vastness of the areas where green stones and *muiraquitãs* once circulated (Boomert, 1987; Moraes et al., 2014; Lima, 2018).

Another line of evidence is the wide distribution and rapid dispersion of some ceramic assemblages when the more intense processes of consolidation of village life began in Amazonia. Neves et al. (2014) write, for example, that around 3000 years ago the Pocó ceramic tradition was already present in the Trombetas river, the lower Tapajós, the central Amazon, the Negro river basin, the Tefé region in the middle Solimões, and probably in many other places that have not yet been surveyed. The recent synthesis of archaeological information presented in the book *Koriabo: from the Caribbean Sea to the Amazon River* (Barreto et al., 2020) shows that, from an archaeological point of view, the contemporary networks of relations mentioned by Guianese ethnology already operated during the pre-colonial period, including in regions outside of the Guiana shield.[14]

[14] As synthesized in the map presented in the opening of the book, while there are still some doubts about the exact chronology, between AD 400 and at least the eighteenth century (Rostain, 2020) a ceramic style known archaeologically as Koriabo circulated throughout a region that encompasses the area between the Guiana shield and at least the Trombetas river. It is present on both sides of the Amazon river, in the estuarine area and, impressively, the Lesser Antilles.

3.4.1 Landscape Domestication and Networks of Relations as Ways to Produce Abundance

The material remains in the historic and archaeological sites we visited point to the existence of both a recent Zo'é occupation, and an old non Zo'é occupation. As well as this evidence, the ethnographic observations we made about their way of life and occupation clearly show that the Zo'é live in a state of abundance in an interfluvial zone (Fig. 3.10).

The paths that connect current villages and the places we visited, the managed resources that occur along them, and the memories and historic connections that the Zo'é weave into these places, show that what we record in the form of material remains is only a very small part of the lived experience of these places and landscapes. In these headwater region, the Zo'é are continuing an ancient occupation and forest management strategy based on seasonal mobility (periods of hunting, opening swidden fields, house construction, açaí, Brazil nut, fishing with timbó poisons, etc.) between different villages and locations of indigenous land, as well as the establishment of new villages in *capoeiras* where previous villages were located. This occupation strategy was characterized by Havt (2001) as "walking in one's own footsteps". As well as the intense occupations by the Zo'é over time, we also observed rock shelters, *terras pretas*, rock engravings, concentrations of certain plant species, and lithic polishing workshops, which are the most visible signals of a landscape also marked by ancient human action (Fig. 3.1 and Table 3.1).

In this scenario, Brazil nut groves are of primordial importance, since Brazil nut is one of the foods most consumed by the Zo'é, and these trees have a long history of relationships with human beings in Amazonia. In addition to consuming Brazil nut, the Zo'é use fibers from its bark to weave hammocks. The Zo'é, however, do

Fig. 3.10 Kujã se pounding açaí fruits in the Iwi'ara campsite. (Photo Raoni Valle/2013)

not plant this species. According to them, it is the agoutis that plant them. Even so, by consuming, discarding and transporting the nuts along land and river paths, and exchanging them with other villages, the Zo'é circulate large quantities of nuts throughout their territory. Although we have no data for the Zo'é case, the data presented by Ribeiro et al. (2014) about Kayapó gathering are interesting when thinking about both the historic relationship between Amerindian occupation and Brazil nut grove formation, and the way that the productive potential of interfluves could have been important in past networks of relations.

At the Caverna da Pedra Pintada site, not too far away from the TIZ, charred remains of Brazil nut are present since the first occupation layers dated to over 10,000 years ago (Shock & Moraes, 2019). Brazil nut was also identified by Shock et al. (2014) in several *terra preta* sites, including the urban area of Santarém, estimated to have been the location of the largest indigenous settlements of the pre-colonial period.

Aside from Brazil nut, botanical data collected in ancient sites in interfluvial areas show that since the beginning of human occupations, people used a wide range of plants such as the açaí, bacaba, patauá, buriti, tucumã and babaçu palms, courbaril, nance, and other species that archaeobotanists are unable to identify (Morcote-Rios et al., 2014, 2017; Shock & Moraes, 2019). Many of these trees continue to be important for current Amazonian populations.

In shell mounds and sites located near waterfalls in southwest Amazonia, archaeobotanical data also show that a significant number of plants were used by the people that constructed these places, including domesticated species such as common bean, squash and rice (Watling et al., 2018; Hilbert et al., 2017). Beans and squash are exotic plants in Amazonia, suggesting that at least 7000 years ago, long-distance networks of relations were responsible for the spread of plants in the region. Botanical remains from ancient sites such as Penã Roja (Morcote-Ríos et al., 1998), and the Llanos de Moxos (Lombardo et al., 2020), as well as genetic data (Clement, 2014) show that the processes of plant management and domestication date back to the beginning of the Holocene. We can thus consider that when village life (supposedly the "floodplain pattern" of occupation) propagated in Amazonia around 3000 years ago (Neves et al., 2014), the more radical processes of transformation, management, dispersal and concentration of plants had already taken place. The basis for polyculture agroforestry strategies evidenced in *terra preta* sites from the period were thus probably established before the formation of these sites (Meazumi et al., 2018; Watling et al., 2020). From 3000 years onwards, with the growth of evidence of villages, what probably happens is the intensification of some of these polyculture systems that allowed for the establishment of larger settlements.

Even in larger sites close to the floodplains, which were supposedly there to benefit from more fertile soils, better potential for intensive agriculture, and a more productive aquatic environment, botanical remains and isotopic data recovered in these contexts show a different pattern. Shock et al. (2014) showed that plants typical of interfluvial regions, such as Brazil nut, were an important part of what was consumed in riverine sites. Analyzing botanical microremains from one of the largest archaeological sites in the central Amazon, Cascon (2010) concluded that, despite being cultivators of maize and manioc, the populations that occupied this region should be recognized as having broad spectrum diets, since palms and other gathered species were well-represented in the analyses.

It is suggestive that such a broad spectrum diet is a result of the diversified exploitation of different environments and not only the floodplain. Interpreting isotopic data from human skeletal material belonging to the Marajoara culture of Marajó Island, on the mouth of the Amazon river, Hermenegildo et al. (2017) concluded that these populations practiced a diversified economy based on the management of wild plants, cultivation, and hunting. These data show how it is not possible to think of the floodplains and *terra firme* as disconnected worlds in the history of human occupation of Amazonia. Isolated occupations observed today are much more a reflection of colonialism, which broke long-distance networks of relations and circumscribed boundaries where before there were paths. The resilience of native populations, even in the face of these territorial impositions and restrictions, is perhaps one of the best examples of what we can call "landscape domestication". Deep knowledge of the domains of landscape is what allows for the continuation of occupation and the slow process of environmental transformation.

In this sense, even in the presence of intensification, the "massive prehistoric fires and forest clearing" expected by Piperno et al. (2015, 2021) as evidence of anthropization are not plausible. Contrary to this idea, Balée (1998) highlights that "landscape is a dialectical phenomenon [...] between those people and the land [...] without necessarily leading to irreversible degradation, biological extinctions, and cultural impoverishment" and points to different types of fire that can produce different types of biodiversity. In a homeostatic system, the idea of "gap producing", contrary to that of "forest cleaner" – which would be closer to *homo devastans* (Balée, 1998) – is more probable.

3.4.2 Numbers for the Amazon as Indigenous Forest

By observing the day-to-day of the Zo'é villages and camps, we can envision, or perhaps imagine, the diverse ways in which the archaeological record is (and must have been) formed. The transformation of the landscape observed throughout the two journeys, especially in the 3 days spent in the camp adjacent to Iwi'ara, gave us a clearer idea of what could have been the historical-ecological trajectory of many indigenous collectives in the tropical forest. In 2013, apart from mapping archaeological sites encountered during the 10 day journey, we also mapped some of the activities that we developed along its 80 km trajectory. We travelled in the company of 22 Zo'é people (16 adults, 8 women and 8 men – 3 of whom were hunters – and 6 children). We spent the night in 3 villages and 2 camps, the latter of which were mounted on the day. We can consider that during this journey, in an area of less than 20 km^2, these activities caused some kind of disturbance that could be considered management. Occupation is even more intense in the region of the Pupurini basin since it is the location of several villages. Some of these villages occupy the limit between the Erepecuru and Cuminapanema watersheds.

From studying the Zo'é, we understand that trekking camps, such as the one we made in Iwi'ara, as well as the dispersal of villages over a broad area of the Erepecuru and Cuminapanema interfluve, suggests that over the 13,000 years of occupational history in the Amazon, it is impossible to think of the existence of areas that have not been significantly altered by human action. In the case of the camps, called *tapaby* by the Zo'é, at Iwi'ara we observed 25 people completely transform an area of 10 × 10 m^2 and disturb everything in the area surrounding it, forming an archaeological campsite. Hypothetically, if 25 people form 6 campsites of 10 × 10 m^2 per year, what would that look like after 13,000 years? In other words, 25 people opening 6 campsites of 10 × 10 m^2 (0.01 ha) per year and disturbing their surroundings over 13,000 years? We would have, then, 78,000 campsites![15] In a scenario of greater population density, we can imagine millions of archaeological sites of small dimensions dispersed throughout Amazonia. However, these numbers must also take into account both the fact that people return to the same places over time (the "walking in one's own footsteps" model) and the possibility that some of these camps will become villages (and then an area of 10 × 10 m^2 would be progressively disturbed, inserting itself in the networks of relationships that connect other places and villages in the territory).

Today, the Zo'é are a small group of around 320 people organized in 18 groups which occupy, in total, 47 villages throughout the year. Even so, they plainly occupy an area of 300,000 hectares. If we imagine a scenario whereby Amazonia was inhabited by 5 million people,[16] even the most conservative estimates point to a long history of interactions between indigenous peoples and the forest. The reader is

---

[15] What would be the potential for identifying these remains with archaeological tools? The majority of surface surveys would not be capable of detecting remains from these activities. With this difficulty in mind, Piperno et al. (2015) state that "if little to no significant forest clearing and agricultural practices with major or minor annual seed and root crops is empirically evidenced, what can be said about ideas that forest species composition was impacted through the systematic removal of non-economic tree species and replacement with edible or other useful fruit trees, resulting in an enrichment of these species and planted or managed orchard-like formations (e.g. Balée, 2013; Erickson, 2008)?". Sub-surface interventions are generally made with test pits spaced over much larger intervals of 100 m or more. McMichael (2012), for instance, selected 13 points in a transect of over 800 km between Porto Velho and Manaus. The AmazonArch project (https://sites.google.com/view/amazonarch/home) has catalogued around 10,000 known archaeological sites in Amazonia, which represent an extremely small portion of potential human activities in the region. If we eliminate rock shelter sites, which are few in number, only 23 archaeological sites possess dimensions that could be compatible with temporary camps by small groups, such as those of the Zo'é. Even so, they appear to be locations where more intensive activities or frequent occupations happened, since studies mention the presence of lithic and ceramic concentrations. These data show that, from the archaeological point of view, we know very little in respect to temporary camps in Amazonia.

[16] Independently of whether we disillusion ourselves with paradise and, like Betty Meggers (1992), consider a maximum Amazonian population density being 1.5–2 million people; or if we think like optimists such as Nigel Smith (1999) and believe that this number could have reached 15 million; or even if we trust in someone who has perhaps most dedicated themselves to studying the subject

probably imagining those 5 million people as a snapshot in time; however, it is necessary to think about how many generations lived in Amazonia before this number was reached. While these numbers are hypothetical, they allow us to reflect upon how much we may be under-estimating what subtle human activities can produce in terms of the formation of the archaeological record and their long-term impact on biodiversity. Archaeology, after all, is demonstrating that humans were present in various parts of Amazonia more than 12,000 years ago (Roosevelt et al., 1996; Magalhães et al., 2019; Morcote-Rios et al., 2017; Pereira & Moraes, 2019).

It is also necessary to exercise caution to not return to the standard model of the "tropical forest culture" (Steward, 1948). While the existence of an "Amazonian substrate" is clear, variations in how Amerindians conceptualize relations with the forest, construct territories, and articulate networks of relations are notable. These variations are of great importance for the study of long-term histories, insofar as different sociocosmologies articulate distinct relationships with the environment (thinking on expansion models of Jê, Arawak, Caribe, Tukano, Yanomami etc.), and these different articulations have direct implications for the formation of the archaeological record. Far from generalizing the Zo'é model, our goal with this exercise was to suggest that even a conservative population estimate can generate notable results in terms of the occupation and transformation of the forest if we think on a scale of thousands of years.

3.5 Towards a Long-Term History of Networks of Relations in the Guianas

As mentioned in the introduction, in this chapter we sought to continue unmaking the images that insist in characterizing the Zo'é as a "Stone Age", "semi-nomadic" people that live in a "paradise" surrounded by "Christianized Karibs". From the material recorded in the field, we made some approximations between the anthropological perspective of networks of relations in the Guianas, the archaeological perspective of deep history, and some assumptions of Historical Ecology. This approximation is important insofar that it presents a more credible, ethnographically-based image of what could have been Amerindian relations and modes of occupation in interfluvial areas of Amazonia.

Although preliminary, the ethnographic and archaeological materials recorded during the journeys point to the long-term persistence of networks of relations in the Erepecuru-Cuminapanema interfluve, the area currently occupied by the Zo'é. From our journey to Iwi'ara and the Erepecuru, we were able to see old occupation areas and listen to what the Zo'é had to say about them. Some of the archaeological sites were places previously-known to the Zo'é (like Iwi'ara, for example), but

such as William Denevan (2014) and opt for 5–10 million, it is consensus that Amazonian population density 3000 years ago was larger than the most conservative estimates.

others were unknown to them (like some places in the Erepecuru). Our visits to the *capoeiras*, in particular, were relevant for evidencing the importance of these places for activating memories of landmark events and the trajectories of people who had the initiative to build new villages in the past. In this way, Iwi'ara was associated with the chiefs Kisiriguhu and Sarakut and, principally, Kunamiju. Pupuruni, in turn, was associated with the story of the death of Wajkyt and to conflicts with *apam* enemies.

These relations are fundamental and are associated exactly to the occupation of the interfluvial zone. Havt's (2001) descriptions of Zo'é modes of conceptualizing the environment and constructing territory, and those of Ribeiro (2020) about Zo'é networks of relations in the present, open windows of interpretation on the Zo'é mode of occupation of the Erepecuru-Cuminapanema interfluve in the past. On the one hand, from an ethnological point of view, Zo'é networks of relations and the complex web of terrestrial and fluvial paths make us think of an analytical approximation between the Zo'é model and the "reticular space" outlined by Albert and Tourneau (2007) for the Yanomami case, in contraposition to "concentric" models that invariably seek to define an "inside" and an "outside".

On the other hand, accounts of events and trajectories of important people, material evidence encountered during the journeys, and the Zo'é mode of occupation, point to a continuity of occupation of the interfluvial zone, reverberating a reflection made by Viveiros de Castro:

> [...] It would be more advised, considering the cultural substrate common to all of Amazonia, to imagine a dynamic of the *gumsa/gumlao* type subjected to conjunctural contractions and expansions, articulating populations of the floodplains and *terra firme* in ecologically and sociopolitically-heterogeneous regional systems [...] It seems increasingly clear that the emergence and persistence of 'simple' or 'complex' social structures – no matter what these characterizations are, evoking the old social evolutionism – cannot be explained by environmental factors taken in abstraction from historical dynamics and structures of large-scale societal interactions, as well as political decision-making processes guided by value systems that respond to much more than extrinsically and objectively defined environmental challenges or problems (2002: 330–331, authors' translation from Portuguese).

In this case, the challenge is to think about the long-term effects of the Zo'é mode of "walking in one's own footsteps" described by Havt (2001). In the case of ethnology, the description of Amerindian networks of relations, have already contributed towards dissolving large analytic dividers (nature/culture; individual/society; myth/history; etc.). The focus on networks of trade and relations enables a reorganization of archaeological classifications, since it is not possible to establish a fixed relationship between "ethnic group", territory, language and material culture over time and space. However, to think of networks and their material correlates in the *longue durée* permits a re-reading of what was perceived, through a simple, evolutionist perspective, as evidence of population migrations and cultural diffusion.

By manner of deepening the argument relative to the pertinence of networks of relations to think about long-term indigenous history, we intend to continue promoting the cross-over between the ethnology, history and archaeology of the Guianas, in dialogue with similar research carried out with indigenous peoples of the region (Cabral, 2014; Queiroz, 2014; Jacome & Xamen Wai Wai, 2020).

Our data also allow us to problematize the great floodplain/*terra firme* divide as an explanation and analytical principal of Amerindian modes of occupation in Amazonia, as relating both to the "standard model" and the "sparse population model". In the perspective of defenders of the sparse population model, the search for evidence of human action still seems to pertain to the idea of biodiversity reduction (massive fires and clearing). However, the data show that the primordial characteristic of the "system" is the production of biodiversity, which leads to a scenario whereby large-scale slash and burn is not the best option for occupation. It is difficult to think of colonialism as an event (the end of prehistory). More plausible would be to think of it as an ongoing process (*longue durée* history). Palaeobotanical remains are important tools for the history of indigenous peoples' past, but they cannot ignore what the data from the "living traces", the landscape and everyday life show. It is thus necessary to understand that the absence or low-density of these practices (slash and burn) in interfluvial zones is not evidence of a land without people.

Finally, we would like to underline, in the wake of Gallois (2016), that some connections presented from the research with the Zo'é suggest a strong continuity between ancient and contemporary indigenous peoples. The ethnographic model of networks of relations, in this sense, by emphasizing concentration and dispersion movements as resulting mainly from political (quality of social relations) factors, far from confirming a "post-contact regression", fruit of colonialism, converges with recent analyses of both historical materials and archaeological contexts in the Amazon.

Acknowledgements We would like to thank the Zo'é people for their company during the trips as well as for the many teachings about forest life in Amazonia in the present and in the past. In the process of producing this text, we would also like to thank Jannifer Watling for the translation and comments on an early version of the text; Guilherme Daltro Siviero for the production of the map in Fig. 3.1; and the editors of this volume for the invitation to publish our research work.

References

Albert, B., & Le Tourneau, F. M. (2007). Ethnogeography and resource use among the Yanomami: Toward a model of reticular space. *Current Anthropology, 48*(4), 584–592.

Andrade, L. M. (1995). Os quilombos da bacia do rio Trombetas: breve histórico. *Revista de Antropologia, 38*(1), 79–99.

Arvello-Jimenez, N., & Biord, H. (1994). The impact of conquest on contemporary indigenous peoples of the Guiana shield: The system of Orinoco regional interdependence. In A. Roosevelt (Ed.), *Amazonian Indians*. The University of Arizona Press.

Assis, R. (2013). Enfim, sós. *Revista Trip*. http://revistatrip.ol.com.br/trip/enfim-sos

Balée, W. (Ed.). (1998). *Advances in historical ecology*. Columbia University Press.

Balée, W. (2013). *Cultural forests of the Amazon: A historical ecology of people and their landscapes*. University of Alabama Press.

Balée, W., & Erickson, C. (2006). *Time and complexity in historical ecology: Studies in the neotropical lowlands*. Columbia University Press.

Balée, W., & Meredith, D. (2020). Applied historical ecology. In *Methods in historical ecology* (pp. 162–170). Routledge.

Barbosa, G. C. (2005). Das trocas de bens. In D. T. Gallois (Org.), *Redes de Relações nas Guianas*. Humanitas/FAPESP.

Barreto, C., Lima, H. P., Rostain, S., & Hofman, C. (Eds.). (2020). *Koriabo, from the Caribbean Sea to the Amazon River* (Vol. 1, 1st ed., 376p.). Museu Paraense Emílio Goeldi.

Boomert, A. (1987). Gifts of the Amazons: Green stone pendants and beads as items of ceremonial exchange in Amazonia and the Caribbean. *Antropológica, 67*, 33–54.

Bush, M. B., McMichael, C. H., Piperno, D. R., Silman, M. R., Barlow, J., Peres, C. A., et al. (2015). Anthropogenic influence on Amazonian forests in pre-history: An ecological perspective. *Journal of Biogeography, 42*(12), 2277–2288.

Butt-Colson, A. (1973). Inter-tribal trade in the Guiana Highland. *Antropológica, 34*, 1–70.

Cabral, M. P. (2014). De cacos, pedras moles e outras marcas: percursos de uma arqueologia não qualificada. *Amazônica – Revista de Antropologia (Online), 6*(2), 314–331.

Cartagenes, R. (2011). Novas tribos, velhas táticas. In B. Ricardo & F. Ricardo (Eds.), *Povos Indígenas no Brasil (2006–2010)*. ISA.

Cascon, L. M. (2010). *Alimentação na Floresta Tropical: um estudo de caso no sítio Hatahara*. Master Thesis, Museu Nacional, Universidade Federal do Rio de Janeiro. Rio de Janeiro.

Clement, C. R. (2014). Crop domestication in the Amazon. In H. Selin (Org.), *Encyclopaedia of the history of science, technology, and medicine in Non-Western cultures* (1st ed., pp. 1–7). Springer Netherlands.

Clement, C. R., & Cassino, M. F. (2018). Landscape domestication and archaeology. *Encyclopedia of Global Archaeology, 7*, 4388–4394.

Clement, C. R., Denevan, W. M., Heckenberger, M. J., Junqueira, A. B., Neves, E. G., Teixeira, W. G., & Woods, W. I. (2015). The domestication of Amazonia before European conquest. *Proceedings of the Royal Society B: Biological Sciences, 282*(1812), 20150813.

Coudreau, O. (1901). *Voyage au Cuminá*. Imprimerie Lahure.

Cruls, G. [1930] (1973). *A Amazônia que eu vi: Óbidos-Tumucumaque*. Livraria José Olympio/Instituto Nacional do Livro.

Denevan, W. M. (2014). Estimating amazonian indian numbers in 1492. *Journal of Latin American Geography, 13*, 207–221.

Dreyfus, S. (1993). Os empreendimentos coloniais e os espaços políticos indígenas no interior da Guiana Ocidental (entre o Orenoco e o Corentino) de 1613 a 1796. In E. Viveiros de Castro, & M. Carneiro da Cunha (Orgs.), *Amazônia: Etnologia e História Indígena*. NHII/USP/FAPESP.

Erickson, C. L. (2008). Amazonia: The historical ecology of a domesticated landscape. In H. Silverman & W. H. Isbell (Eds.), *The Handbook of South American Archaeology* (pp. 157–183). Springer.

Farage, N. (1991). *As Muralhas do Sertão: os Povos Indígenas no Rio Branco e a Colonização*. Paz e Terra/ANPOCS.

Frikel, P. (1958). Classificação linguística-etnológica das tribos indígenas do Pará setentrional e zonas adjacentes. *São Paulo: Revista de Antropologia, 6*(2), 113–189.

Funes, E. (1995). *Nasci nas matas, nunca tive senhor: história e memória dos mocambos do baixo Amazonas*. PhD Thesis (História Social), PPGHS/USP.

Gallois, D. T. (1986). *Migração, Guerra e Comércio: os Waiãpi na Guiana*. FFLCH/USP.

Gallois, D. T. (2005). Introdução: percursos de uma pesquisa temática. In D. T. Gallois (Org.), *Redes de Relações nas Guianas*. Humanitas, FAPESP.

Gallois, D. T. (2015). Sobre alguns modos Zo'é de fazer coletivos e lideranças. In D. Fajardo & L. M. M. Andrade (Eds.), *Entre águas bravas e mansas: índios e quilombolas em Oriximiná*. CPI-SP/Iepé.

Gallois, D. T. (2016). Algumas aproximações entre etnologia, história e arqueologia. *Teoria & Sociedade, 24*(2), 17.

Gallois, D. T., & Havt, N. B. (1998). *Relatório Circunstanciado de Identificação e Delimitação da Terra Indígena Zo'é*. DAF/FUNAI.

Havt, N. B. (2001). *Representações do Ambiente e da Territorialidade entre os Zo'é/PA*. Master Thesis (Antropologia Social), São Paulo, PPGAS/USP.

Heckenberger, M. (2005). *The ecology of power: Culture, place, and personhood in the southern Amazon, AD 1000–2000*. Routledge.

Hermenegildo, T., O'Connell, T. C., Guapindaia, V. L., & Neves, E. G. (2017). New evidence for subsistence strategies of late pre-colonial societies of the mouth of the Amazon based on carbon and nitrogen isotopic data. *Quaternary International, 448*, 139–149.

Hilbert, P. P. (1982). Pottery from the Cumina river, Brazil and its affiliations with the Koriabo phase of Guyana. *Journal of the Walter Roth Museum of Archaeology and Anthropology, 5*(2), 74–81.

Hilbert, L., Neves, E. G., Pugliese, F., Whitney, B. S., Shock, M., Veasey, E., et al. (2017). Evidence for mid-Holocene rice domestication in the Americas. *Nature Ecology & Evolution, 1*(11), 1693–1698.

Howard, C. (1993). Pawana: a farsa dos visitantes entre os Waiwai da Amazônia Setentrional. In E. Viveiros de Castro, & M. Carneiro da Cunha (Orgs.), *Amazônia: Etnologia e História Indígena*. NHII/USP/FAPESP.

Jácome, C., & Xamen Waiwai, J. (2020). A paisagem e as cerâmicas arqueológicas na bacia Trombetas: uma discussão da Arqueologia Karaiwa e Wai Wai. *Boletim do Museu Paraense Emílio Goeldi. Ciências Humanas, 15*(3). https://doi.org/10.1590/2178-2547-BGOELDI-2019-0140

Junqueira, A. B., Levis, C., Bongers, F., Peña-Claros, M., Clement, C. R., Costa, F., & Ter Steege, H. (2017). Response to comment on "Persistent effects of pre-Columbian plant domestication on Amazonian forest composition". *Science, 358*(6361), 925–931.

Levis, C., Costa, F. R., Bongers, F., Peña-Claros, M., Clement, C. R., Junqueira, A. B., et al. (2017). Persistent effects of pre-Columbian plant domestication on Amazonian forest composition. *Science, 355*(6328), 925–993.

Lévi-Strauss, C. [1991] (1993). *História de Lince*. Cia. das Letras.

Lima, A. M. A. (2018). A Ecologia de Assentamentos, Interações Sociais Ameríndias e o Contexto Geográfico dos Muiraquitãs no Baixo Amazonas. *Cadernos do Lepaarq (UFPEL), XV*, 121–141.

Lombardo, U., Iriarte, J., Hilbert, L., Ruiz-Pérez, J., Capriles, J. M., & Veit, H. (2020). Early Holocene crop cultivation and landscape modification in Amazonia. *Nature, 581*(7807), 190–193.

Maezumi, S. Y., Alves, D., Robinson, M., Gregorio de Souza, J., Levis, C., Barnett, R. L., et al. (2018). The legacy of 4,500 years of polyculture agroforestry in the eastern Amazon. *Nature Plants, 4*(8), 540–547. https://doi.org/10.1038/s41477-018-0205-y

Magalhães, M. P., Lima, P. G. C., Santos, R. S., Maia, R. R., Schmidt, M., Barbosa, C. A. P., & Aires da Fonseca, J. (2019). O Holoceno inferior e a antropogênese amazônica na longa história indígena da Amazônia oriental (Carajás, Pará, Brasil). *Boletim do Museu Paraense Emílio Goeldi. Ciências Humanas, Belém, 14*(2), 291–325, maio/ago. https://doi.org/10.1590/1981.81222019000200004

McMichael, C., Piperno, D., Bush, M., Silman, M., Zimmerman, A., Raczka, M., & Lobato, L. (2012). Sparse pre-Columbian human habitation in western Amazonia. *Science, 336*, 1429–1431. https://doi.org/10.1126/science.1219982

McMichael, C., Palace, M., Bush, M., Braswell, B., Hagen, S., Neves, E. G., et al. (2014). Predicting pre-Columbian anthropogenic soils in Amazonia. *Proceedings of the Royal Society B: Biological Sciences, 281*(1777), 20132475.

McMichael, C. N., Matthews-Bird, F., Farfan-Rios, W., & Feeley, K. J. (2017a). Ancient human disturbances may be skewing our understanding of Amazonian forests. *Proceedings of the National Academy of Sciences, 114*(3), 522–527.

McMichael, C. H., Feeley, K. J., Dick, C. W., Piperno, D. R., & Bush, M. B. (2017b). Comment on "Persistent effects of pre-Columbian plant domestication on Amazonian forest composition". *Science, 358*(6361). https://doi.org/10.1126/science.aam9035pmid:28473586
Meggers, B. (1971). *Amazonia: Man and culture in a counterfeit paradise*. Aldine.
Meggers, B. (1992). Prehistoric population density in the Amazon basin. In *Disease and demography in the Americas* (pp. 197–205). Smithsonian Institution Press.
Métraux, A. [1928] (1979). *A religião dos Tupinambás*. Edusp.
Moraes, C. P., Lima, A. M. A., & Santos, R. A. (2014). Os artesãos das Amazonas: a diversidade da indústria lítica dos Tapajó e o muiraquitã. In S. Rostain (Org.), *Antes de Orellana – Actas del 3er Encuentro Internacional de Arqueología Amazónica* (1st ed., pp. 133–140). Instituto Francés de Estudios Andinos.
Morcote-Rios, G., Cabrera, C., Mahecha, D., Franky, C., & Cavelier, I. (1998). Las palmas entre los grupos cazadores recolectores de la Amazonia colombiana. *Caldasia, 20*, 57–74.
Morcote-Ríos, G., Aceituno Bocanegra, F. J., & Sicard, T. L. (2014). Recolectores del Holoceno Temprano en la Floresta Amazónica Colombiana. In S. Rostain (Ed.), *Antes de Orellana. Actas del 3er Encuentro Internacional de Arqueología Amazónica* (pp. 39–50). Instituto Francés de Estudios Andinos.
Morcote-Ríos, G., Mahecha, D., & Franky, C. (2017). Recorrido en el tiempo: 12000 años de ocupación de la Amazonia. In Universidad Nacional de Colômbia (Ed.), *Universidad y territorio* (Vol. 5, t. 1, pp. 66–93). Universidad Nacional de Colombia.
Neves, E. G. (2012). *Sob os tempos do equinócio: oito mil anos de história na Amazônia Central (6.500 AC – 1.500 DC)*. Universidade de São Paulo.
Neves, E. G., Guapindaia, V. L., Lima, H. P., Costa, B. L., & Gomes, J. (2014). A tradição Pocó-Açutuba e os primeiros sinais visíveis de modificações de paisagens na calha do Amazonas. In *Amazonía: Memorias de las Conferencias Magistrales del 3er Encuentro Internacional de Arqueología Amazónica* (pp. 137–158). Ekseption Publicidad.
Nicolino, J. R. S. (1946). *Diário das três viagens (1877 – 1878 – 1882)*. Imprensa Nacional.
Nimuendaju, C. (1944). *Mapa etno-histórico do Brasil e regiões adjacentes*. IBGE.
Overing, J. (1983/1984). Elementary structures of reciprocity: A comparative note on guianese, central brazilian and north-west amazon socio-political thought. *Antropológica, 59*, 331–448.
Pereira, E. (2004). *Arte rupestre na Amazônia: Pará*. Unesp.
Pereira, E., & Moraes, C. P. (2019). A cronologia das pinturas rupestres da Caverna da Pedra Pintada, Monte Alegre, Pará: revisão histórica e novos dados. *Boletim do Museu Paraense Emílio Goeldi. Ciências Humanas, 14*(2), 327–342.
Perrone-Moisés, B. (2006). Notas sobre uma certa confederação guianense. [Paper presentation] *Colóquio Guiana Ameríndia. História e Etnologia*. Belém-PA.
Perrone-Moisés, B. (2015). *Festa e Guerra*. Livre Docência Thesis (Antropologia Social), PPGAS/USP, São Paulo.
Perrone-Moisés, B., & Sztutman, R. (2010). Notícias de uma certa confederação tamoio. *Mana, 16*(2), 401–433.
Piperno, D., McMichael, C., & Bush, M. (2015). Amazonia and the Anthropocene: What was the spatial extent and intensity of human landscape modification in the Amazon Basin at the end of prehistory? *The Holocene, 25*(10), 1588–1597.
Piperno, D. R., McMichael, C., & Bush, M. B. (2017). Further evidence for localized, short-term anthropogenic forest alterations across pre-Columbian Amazonia. *Proceedings of the National Academy of Sciences, 114*(21), E4118–E4119.
Piperno, D. R., McMichael, C. H., Pitman, N. C., Andino, J. E. G., Paredes, M. R., Heijink, B. M., & Torres-Montenegro, L. A. (2021). A 5,000-year vegetation and fire history for tierra firme forests in the Medio Putumayo-Algodón watersheds, northeastern Peru. *Proceedings of the National Academy of Sciences, 118*(40), e2022213118.
Porro, A. (1985). Mercadorias e rotas de comércio intertribal na Amazônia. *Revista do Museu Paulista, 30*, 7–12.
Queiroz, R. C. (2014). Olhares e perspectivas que fabricam a diversidade do passado e do presente: por uma arqueologia etnográfica das bacias dos rios Trombetas e Nhamundá. *Anuário Antropológico, 39*(2), 161–200.

Ribeiro, F. A. N. (2020). *Encontros Zo'é nas Guianas*. PhD thesis (Antropologia Social), São Paulo, PPGAS/USP.
Ribeiro, M. B. N., Jerozolimski, A., de Robert, P., & Magnusson, W. E. (2014). Brazil nut stock and harvesting at different spatial scales in southeastern Amazonia. *Forest Ecology and Management, 319*, 67–74.
Rivière, P. (1984). *Individual and society in Guiana*. Cambridge University Press.
Rivière, P., Grupioni, D. F., Gallois, D. T., Barbosa, G., Sztutman, R., & Pateo, R. D. D. (2007). A propósito de Redes de relações nas Guianas. *Mana, 13*(1), 251–273.
Rondon, C. (1953). *Índios do Brasil. vol. III*. Conselho Nacional de Proteção aos Índios. Publicação n° 99.
Roosevelt, A. (1989). Resource management in Amazonia before the conquest: Beyond ethnographic projection. In *Advances in economic botany* (pp. 30–62). Botanical Garden Press.
Roosevelt, A. C., Da Costa, M. L., Machado, C. L., Michab, M., Mercier, N., Valladas, H., & Schick, K. (1996). Paleoindian cave dwellers in the Amazon: The peopling of the Americas. *Science, 272*(5260), 373–384.
Rostain, S. (2020). Koriabo pottery in the Guianas. In C. Barreto, et al. (Org.), *Koriabo, from the Caribbean Sea to the Amazon River* (Vol. 1, 1st ed., pp. 55–80). Museu Paraense Emílio Goeldi.
Salgado, S., & Serva, L. (2020). Zo'é se isolam para fugir da Covid. In *Folha de São Paulo* (pp. 1–10). Caderno Especial Sebastião Salgado na Amazônia. Disponível em https://www.povosindigenas.org.br/es/Not%C3%ADcias?id=208105
Shock, M. P., & Moraes, C. P. (2019). A floresta é o domus: a importância das evidências arqueobotânicas e arqueológicas das ocupações humanas amazônicas na transição Pleistoceno/Holoceno. *Boletim do Museu Paraense Emílio Goeldi. Ciências Humanas, 14*(2), 263–289.
Shock, M. P., Moraes, C. P., Belletti, J. S., Lima, M. N., Silva, F. M., Lima, L. T., Cassino, M. F., & Lima, A. M. A. (2014). Initial contributions of charred plant remains from archaeological sites in the Amazon to reconstructions of historical ecology. In *3er Encuentro Internacional de Arqueología Amazónica, Quito. Antes de Orellana* (pp. 291–296). Artes Gráficas Señal.
Smith, N. J. H. (1999). *The Amazon River forest: A natural history of plants, animals, and people*. Oxford University Press.
Steward, J. H. (1948). *Handbook of South American Indians, Vol. III. The tropical forest tribes* (p. 986). Smithsonian Institution, Bureau of American Ethnology, Bulletin 143.
The Sunday Telegraph. (2008). *Amazon Grace*.
Thomas, D. (1982). *Order without government*. University of Illinois Press.
Valle, R. M., Echevarría López, G. T., Tenório Tuyuka, P. H. P., & Munduruku, J. S. (2018). What is anthropogenic? On the cultural aetiology of geo-situated visual imagery in Indigenous Amazonia. *Rock Art Research, 35*(2), 123–144.
Viveiros de Castro, E. (1987). Sociedades minimalistas: a propósito de um livro de Peter Rivière. *Anuário antropológico, 12*(1–2), 265–282.
Viveiros de Castro, E. (2002). *A inconstância da alma selvagem*. Cosac Naify.
Watling, J., Iriarte, J., Mayle, F. E., Schaan, D., Pessenda, L. C., Loader, N. J., et al. (2017a). Impact of pre-Columbian "geoglyph" builders on Amazonian forests. *Proceedings of the National Academy of Sciences, 114*(8), 1868–1873.
Watling, J., Iriarte, J., Mayle, F. E., Schaan, D., Pessenda, L. C., Loader, N. J., et al. (2017b). Reply to Piperno et al.: It is too soon to argue for localized, short-term human impacts in interfluvial Amazonia. *Proceedings of the National Academy of Sciences, 114*(21), E4120–E4121.
Watling, J., Shock, M. P., Mongeló, G. Z., Almeida, F. O., Kater, T., De Oliveira, P. E., & Neves, E. G. (2018). Direct archaeological evidence for Southwestern Amazonia as an early plant domestication and food production centre. *PLoS One, 13*(7), e0199868.
Watling, J., Almeida, F., Kater, T., Zuse, S., Shock, M. P., Mongeló, G., et al. (2020). Arqueobotânica de ocupações ceramistas na Cachoeira do Teotônio. *Boletim do Museu Paraense Emílio Goeldi. Ciências Humanas, 15*(2), e20190075.

Chapter 4
Historical Ecology in Amazonia

Stéphen Rostain and Doyle McKey

Abstract Historical ecology is at the heart of the ongoing debate about the degree to which pre-Columbian humans transformed Amazonian ecosystems. For more than 30 years, specialists in the Amazon have been studying ancient human impacts on Amazonian forests and savannas. People and their environment must be studied from the point of view of the dynamics of their interaction, taking into account historical episodes, cultural practices and livelihood strategies. The basic condition for the success of this type of research is effective interdisciplinary cooperation. Four examples of recent themes in Amazonian historical ecology are presented here: domesticated plants and landscapes in forested Amazonia; the geoglyphs of western Amazonia; domestication of floodplain savanna landscapes; and anthroposols of forested Amazonia.

Keywords Amazonia · Archaeology · Domestication · Historical ecology · Landscape · Plant

For most of the twentieth century, the Amazon was viewed as a uniform, pristine rainforest in which the indigenous people had always led a modest life subject to the whims of Nature. According to the theory of cultural ecology, defended among others by Julian Steward (1948), the geography, perceived a priori as hostile and unfavorable, placed strict limits on the degree of culture and the level of human progress that could be attained. On the basis of this neo-evolutionary vision, a simplistic distinction was made between *terra firme* and *várzea* (Meggers, 1996[1971]).

S. Rostain (✉)
National Center for Scientific Research (CNRS), Umr 8096 "Archéologie des Amériques", Paris, France
e-mail: stephen.rostain@cnrs.fr

D. McKey
CEFE, Univ Montpellier, CNRS, EPHE, IRD, Montpellier, France
e-mail: doyle.mckey@cefe.cnrs.fr

A. C. Colonese, R. G. Milheira (eds.), *Historical Ecology and Landscape Archaeology in Lowland South America*, Interdisciplinary Contributions to Archaeology, https://doi.org/10.1007/978-3-031-32284-6_4

The former corresponded to the inter-riverine forest areas – that is to say, most of the Amazon – whose poor soils were supposed to be of limited agricultural potential and thus incapable of sustaining development of large and complex human societies (Meggers also later [Meggers, 2003] extended her argument to include Amazonian savannas as areas of limited agricultural potential). The second comprised the alluvial plains along the Amazon and its main tributaries, plains that, thanks to the annual contribution of rich silts torn by rainfall from the slopes of the Andes during high waters, would have favored the appearance of productive agriculture, and the birth of more complex societies.

This Manichean and reductionist vision is no longer defended. The reality is more subtle and diverse. First, the variety of Amazonian landscapes is much greater than can be subsumed by terms such as *terra firme*, *várzea,* and savanna (Rostain, 2016). Second, whereas *várzea* soils may be more fertile, they are also prone to crop-destroying floods, and societies that use their abundant aquatic resources locate many of their farms on *terra firme* nearby (Denevan, 1996). Third, Indigenous populations invented a range of solutions to procure and cultivate useful plants in what may at first glance seem to be inhospitable environments. It is this capacity of humans to shape the properties of environments, rather than being 'determined' by traits of environments impervious to transformation by them (Arroyo-Kalin, 2016), that has been taken into account since the last two decades of the twentieth century, radically changing our view of this very special tropical environment. Historical ecology is at the heart of the ongoing debate on the degree to which pre-Columbian occupants transformed Amazonian ecosystems (Fig. 4.1). Indeed, for more than 30 years, Amazon scholars have been examining ancient human impact and pre-Columbian anthropogenic modification of Amazonian environments (Balée & Erickson, 2006). William Balée (1998, 2013) argues that people and their environment must be studied from the perspective of the dynamics of their interaction, taking into account historical episodes, cultural practices, and livelihood strategies. The basic requirement for success in this type of research is effective interdisciplinary work (Odonne & Molino, 2021).

A central concept of historical ecology is that pre-modern human land use can have long-lasting effects – legacy effects – on plant communities, soils, and ecosystem functioning. Åkesson et al. (2021) distinguish between direct ecological legacies, that is, those that are intended products of manipulation of the environment, and indirect legacies, that is, those that are unintended products of manipulation. This distinction is subtle but important. As we will see below, both kinds of legacies can have positive impacts on the capacity of the environment to support human activities. Although these positive impacts have attracted more attention from historical ecologists, legacies may also affect ecosystem functioning without affecting humans. Dissection of these legacies continues to raise important questions about the ways in which human and landscape histories reciprocally affect each other. These questions define a rich field of research about the ways in which human and landscape histories intertwine beyond the temporal scale of a few human generations (Fig. 4.2).

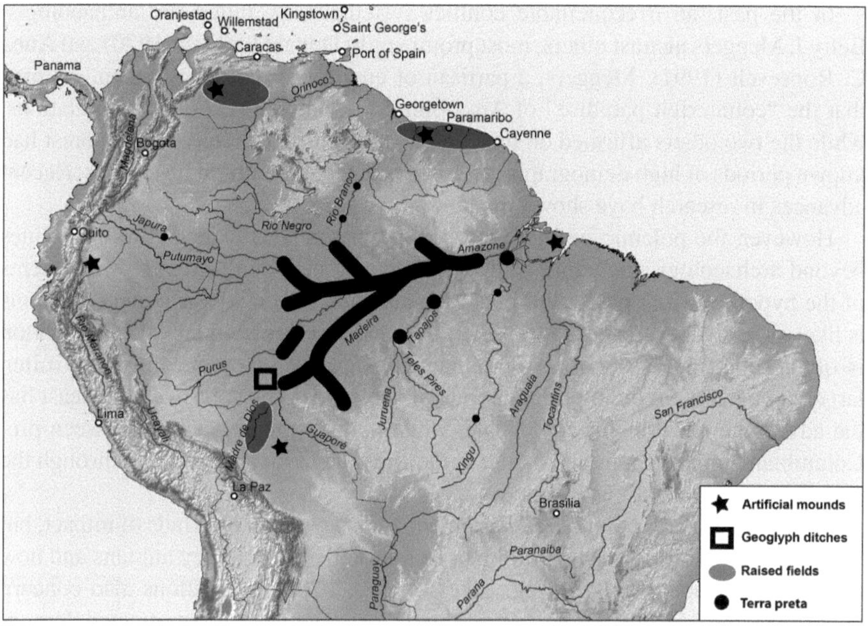

Fig. 4.1 Map of some remarkable archaeological sites in Amazonia. (© Drawing S. Rostain)

Fig. 4.2 Kichwa woman going to her "*chacra*", a cultivated field in the forest, Yasuni National Park, Ecuador. (© Photo S. Rostain)

In the past, an irreconcilable conflict systematically pitted the archaeologist Betty J. Meggers against others, most prominently Donald Lathrap (1970) and Anna C. Roosevelt (1991). Meggers, a partisan of environmental determinism, claimed that the "counterfeit paradise" of Amazonia could not support dense populations, while the two others affirmed on the contrary that this evergreen tropical forest had known periods of high demography and essential techno-cultural inventions. Recent advances in research have shown the latter to be correct.

However, the polemic has not died down, but has been extended to disciplines beyond archaeology, including ecology. Thus, any new publication by proponents of the hypothesis that pre-Columbian humans transformed Amazonian ecosystems is likely to provoke a response from skeptics, and vice versa. In fact, this opposition is often about matters of form or semantics, since its actors are in reality often largely in agreement on important points of substance. The controversy at least has the advantage of enriching the debate, while making the interaction between pre-Columbian human groups and nature in the Amazon particularly visible through the abundance of publications it generates.

Today, the biggest point of real disagreement is not about the kinds of impact, but on how much of Amazonia was more or less strongly impacted by humans and how long these impacts persisted (e.g., Åkesson et al., 2021). Questions also concern how the reciprocal interactions between humans and environment affected demography (e.g., Arroyo-Kalin & Riris, 2020; Flores & Levis, 2021). Interpretations are now based on quantified data ranging from macroscopic to microscopic observation, using advanced techniques. However, the virulence of the debates has not lost any of its intensity. We continue to kill each other with insults, bird names and venomous diatribes uttered at conferences and in the corridors, because we hold back in the publications. Sometimes, science in the Amazon is like an ancient circus game. Mirmilions of archaeology, retiaries of ecology and secutors of botany confront each other without mercy (usually stopping short of bloodshed) in the arenas of ancient tropicality. These intellectual jousts, often futile, have intensified in the last 13 years because it is really with the new millennium that historical ecology and interdisciplinarity have become popular among, and indeed imposed upon, scientists of Amazonia.

Yet alternative paths are possible (Smith, 2013). In rejecting neo-evolutionism, do we run the risk of throwing the baby out with the bathwater? McKey et al. (2016) propose combining elements of historical ecology with others from cultural ecology to shape an intermediate approach that is more coherent than the sometimes cartoonish versions of each of these "half-approaches". Drawing on ecological and archaeological data, they argue that environmental constraints – and opportunities – help shape the contours, limits, and diversity of human cultural adaptation. Thus, concepts such as "cultural niche construction" marry environmental influence and human agency, without granting primacy to either.

This chapter follows a logical thread:

1. Indigenous Amazonians have had lasting effects up to the present day. This is the essence of historical ecology: understanding the ecology of today's ecosystems and environments through knowledge of their history. These legacy effects are: on

plant genomes (domestication); on vegetation composition (landscape domestication); on soils (Amazonian Dark Earths, raised fields); and on ecosystem functioning.

2. These legacy effects often leave a positive ecological inheritance for subsequent human inhabitants. The positive effects of domesticated plants, domesticated forests, and Amazonian Dark Earths for subsequent human populations last for centuries or millennia, constituting a useful ecological inheritance transferred across generations. Although some would apply terms such as "cultural niche construction" or "landesque capital" only when the positive effects are intentional (e.g., Arroyo-Kalin. 2016), some of the most persistent positive legacy effects may have originated as unintended consequences of human actions. They nonetheless have similar beneficial effects on the quality of the environment for future generations.

3. Not all legacy effects are positive for humans. Some are at best neutral. Pre-Columbian raised fields offer an example: the physical vestiges of raised fields remain, but nutrient enrichment of their soils was only temporary. They continue to shape ecosystem functioning (communities of soil animals and of plants), but not in ways that affect humans.

4. The example of vestiges of raised fields shows how important it is to fully integrate ecology and archaeology to interpret landscape history.

5. We are now beginning to be able to tie the legacy effects of the alteration of plant genomes, plant communities, soils, and landscapes to human demography in pre-Columbian Amazonia.

4.1 Domesticated Plants and Landscapes in Forested Amazonia

The last two decades have seen the sharpening of the dispute around the intensity of the transformation of Amazonian ecosystems by pre-Columbian peoples (Neves & Heckenberger, 2019; Clement et al., 2021) (Fig. 4.3). Charles Clement and colleagues do not hesitate to advance the term "domesticated landscape". This formula, which has been as seminal as it has been controversial, is based on the following assumption: the immense variety of socio-cultural and historical processes that participated in shaping the Amazon during the Holocene allows this immense area to be ranked among the world's biomes that have been the arena of great anthropogenic impacts (Neves & Rostain, 2012).

The last few years have brought solid elements of support for this proposition. Long designated as the debtor of innovations that originated in the Andes (Meggers, 1996[1971]), Amazonia is in fact proving to be creative. At least 83 plant species have been to some extent domesticated there (Clement et al., 2015), in the strict sense of having undergone evolutionary change mediated by human management, making this region one of the world's most important centers of plant domestication (although 'centers' is somewhat misleading, as different plants were domesticated in various parts of this vast region [Clement et al., 2010]). These domesticates

Fig. 4.3 The Amazonian forest is much less "virgin" than it seems at first glance. (© Photo S. Rostain)

include manioc (also known as cassava) (*Manihot esculenta*, Euphorbiaceae), chilies (*Capsicum* spp., Solanaceae), 10,000–8000 years ago, and later pineapple (*Ananas comosus*, Bromeliaceae), peach palm (*Bactris gasipaes*, Arecaceae), cootie (*Bixa orellana*, Bixaceae), cocoa (*Theobroma cacao*, Malvaceae), tobacco (*Nicotiana tabacum*, Solanaceae), peanuts (*Arachis hypogaea*, Fabaceae), and many others (Clement, 1999).

Domestication was a slow process of coevolution between plants and humans based on observations and manipulations of plants and of their environments. In Amazonia, this process began probably 10,000 years ago (Clement et al., 2021). As Lévi-Strauss already recognized over 70 years ago (Lévi-Strauss, 1948: 465), the mutually beneficial relationships between plants and humans emerging from this process form a continuum, ranging from wild plants that were favored by silvicultural systems, to incipient domesticates (cultivation and dispersal by humans, but with – so far – little or no evidence for evolutionary change), to full domesticates. Clement et al. (2021) estimate that in the South American lowlands, at least 6261 native plant species have been used, of which 742 are managed outside and inside food production systems, 600 are cultivated, and at least 45 are fully domesticated. Whereas cultivation and domestication began perhaps 10,000 years ago, elaborate agricultural systems began to emerge in Amazonia only much later, around 4000 BP (Neves, 2007; Clement et al., 2015). Throughout this period, and up to the present, indigenous Amazonians continued to rely heavily on incipient domesticates and wild plants, managed in food production systems or favored by silvicultural practices (Neves & Heckenberger, 2019; Iriarte et al., 2020; Clement et al., 2021).

Although humans may have transformed the genomes of only a relatively small number of plant species – but so far few "incipient" domesticates have been investigated for genetic signatures of human management – they unquestionably altered the composition of plant communities, usually in ways that increased their capacity to produce food and other resources for humans (Flores & Levis, 2021). This is one facet of what Clement and colleagues (Clement et al., 2015, 2021; Levis et al., 2017, 2018; WinklerPrins & Levis, 2020) term landscape domestication. Levis et al. (2018) extensively document how indigenous Amazonians remove non-useful plants, protect useful plants and attract their non-human seed dispersers, transport useful plants to new sites, and manage landscapes with fire and by other means. The effects of these actions persist to the present day. What proportion of Amazonian forests was thus profoundly altered by humans is still a matter of debate, because vast expanses are still archaeologically unexplored (Iriarte et al., 2020). Such 'domesticated forests' may be very widespread. Balée (2013) estimated that around 12% of forests in Brazilian Amazonia have been so altered. In a basinwide survey of the Amazon tree flora in 1170 forest plots, ter Steege et al. (2013) found that of the 227 hyperdominant species that accounted for half of all trees, many are species heavily used by indigenous Amazonians. They considered widespread pre-1492 cultivation as one plausible explanation to account for this. Levis et al. (2017) marshaled arguments supporting this hypothesis. However, McMichael et al. (2017) raised methodological concerns, including the under-estimation of post-1492 impact on forest composition and the fact that a disproportionate number of plots was located near archaeological sites. Åkesson et al. (2021) question the pertinence of inferring legacy effects from the composition of current vegetation without empirical data on past vegetation change. While the existence of 'domesticated forests' appears solidly established, their extent is still debated. Recent work, integrating evidence from archaeology, paleoecology, ethnography, and studies of modern vegetation (Maezumi et al., 2018), supports the hypothesis that they were indeed widespread, at least in eastern Amazonia. The following example traces a similar debate for southwestern Amazonia.

4.2 The Geoglyphs of Western Amazonia

One of the collateral effects of the drastic deforestation that the Amazon has undergone in recent decades, especially in its southwestern part, has been to reveal countless archaeological sites. For example, in the state of Acre, Brazil, the opening of the vegetation has brought to light hundreds of geometric ditches forming squares or circles and called "geoglyphs" and spread over an expanse of nearly 13,000 km^2 (Fig. 4.4). Many more perhaps remain to be discovered under uncleared forest. The diameter of the geoglyphs varies from 100 to 300 meters and the ditches reach 11 meters in width and 4 meters in depth (Schaan, in Rostain & Jaimes Betancourt, 2017). In addition, some sites have up to six closed ditches. Archaeological excavations have revealed that they range in age from 2000 to 650 years calibrated BP (Schaan et al., 2012), although it is possible that some were built as early as 3500–3000 years BP. The almost complete absence of archaeological remains

Fig. 4.4 Geoglyph with geometric ditches from the site of Fazenda Colorada, Acre State, Brazil. (© Photo M. Pärssinen)

argues for a ritual function; they are interpreted as occasional gathering places for collective ceremonies.

Recent work has combined the efforts of archaeologists, archaeobotanists, paleoecologists, botanists, anthropologists, and geographers to reconstruct the environmental history and impact of human groups in the area over millennia (Watling et al., 2015, 2017a). The hypothesis of these authors at the outset was that the geoglyph-builders had deforested the area in pre-Columbian times. Under this scenario, the regrowth of the forest – until it was cut by modern settlers – would indicate a surprising degree of resilience.

What they found suggested a very different scenario. Analyses of phytoliths, charcoal, and stable carbon isotopes collected from dated levels of geoglyph sites, showed that the region had indeed been forested for over 6000 years, bearing forest dominated by bamboo (*Guadua* sp.). However, contrary to the original hypothesis, the geoglyph-builders did not conduct large-scale deforestation. They were living in a forested landscape that had already been managed by humans for hundreds of years or more. Even before the geoglyphs were built, areas of forest had been cleared at least twice, around 4000 and 3600 BP. The geoglyph builders opened only small clearings for limited periods, meaning that the sites were hidden in the forest and not visible to each other, which is, as the authors note, a surprising conclusion. The authors make the intriguing suggestion that clearance of small areas may have been facilitated by the life cycle of the bamboo, patches of which flower, fruit and die about every 27–28 years (de Carvalho et al., 2013). Forest clearance may not have required the burning of trees. The forest that regrew in cleared areas was dominated for millennia by palm trees, their occurrence beginning to decline only after the abandonment of the geoglyph sites, around 650 years ago.

In sum, human management of Acre's vegetation cover is attested to for at least 4400 years. Nevertheless, the period of geoglyph construction was not accompanied by large-scale intensive deforestation over long periods of time, but only by occasional spot clearing and the practice of agroforestry, including the cultivation of maize (*Zea mays*) and squash (*Cucurbita* sp.). Thus, and contrary to the initial hypothesis of extensive tree cutting and the creation of huge open spaces, the geoglyph builders conducted their rites in isolated sites in the forest. An auxiliary message of the study is that nothing in the region's pre-Columbian history approached the massive scale of present-day deforestation, and that the forest's resilience to such an attack is by no means assured.

As in other cases, this interpretation of rather intensive and large-scale human management of 'domesticated forests' has met with skepticism from some quarters. Piperno et al. (2017) regard the data as being more consistent with short-term, localized management near the geoglyph sites. In their reply, Watling et al. (2017b) insist on how patterns in their palm-phytolith data support their conclusions and fault the emphasis by Piperno et al. (2017) on soil charcoal, arguing that indigenous Amazonians augmented productivity of agroforests by means more subtle than whole-scale burning.

4.3 Domestication of Floodplain Savanna Landscapes

Forested Amazonian landscapes were domesticated through changes in vegetation composition (above sections) and by the formation of Amazonian Dark Earths (see below). Indigenous Amazonians also domesticated the floodplain savanna landscapes of Bolivia's Llanos de Mojos, by hydrological engineering (construction of various earthworks) and controlling fire regimes (Duncan et al., 2021). These landscapes were also a focus of domestication of plants, grown both on human-created forest islands (Lombardo et al., 2020) and in the seasonal floodplains themselves (Hilbert et al., 2017). At one of the best-studied sites, from the mid to late Holocene, plant management practices were characterized by increasing diversification, rather than intensification (Furquim et al., 2021).

Like their forest-dwelling counterparts, the pre-Columbian inhabitants of floodplain savannas invested vast amounts of time and labor in domesticating these landscapes. However, the legacy effects of their activities on today's vegetation and soils appear modest compared to those in forested Amazonia. Unlike domesticated forests, there is little evidence that pre-Columbian humans caused persistent alterations in the composition of plant communities. This is probably because the herbaceous and shrubby plants that dominate savanna vegetation are shorter-lived than forest trees, so that human-induced changes in species composition are effaced more rapidly by succession. Also, unlike the dark-earth Anthroposols of forested Amazonia, which are more fertile than adjacent soils even centuries after abandonment (Glaser & Birk, 2012), there is no evidence that soils in vestiges of pre-Columbian agricultural raised fields in savannas are enriched compared to adjacent unmanaged soils. While some have considered this to indicate the absence of fertility-management practices in raised fields (Boixadera et al., 2019), studies in present-day wetland raised-field agriculture in Africa suggest an alternative explanation. There, large amounts of plant biomass are composted in raised fields, resulting in a marked, but very short-lived, increase in nutrient availability as organic matter is quickly decomposed (Rodrigues et al., 2020). In contrast, the persistence of nutrient-retaining properties of dark-earth anthroposols in forested Amazonia is conferred by a very stable form of organic matter, charcoal (Glaser & Birk, 2012).

Present-day plant communities and soils in landscapes featuring vestiges of raised fields thus show no conspicuous direct legacy effects, i.e., the intentional results of manipulation. However, these landscapes offer striking examples of indirect legacy effects (Fig. 4.5). Although these unintended products of manipulation do not increase the landscape's usefulness to humans, they impact ecosystem functioning in important ways. With little consequence for human use of ecosystems, these effects are understandably less interesting to archaeologists and anthropologists than are legacy effects that confer a positive ecological inheritance across generations. Nevertheless, understanding them is key to interpreting the historical ecology of floodplain savannas.

These indirect legacy effects result from unique features of floodplain savanna ecology. Soil animals require air to breathe, but during the high-water season, soils are waterlogged. Soil engineer organisms – earthworms, termites, and ants – that can build mounds elevated above the water surface thus have an important advantage, creating habitats that enable them to survive. Their mound-building activities also benefit other organisms, including plants, whose roots then help stabilize mounds. There is thus continual reciprocal positive feedback between organisms and topography. It is therefore not surprising that floodplain savannas host many examples of natural earth-mound landscapes (McKey et al., 2014; López Mazz et al., 2016; Zangerlé et al., 2016). Mound-building termites and ants can forage throughout the savanna during low-water periods, while their colonies find refuge in the mounds during high water. Earthworms, on the other hand, retreat deep in the soil during the dry season. During high water, they forage in the waterlogged soil, building mounds to which they periodically come to respire, at the same time

Fig. 4.5 Pre-Columbian raised fields of Diamant, west of Kourou, French Guiana. (© Photo S. Rostain)

depositing their excrement and thereby adding soil to the mounds. Among the most spectacular of these natural earth-mound landscapes are the surales in the floodplain savannas of the Orinoco Basin in Colombia and Venezuela (Fig. 4.6), composed of thousands of regularly spaced mounds, most often 1–2 m in diameter. Marston Bates (1948: 566–568) described them as follows: "The surales present a reticulate pattern of deep ditches surrounding mounds a meter or two in diameter; the top of the mound is a meter or more above the bottom of the surrounding ditch. … The reticulate ditching is like the pattern formed by the drying of a gigantic mud flat. … Well-developed sural country is difficult to traverse. If you are on foot, you have to decide whether to follow the endless twistings of the boggy ditches or to jump from mound to mound, both awkward expedients. If you are mounted, the animal has to make the same decision and generally ends up in complete frustration: I have heard stories of man and mule firmly stuck in a narrow, deep ditch between two sural mounds."

Fig. 4.6 Landscape of *surales* in the llanos of Colombia. (Above, images taken with a Pixy™ drone; below, ground view) (© Photo D. Renard & D. McKey)

Field studies of the surales by Zangerlé et al. (2016) indicate that the mounds are in large part the accumulated excreta of large earthworms. They begin as small mounds of modest diameter (a few dm to less than 1 m), apparently built by individual earthworms, that then merge into larger mounds, with multiple peaks, resulting from the joint actions of numerous earthworms. The multiple-peaked mounds grow and become more homogeneous in shape, becoming in some sites imposing mounds of up to 4.5 meters in diameter and a maximum of 2 meters high, from the top to the bottom of the ditch.

What do these natural earth-mound landscapes have to do with pre-Columbian raised fields? In fact, once they were abandoned, raised fields in floodplain savannas "plugged into" the same reciprocal feedbacks between topography and organisms that drive the formation of natural earth-mound landscapes. Abandoned raised fields are islands of well-drained soils in which activities of earthworms, termites, ants, and plants are all concentrated. In landscapes bearing vestiges of pre-Columbian raised fields in coastal savannas of French Guiana, where there are large complexes of such vestiges (Rostain, 2010, 2012, 2015), ecological studies showed the importance of these feedbacks (McKey et al., 2010). Nests of social insects are located exclusively on these mounds, because any nest initiated in the flooded areas between the mounds does not survive. Social insects such as fungus-growing leaf-cutter ants continuously transport organic matter into their nests. In addition, during nest building and maintenance, excavated material is transported to the top of the mound. Earthworms foraging over the flooded savanna come to the mounds to respire, depositing their excrements at the same time. The organic and mineral matter that all of these animals bring to the mounds compensates for loss by erosion. In addition to transporting organic and mineral matter to the mounds, the biogenic structures made by these animals stabilize the mound soils, reducing erosion. The topographic heterogeneity initially created by humans thus set in motion feedbacks that drive ecosystem functioning today, producing species-rich soil invertebrate communities (Renard et al., 2013).

Such indirect legacy effects on ecosystem functioning did not make floodplain savannas more useful to humans; their importance to archaeologists lies elsewhere. By "re-engineering" raised fields, natural engineers help preserve the physical legacy of human mound-building activities (McKey et al., 2010; Rostain & McKey, 2015a, 2015b). Without them, erosion may have left few pre-Columbian raised fields for archaeologists to study. It would also behoove archaeologists to learn more about natural earth-mound landscapes, to avoid confusing them with vestiges of raised fields (Fig. 4.7) (McKey et al., 2014). Landscapes that look similar may have dramatically different origins and histories: one explanation does not fit all. This example shows how important it is for historical ecology to integrate

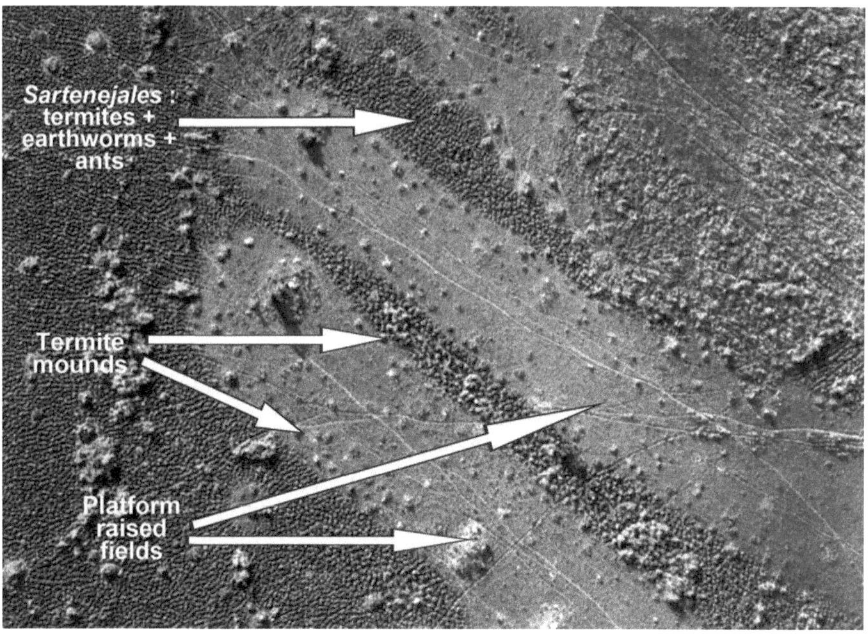

Fig. 4.7 Mixed natural and cultural landscapes of the tropics: pre-Columbian raised fields and natural mounds in the flooded savanna of the Llanos de Mojos, Bolivia. (© Photo D. Renard with the Pixy™ drone)

research from all its parent disciplines. Finally, natural earth-mound landscapes may have served as an "ideation template" (Finney, 2012) imitated by human builders of mounds, including raised fields (Cunha et al., 2016).

4.4 Anthroposols of Forested Amazonia

In addition to domesticating plants and forests, indigenous Amazonians also transformed the properties of soils of forested Amazonia in ways that made them more favorable for agriculture, even up to the present day. No theme in Amazonian historical ecology has seen more debate about whether this ecological inheritance conferred by the activities of pre-Columbian people resulted from the purposeful creation of agricultural soils, or were the unintended, but felicitous consequences of long-term human occupation. Following this debate requires taking into account the variation observed in these anthroposols, which are collectively termed Amazonian Dark Earths (ADEs). Their dark color owes to their content of pyrogenic carbon (fine particles of charcoal), and this highly stable form of organic matter appears to be key to their favorable nutrient-retention properties (Glaser & Birk, 2012). They have been long divided by researchers, indigenous communities, and local farmers into two groups, *terra preta* and *terra mulata*. *Terra preta* (meaning "black soil" in Portuguese) soils are dark, containing large amounts of pyrogenic carbon, abundant artifacts such as ceramics, and faunal bones. Robinson et al. (2021) and Iriarte et al. (2020) refer to them as ADEs (in a stricter sense), distinguishing them from *terra mulata* ("brown soil"), which they name "Amazonian Brown Earths" (ABEs). ABEs contain less pyrogenic carbon and lack cultural artifacts and faunal bones, and usually cover larger expanses. Some consider *terra preta* and ABEs as simply two extremes of a continuum (Fraser et al., 2011), reflecting variation in intensity and type of activities. However, Robinson et al. (2021) consider the differences between them to reflect fundamental differences in how soils were managed and formed.

Hypotheses about the origin(s) of these anthroposols feature a mix of purposeful and unintentional effects. According to a widely accepted hypothesis (the "midden model" Kämpf et al., 2003), *terra preta* arose as a by-product of waste accumulation near human settlements (Fig. 4.8). Humans quickly observed the salubrious effects of the soils resulting from habitation sites and used them to grow nutrient-demanding crops in home gardens, or repurposed them as fields (Robinson et al., 2021). More controversial is the genesis of ABEs. Some hold that the same actions gave rise to both: as locations of villages and their surrounding territories shifted over time, areas nearest the long-term residential-area core developed classic *terra preta*, while areas further away developed ABE soils with less marked anthropic influence (Fraser et al., 2011). An alternative hypothesis (the "agricultural model" [Woods & McCann, 1999]) holds that in contrast to *terra preta*, ABEs were the results of intentional long-term soil management practices for agriculture, such as "slash-and-char" "cool burning" that generated less ash and more charcoal. Robinson et al. (2021) marshal geoarchaeological data supporting this hypothesis.

Fig. 4.8 Dark-earth Anthroposol (ADE) full of pottery sherds in an archaeological site of the Lower Amazon, Brazil. (© Photo H. Lima)

They also point out that knowledge about distribution and properties of ABEs is even spottier than for *terra preta* and that their extent is likely grossly underestimated, as they are often not recognized as archaeological sites.

Regardless of the relative roles of intentionality and fortuitous by-products in the genesis of these anthroposols, their agricultural value is commonly exploited by present-day Amazonian cultivators, both Indigenous people (who continue to create them [Schmidt, 2013]) and *caboclos* (Fraser & Clement, 2008). Surprisingly, however, until recently there was no strong evidence that dark-earth Anthroposols were in fact intensively used by pre-Columbian cultivators. Iriarte et al. (2020) have now begun to fill the gaps in our knowledge of the land-use system these soils supported. Their results from terrestrial phytolith and charcoal records from profiles in 14 ADE and nine ABE sites from across Amazonia, as well as pollen-charcoal records from two lakes, show that both annual and perennial crops were cultivated on these soils, in systems interpreted as polyculture agroforestry "characterised by a mosaic of patches in different stages of succession, forming a complex landscape that transitions from forest to field and back to forest again" (Iriarte et al., 2020: 19).

Evidence now also links the creation and use in agriculture of these transformed, enriched soils, which must have allowed greater production per unit area, to the spread of farmers associated with the ancestors of major linguistic families (Iriarte et al., 2020). Evidence further links the creation of these soils and their intensive use in agriculture to demographic expansion in Amazonia (Arroyo-Kalin & Riris, 2020). Increased food production beginning in the mid-Holocene led to increases in the pace and extent of landscape modification. These transformations culminated in the late Holocene, when the distribution of radiocarbon dates indicated that human populations in Amazonia had reached their maximum demographic expansion (Arroyo-Kalin & Riris, 2020).

Data presented by Arroyo-Kalin & Riris (2020) show that the growth of the human population of Amazonia over the 1700 years prior to European colonisation adheres to a logistic model of demographic growth. They raise the intriguing question of whether Amazonian populations in the late Holocene had reached environmental carrying capacity before the European conquest. This question is a striking illustration of how our conception of the roles of environment and human agency in shaping Amazonian societies have changed over time. 'Carrying capacity' is a notion assoiated with neo-evolutionist cultural ecology, a tradition largely rejected by early historical ecologists working in Amazonia in favor of approaches that grant primacy to human agency. The emphasis on human agency spurred extensive research on how humans modify environments to suit their needs. The question posed by Arroyo-Kalin and Riris includes the realization that while human niche-constructing activities extend the possibilities offered by environment, they do not altogether remove environmental limits. Archaeology needs ecology, and ecology needs archaeology (Briggs et al., 2006). Historical ecology needs both.

4.5 Conclusions

Adopting historical ecology as an approach obliges one to abandon the observation of an archaeological snapshot, a frozen capture of a past moment, in order to try to understand a long-term evolutionary process. Historical ecology opens a privileged way to understanding the dynamics of landscapes. In short, historical ecology has opened a scientific Pandora's box in Amazonia, mobilizing a host of methodologies and disciplines, and releasing a flood of inferences, hypotheses and innovations. A recent paper (Silva et al., 2021) attempted to revive a hypothesis for the geogenic (rather than anthropogenic) origin of ADEs. They argued that the fertility of ADEs results primarily from fluvial deposition, and that these soils were simply used by pre-Columbian humans. In responding to this paper, Lombardo et al. (2021) point out that this hypothesis was laid to rest by Smith (1980), but also present the further evidence accumulated for an anthropogenic origin in the last 40 years. Lombardo and co-authors illustrate the richness of the methods mobilized in Amazonian historical ecology and the power of the results this approach has produced. Judging by the rapid advances over the past few years reviewed in this chapter, the flood of new information does not appear ready to dry up.

Aknowledgments DM's research on this topic has been funded by a fellowship from the Institut Universitaire de France and by grants from the CNRS (Institute of Ecology and Environment; Mission for Interdisciplinarity) and from the TOSCA committee (Terre Solide, Océan, Surfaces Continentales, Atmosphère) of the CNES (French National Center for Space Research), FLOODSCAPE project. SR's archaeological research on earthmounds in Ecuador was funded by the French Ministry of Foreign Affairs and that on raised fields in French Guiana was funded by the French Ministry of Culture. Research conducted jointly by McKey and Rostain on vestiges of raised fields in French Guiana was funded by a grant from the CNRS programme "Amazonie". We thank Drs. Eduardo Neves, Martti Pärssinen, and Delphine Renard for permission to use their photographs.

References

Åkesson, C. M., McMichael, C. N., Raczka, M. F., Huisman, S. N., Palmeira, M., Vogel, J., Neill, D., Veizaj, J., & Bush, M. B. (2021). Long-term ecological legacies in western Amazonia. *Journal of Ecology, 109*(1), 432–446.

Arroyo-Kalin, M. (2016). Landscaping, landscape legacies, and Landesque Capital in pre-Columbian Amazonia. In C. Isendahl & D. Stump (Eds.), *The Oxford handbook of historical ecology and applied archaeology* (Vol. 29, pp. 1–24). Oxford University Press. https://doi.org/10.1093/oxfordhb/9780199672691.013.16

Arroyo-Kalin, M., & Riris, P. (2020). Did Amazonian pre-Columbian populations reach carrying capacity during the late Holocene? *Philosophical Transactions of the Royal Society B, 376*, 20190715.

Balée, W. (1998). History ecology: Premises and postulates. In W. Balée (Ed.), *Advances in historical ecology* (pp. 13–29). Columbia University Press.

Balée, W. (2013). *Cultural forests of the Amazon: A historical ecology of people and their landscapes, Tuscaloosa*. University of Alabama Press.

Balée, W., & Erickson, C. L. (Eds.). (2006). *Time and complexity in historical ecology*. Columbia University Press.

Bates, M. (1948). Climate and vegetation in the Villavicencio region of eastern Colombia. *Geographical Review, 38*, 555–574. https://doi.org/10.2307/211443

Boixadera, J., Esteban, I., Albert, R. M., & Poch, R. M. (2019). Anthropogenic soils from llanos de Moxos (Bolivia): Soils from pre-Columbian raised fields. *Catena, 172*, 21–39.

Briggs, J. M., Spielmann, K. A., Schaafsma, H., Kintigh, K. W., Kruse, M., Morehouse, K., & Schollmeyer, K. (2006). Why ecology needs archaeologists and archaeology needs ecologists. *Frontiers in Ecology and the Environment, 4*(4), 180–188.

Clement, C. R. (1999). 1492 and the loss of Amazonian crop genetic resources. I. the relation between domestication and human population decline. *Economic Botany, 53-2*, 188–202.

Clement, C. R., de Cristo-Araújo, M., d'Eeckenbrugge, G. C., Alves Pereira, A., & Picanço-Rodrigues, D. (2010). Origin and domestication of native Amazonian crops. *Diversity, 2*, 72–106.

Clement, C. R., Denevan, W. M., Heckenberger, M. J., Braga Junqueira, A., Neves, E., & G., Teixeira, W. G. & Woods, W. I. (2015). The domestication of Amazonia before European conquest. *Proceedings of the Royal Society B: Biological Sciences, 282*, 1812. https://doi.org/10.1098/rspb.2015.0813

Clement, C. R., Casas, A., Parra-Rondinel, F. A., Levis, C., Peroni, N., Hanazaki, N., Cortés-Zárraga, L., Rangel-Landa, S., Palhares Alves, R., Ferreira, M. J., Franco Cassino, M., Deambrozi Coelho, S., Cruz-Soriano, A., Pancorbo-Olivera, M., Blancas, J., Martínez-Ballesté, A., Lemes, G., Lotero-Velásquez, E., Mutti Bertin, V., & Mazzochini, G. G. (2021). Disentangling domestication from food production systems in the Neotropics. *Quaternary, 4*(1), 4.

Cunha, L., Brown, G. G., Stanton, D. W., Da Silva, E., Hansel, F. A., Jorge, G., McKey, D., Vidal-Torrado, P., Macedo, R. S., Velasquez, E., & James, S. W. (2016). Soil animals and pedogenesis: The role of earthworms in anthropogenic soils. *Soil Science, 181*(3/4), 110–125.

de Carvalho, A. L., Nelson, B. W., Bianchini, M. C., Plagnol, D., Kuplich, T. M., & Daly, D. C. (2013). Bamboo-dominated forests of the Southwest Amazon: Detection, spatial extent, life cycle length and flowering waves. *PLoS One, 8*(1), e54852.

Denevan, W. M. (1996). A bluff model of riverine settlement in prehistoric Amazonia. *Annals of the Association of American Geographers, 86*(4), 654–681.

Duncan, N. A., Loughlin, N. J., Walker, J. H., Hocking, E. P., & Whitney, B. S. (2021). Pre-Columbian fire management and control of climate-driven floodwaters over 3,500 years in southwestern Amazonia. *Proceedings of the National Academy of Sciences., 40*, 202022206. https://doi.org/10.1073/pnas.2022206118

Finney, F. A. (2012). The forgotten natural prairie mounds of the upper Midwest: Their abundance, distribution, origin, and archaeological implications. In J. L. Horwath Burnham & D. L. Johnson (Eds.), *Mima mounds: The case for polygenesis and bioturbation: Geological society of America* (Special Paper) (Vol. 490, pp. 85–133). https://doi.org/10.1130/2012.2490(05)

Flores, B. M., & Levis, C. (2021). Human-food feedback in tropical forests. *Science, 372*(6547), 1146–1147.

Fraser, J. A., & Clement, C. R. (2008). "Dark earths and manioc cultivation in Central Amazonia: A window on pre-Columbian agricultural systems?", *Boletim do Museu Paraense Emílio Goeldi. Ciências Humanas, 3*, 175–194.

Fraser, J., Teixeira, W., Falcão, N., Woods, W., Lehmann, J., & Junqueira, A. B. (2011). Anthropogenic soils in the Central Amazon: From categories to a continuum. *Area, 43*(3), 264–273.

Furquim, L. P., Watling, J., Hilbert, L. M., Shock, M. P., Prestes-Carneiro, G., Calo, C. M., Py-Daniel, A. R., Brandão, K., Pugliese, F., Zimpel, C. A., & Silva, C. A. D. (2021). Facing change through diversity: Resilience and diversification of plant management strategies during the mid to late Holocene transition at the Monte Castelo Shellmound, SW Amazonia. *Quaternary, 4*(1), 8.

Glaser, B., & Birk, J. J. (2012). State of the scientific knowledge on properties and genesis of anthropogenic dark earths in Central Amazonia (terra preta de Índio). *Geochimica et Cosmochimica Acta, 82*, 39–51.

Hilbert, L., Neves, E. G., Pugliese, F., Whitney, B. S., Shock, M., Veasey, E., Zimpel, C. A., & Iriarte, J. (2017). Evidence for mid-Holocene rice domestication in the Americas. *Nature Ecology & Evolution, 1*, 1693e1698.

Iriarte, J., Elliott, S., Maezumi, S. Y., Alves, D., Gonda, R., Robinson, M., de Souza, J. G., Watling, J., & Handley, J. (2020). The origins of Amazonian landscapes: Plant cultivation, domestication and the spread of food production in tropical South America. *Quaternary Science Reviews, 248*, 106582.

Kämpf, N., Woods, W. I., Sombroek, W. G., Kern, D. C., & Cunha, T. J. F. (2003). Classification of Amazonian dark earths and other ancient anthropic soils. In J. Lehmann, D. Kern, B. Glaser, & W. Woods (Eds.), *Amazonian dark earths. Origin, properties and management* (pp. 77–102). Kluwer.

Lathrap, D. W. (1970). *The upper Amazon*. Thames & Hudson.

Leonor, R., Tobias, S., Carine, B. M., Bernard, G. B., Isabelle, B., Marion, C., Stéphen, R., Joseph, Y, & Doyle, M. (2020). *A Congo Basin ethnographic analogue of pre-Columbian Amazonian raised fields shows the ephemeral legacy of organic matter management Abstract Scientific Reports, 10*(1), https://doi.org/10.1038/s41598-020-67467-8

Levis, C., et al. (2017). Persistent effects of pre-Columbian plant domestication on Amazonian forest composition. *Science, 355*(6328), 925–931. https://doi.org/10.1126/science.aal0157

Levis, C., Flores, B. M., Moreira, P. A., Luize, B. G., Alves, R. P., Franco-Moraes, J., Lins, J., Konings, E., Peña-Claros, M., Bongers, F., & Costa, F. R. (2018). How people domesticated Amazonian forests. *Frontiers in Ecology and Evolution, 5*, 171.

Lévi-Strauss, C. (1948). The use of wild plants in tropical South America. In J. Steward (Ed.), *Handbook of south American Indians* (Vol. 6, pp. 465–486). Bureau of American Ethnology.

Lombardo, U., et al. (2021). Evidence confirms an anthropic origin of Amazonian dark earths. *Nature Communications, 13*(1), 3444.

Lombardo, U., Iriarte, J., Hilbert, L., Ruiz-Perez, J., Capriles, J. M., & Veit, H. (2020). Early Holocene crop cultivation and landscape modification in Amazonia. *Nature, 581*, 190e193.

López Mazz, J., Rostain, S., & McKey, D. B. (2016). Cerritos, tolas, tesos, camellones y otros montículos de las tierras bajas de Sudamérica. *Revista de Arqueologia, 29-1*, 86–113. www.revista.sabnet.com.br/revista/index.php/SAB/.../419/

Maezumi, S. Y., Alves, D., Robinson, M., de Souza, J. G., Levis, C., Barnett, R. L., Almeida de Oliveira, E., Urrego, D., Schaan, D., & Iriarte, J. (2018). The legacy of 4,500 years of polyculture agroforestry in the eastern Amazon. *Nature plants, 4*(8), 540–547.

McKey, D. B., Rostain, S., Iriarte, J., Glaser, B., Birk, J. J., Holst, I., & Renard, D. (2010). Pre-Columbian agricultural landscapes, ecosystem engineers and self-organized patchiness in Amazonia. *Proceedings of the National Academy of Sciences USA, 107-17*, 7823–7828. https://doi.org/10.1073/pnas.0908925107

McKey, D. B., Durécu, M., Solibiéda, A., Raimond, C., Montoya, K. L. A., Iriarte, J., Renard, D., Suarez Jimenez, L. E., Rostain, S., & Zangerlé, A. (2014). New approaches to pre-Columbian raised-field agriculture: Ecology of seasonally flooded savannas, and living raised fields in Africa, as windows on the past and the future. In S. Rostain (Ed.), *Amazonía. Memorias de las conferencias magistrales del 3er Encuentro Internacional de Arqueología Amazónica* (pp. 91–136). MCCTH, SENESCYT, 3EIAA (Ministerio Coordinador de Conocimiento y Talento Humano, Secretaría de Educación Superior, Ciencia, Tecnología e Innovación, 3e International Congress of Amazonian Archaeology).

McKey, D. B., Durécu, M., Pouilly, M., Béarez, P., Ovando, A., Kalebe, M., & Huchzermeyer, C. F. (2016). Present-day African analogue of a pre-European Amazonian floodplain fishery shows convergence in cultural niche construction. *Proceedings of the National Academy of Sciences USA, 113-52*, 14938–14943. https://doi.org/10.1073/pnas.1613169114

McMichael, C. H., Feeley, K. J., Dick, C. W., Piperno, D. R., & Bush, M. B. (2017). Comment on "persistent effects of pre-Columbian plant domestication on Amazonian forest composition". *Science, 358*(6361).

Meggers, B. J., (1996[1971]), *Amazonia. Man and culture in a counterfeit paradise* (revised edition). Smithsonian Institution Press.

Meggers, B. J. (2003). Natural versus anthropogenic sources of Amazonian biodiversity: The continuing quest for El Dorado. In G. A. Bradshaw & P. A. Marquet (Eds.), *How landscapes change* (pp. 89–107). Springer.

Neves, E. G. (2007). El Formativo que nunca terminé. La larga historia de la estabilidad en las ocupaciones humanas de la Amazonía Central. *Boletín de Arqueología PUCP, 11*, 117–142.

Neves, E. G., & Heckenberger, M. J. (2019). The call of the wild: Rethinking food production in ancient Amazonia. *Annual Review of Anthropology, 48*, 371–388.
Neves, E. G., & Rostain, S. (2012). Diversité linguistique et agrobiologique dans le passé amazonien. In N. Schlanger & A.-C. Taylor (Eds.), *L'archéologie des autres* (pp. 119–136). Perspectives archéologiques et anthropologiques, La Découverte.
Odonne, G., & Molino, J.-F. (Eds.). (2021). *Methods in historical ecology*. Insights from Amazonia.
Piperno, D. R., McMichael, C., & Bush, M. B. (2017). Further evidence for localized, short-term anthropogenic forest alterations across pre-Columbian Amazonia. *Proceedings of the National Academy of Sciences, 114*(21), E4118–E4119.
Renard, D., Birk, J. J., Zangerlé, A., Lavelle, P., Glaser, B., Blatrix, R., & McKey, D. B. (2013). Ancient human agricultural practices can promote activities of contemporary non-human soil ecosystem engineers: A case study in coastal savannas of French Guiana. *Soil Biology and Biochemistry, 62*, 46–56. https://doi.org/10.1073/pnas.0908925107
Robinson, M., Jaimes-Betancourt, C., Elliott, S., Maezumi, S. Y., Hilbert, L., Alves, D., de Souza, J. G., & Iriarte, J. (2021). Anthropogenic soil and settlement organisation in the Bolivian Amazon. *Geoarchaeology, 36*(3), 388–403.
Rodrigues, L., Sprafke, T., Bokatola Moyikola, C., Barthès, B. G., Bertrand, I., Comptour, M., Rostain, S., Yoka, J., & McKey, D. (2020). A Congo Basin ethnographic analogue of pre-Columbian Amazonian raised fields shows the ephemeral legacy of organic matter management. *Scientific Reports, 10*(1), 10851. https://doi.org/10.1038/s41598-020-67467-8
Roosevelt, A. C. (1991). *Moundbuilders of the Amazon: Geophysical archaeology on Marajo Island*. Academic.
Rostain, S. (2010). Pre-Columbian earthworks in coastal Amazonia. *Diversity, 2*(3), 353–369.
Rostain, S. (2012). *Islands in the rainforest. Landscape management in pre-Columbian Amazonia*. Left Coast Press, Walnut Creek.
Rostain, S. (Ed.), (2015). Archéologie de l'Amazonie. *Les premiers habitants de la Guyane côtière*, BAR International Series 2758, Paris Monographs in American Archaeology 44, Archaeopress.
Rostain, S. (2016). *Amazonie. Un jardin sauvage ou une forêt domestiquée*. Essai d'écologie historique.
Rostain, S., & Betancourt, C. J. (Eds.), (2017). *Las Siete Maravillas de la Amazonia precolombina*, IV Encuentro Internacional de Arqueología Amazónica, Bonner Altamerika-Sammlung un Studien. La Paz (Bolivie), Plural Editores.
Rostain, S., & McKey, D. B. (2015a). Les paysages de champs surélevés de Guyane française: un patrimoine bioculturel menace. *Revue d'Ethnoécologie, 7*. https://doi.org/10.4000/ethnoecologie.2193
Rostain, S., & McKey, D. B. (2015b). Des fourmis et des hommes. Les constructeurs de paysage d'Amazonie. *Les carnets du paysage, 27*, 128–141.
Schaan, D., Parssinen, M., Saunaluoma, S., Ranzi, A., Bueno, M., & Barbosa, A. (2012). New radiometric dates for Precolumbian (2000–700 BP) earthworks in western Amazonia, Brazil. *Journal of Field Archaeology, 37*, 132–142. https://doi.org/10.1179/0093469012Z.00000000012
Schmidt, M. J. (2013). Amazonian dark earths: Pathways to sustainable development in tropical rainforests? *Boletim do Museu Paraense Emílio Goeldi Ciências Humanas, 8*(1), 11–38.
Silva, L. C. R., et al. (2021). A new hypothesis for the origin of Amazonian dark earths. *Nature Communications, 12*, 127. https://doi.org/10.1038/s41467-020-20184-2
Smith, N. J. H. (1980). Anthrosols and human carrying capacity in Amazonia. *Annals of the Association of American Geographers, 70*, 553–566.
Smith, E. A. (2013). Agency and adaptation: New directions in evolutionary anthropology. *Annual Review of Anthropology, 42*, 103–120. https://doi.org/10.1146/annurev-anthro-092412-155447
Steward, J. (1948). Handbook of South American Indians. In *The Tropical Forest Tribes* (Vol. 3). Smithsonian Institution, Bureau of American Ethnology, bulletin 143.
Ter Steege, H., et al. (2013). Hyperdominance in the Amazonian tree flora. *Science, 342*(6156), 325–333.

Watling, J., Saunaluoma, S., Pärssinen, M., & Schaan, D. (2015). Subsistence practices among earthwork builders: Phytolith evidence from archaeological sites in the southwest Amazonian interfluves. *Journal of Archaeological Science: Reports, 4*, 541–551. https://doi.org/10.1016/j.jasrep.2015.10.014

Watling, J., Iriarte, J., Mayle, F., Schaan, D., Pessenda, L., Loader, N. F., Street-Perrott, A., Dickau, R. E., Damasceno, A., & Ranzi, A. (2017a). Impact of pre-Columbian "geoglyph" builders on Amazonian forests. *Proceedings of the National Academy of Sciences USA, 114-8*, 1868–1873. https://doi.org/10.1073/pnas.1614359114

Watling, J., Iriarte, J., Mayle, F. E., Schaan, D., Pessenda, L. C., Loader, N. J., Street-Perrott, F. A., Dickau, R. E., Damasceno, A., & Ranzi, A. (2017b). Reply to Piperno et al.: It is too soon to argue for localized, short-term human impacts in interfluvial Amazonia. *Proceedings of the National Academy of Sciences, 114*(21), E4120–E4121.

WinklerPrins, A. M. A., & Levis, C. (2020). Reframing pre-European Amazonia through an Anthropocene lens. *Annals of the American Association of Geographers, 111*(3), 858–868. https://doi.org/10.1080/24694452.2020.1843996

Woods, W. I., & McCann, J. M. (1999). The anthropogenic origin and persistence of Amazonian dark earths. In *Yearbook, conference of Latin Americanist geographers* (Vol. 25, pp. 7–14). University of Texas Press.

Zangerlé, A., Renard, D., Iriarte, J., Suarez Jimenez, L. E., Montoya, K. L. A., Juilleret, J., & McKey, D. B. (2016). The *surales*, self-organized earth-mound landscapes made by earthworms in a seasonal tropical wetland. *PLoS One, 11*(5), e0154269. https://doi.org/10.1371/journal.pone.0154269

Chapter 5
Building Databases for the Generation of Archeological Landscape Models in Santa Catarina (Brazil)

Mirian Carbonera and Daniel Loponte

Abstract In this study we investigate the biases that affect the database of the archaeological record for certain areas of the state of Santa Catarina (Brazil). For this purpose, we describe the development of a project aimed to identify archaeological sites, which allowed the analysis of the main biases that affect the construction of reliable archaeological information regarding the distribution of archaeological sites. These major biases include destruction of sites, lack of systematic surveys, and duplicate, incomplete, or erroneous records. Beyond these problems, the identification of the different sources of error becomes an essential source of information to model archaeological landscapes in a more prudent way, considering the potential biases in the distribution of the known record.

Keywords South Brazil · Landscape archaeology · Archaeological databases · Heritage conservation

5.1 Introduction

In this chapter we discuss different biases affecting the databases used to construct the distribution of the archaeological record. To do this, we present some results and reflections derived from the field work within the PRESASC program carried out by the authors in the state of Santa Catarina (Brazil), which allowed the construction of a new database of the archaeological record for some areas of the state. Then, we

---

M. Carbonera (✉)
Programa de Pós-Graduação em Ciências Ambientais e Centro de Memória do Oeste de Santa Catarina (CEOM) – Unochapecó, Chapecó, Brazil
e-mail: mirianc@unochapeco.edu.br

D. Loponte
Consejo Nacional de Investigaciones Científicas y Técnicas (CONICET). Instituto Nacional de Antropología y Pensamiento Latinoamericano (INAPL), Buenos Aires, Argentina
e-mail: dashtown@gmail.com

A. C. Colonese, R. G. Milheira (eds.), *Historical Ecology and Landscape Archaeology in Lowland South America*, Interdisciplinary Contributions to Archaeology, https://doi.org/10.1007/978-3-031-32284-6_5

identified several biases that impact these records to finally consider in the discussion how these bias could affect the construction of archaeological landscape models.

5.2 The PRESASC Program

The PRESASC project was a program to locate and assess the state of preservation of 533 known archaeological sites distributed in 47 counties along four major geographic areas in the state of Santa Catarina (Fig. 5.1). This ambitious plan was an initiative of the *Instituto do Patrimônio Histórico e Artístico Nacional* (IPHAN)[1] ("National Historical and Artistic Heritage Institute"), through its 11th Regional Superintendency located in Santa Catarina, led by Liliane Janine Nizzola, and it was created as a result of a Conduct Adjustment Term (TAC in Portuguese), which is an extrajudicial instrument for conflict resolution. The TAC was generated after an energy company caused presumed damage to archaeological goods, violating Federal Law 3924 of 1961. The repair agreement between the company and the IPHAN included a compensation payment to the state. The IPHAN allocated these

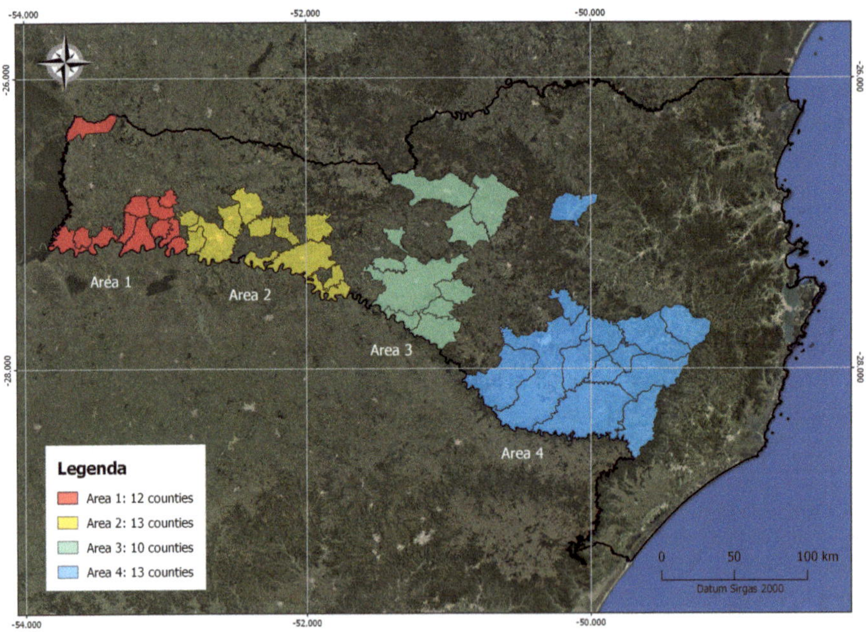

Fig. 5.1 Research areas included in the program

[1] IPHAN is a federal agency (Ministry of Tourism) with 27 regional offices along the country, which are responsible for the preservation of the Brazilian cultural heritage, including the management of the archaeological heritage.

financial resources to improve its databases through the accurate location of archaeological sites, the evaluation of their state of preservation, and the marking of 30 sites of great interest for local archeology or for their pedagogical capacity through allusive plaques. To operationalize the TAC, the company conducted a bid for prices and background evaluation of the investigation teams that were invited to quote. As a result, the Universidade Comunitária da Região de Chapecó (UNOCHAPECÓ) and the Centro de Memória do Oeste de Santa Catarina (CEOM) were chosen to carry out the program. After that, a survey project was sent by the CEOM to IPHAN, which was evaluated and approved. The respective field work permits were later published in the official journal of the federal state prior to starting the surveys.

The sites included in the project were identified between the 1960s and 2000s during the first stages of the archaeological investigations of South Brazil, while others were recorded during different unpublished CRM projects beginning in the 1980s. Most of them were generically located and briefly described in the original studies, and with these high levels of uncertainty in terms of location and preservation status, they were transcribed into the official databases. The most common information of the location of the sites included the name of the landowner, or the name of the rural property, and a brief generic location (*e.g.* "near the river..."; "400 to 500 meters approximately from the small waterfall to the north…"). In farms of hundreds of hectares, or large areas covered by natural and dense vegetation, this information had a different degree of precision and usefulness. Generally speaking, this type of information allowed sites to be identified quickly if the owner who originally identified the site was available for such a task, or if the location information was transmitted to another person in an accurate manner. The reality is that most of the original owner had passed away, added to the fact that many of these properties have completely changed their appearance since the sites had been recorded. Some of these lands went from forest and small-scale agricultural production or cattle exploitation, to intensive agricultural production, changing deeply the landscape. Others have been divided for the construction of condos and urban expansion. A large number of these properties have also been sold since then, changing their names and the owners, or divided between different heirs of the original owner, or divided between heirs of new owners, hindering the location of properties and specific areas of the sites. If this information were somewhat complex to elucidate, it then became impossible to identify the areas of some sites generically located, such as "two pit-houses in Coxilha Rica" being Coxilha Rica a region of hundreds of km^2, or "…on the south face of the hill", being difficult to identify "the hill" among several existing in a property (often covered by dense vegetation), and where no memory of the current owners of that record persists. This type of very generic location often constituted an insurmountable problem for site identification. Finally, a small fraction of the sites identified in recent CRM projects (from the 1990s onwards) included geographical coordinates obtained by GPS. However, some of them feature several issues such as incomplete coordinates or coordinates falling outside the region, the State and even outside Brazil.

The PRESASC encompasses four areas divided into two well-defined regions. The first corresponds to the lowlands of the western region (~ 400 to 600 masl) of the Upper Uruguay River valley (Santa Catarina State, south Brazil). Most of the sites to be rediscovered were open-air sites of horticulturalists grouped in the Itararé-Taquara and Guarani[2] archaeological units, and a small fraction of sites belong to pre-pottery hunter-gatherers. All these sites are located in Areas 1 and 2 of Fig. 5.1. The first surveys in both areas were done by the Jesuits Balduíno Rambo and Pedro Ignácio Schmitz from the 1940s and 1950s respectively (Schmitz, 1957), followed by Jesuit João Alfredo Rohr, who explored and recorded 53 sites (Rohr, 1966, 1984). In the 1960s and 1970s, Walter Piazza's survey within the National Archaeological Research Program (PRONAPA) recorded numerous sites, covering the areas between the Peperi-Guaçu and Irani rivers, and later between the Irani and Peixe rivers, which included the Piratuba, Ipira, Itá and Concórdia counties (Piazza, 1969, 1971). At the same time, María José Reis, during the 1970s, surveyed the same region in neighboring counties such as Chapecó, São Carlos, Palmitos, Concórdia, Ipumirim and Pinhalzinho (Reis 1980[2007]), also recording a great quantity of archaeological sites. Other researches were carried out by recent CRM projects related to the construction of several dams (Foz do Chapecó, Itá, and Machadinho), and to environmental licensing projects (Campos, 2018; De Masi, 2012; Goulart, 1985, 1995; PUCRS, 2001).

The second region corresponds to the Santa Catarina Plateau, in the highlands (~ 700 to 1800 masl) of the eastern section of the Upper Uruguay valley (Areas 3 and 4 of Fig. 5.1), which comprises from Campos Novos to Urubici and Bom Jardim da Serra counties. The records here include open sites, pit-houses, mortuary mounds, platforms, ring-enclosures often with central mounds, rock engravings and funerary caves. Most of these sites belong to the Itararé-Taquara archaeological unit and a few sites to pre-pottery hunter-gatherers. The first investigation in this area was carried out by Dr. Jorge C. Bleyer, a German physician who settled in Lages city and published his data in the first decades of the twentieth century, which served in some cases for subsequent investigations. Between the 1960s and the 1970s, several researchers surveyed the region, recording numerous sites in Lages, Sao Joaquin, Urubici, Bom Jardim da Serra, and Bom Retiro counties (Miller, 1969; Piazza, 1966; Reis, 1980; Rohr, 1971, 1984). Subsequent work in this same region, among the Pelotas and Canoas river basins, were carried out by De Masi (2005) and PUCRS (1989), whose main researchers were Guilherme Naue, José J. P. Brochado and Fernando La Salvia. Most of the sites to be evaluated were identified by the projects and researchers mentioned above. In these same regions, other projects have been developed, most of them of recent date, which were left out of the specific interest of the PRESASC program.

[2] The names of the archaeological units and the geographic locations are expressed in Portuguese.

5.3 Theoretical and Methodological Issues

The PRESAC program included many activities that are directly related to the construction of solid archaeological databases, but also to the preservation of the record over time within the complex socioeconomic dynamics of the area. We will briefly point out all these aspects for a better understanding of the program.

5.3.1 Survey Design

The field work to rediscover the sites were designed using different kinds of information. Among them, the files of each site provided by the IPHAN, which generally mirror the basic data provided by the original researchers. In a few original reports we found some additional data for the localization of the sites (*e.g.* sketches). All this information was organized by county, and shared with the authorities of each municipality in order to identify the properties, tasks which were mainly derived to the departments of agriculture, who have the location of each rural property (landowner, location, property transfer processes by sale or inheritance). Thus, prior to the field work, we located many of the properties and the areas to be visited in a satellite database, which were analyzed by different layers of satellite images (Landsast; Google Earth; Landviewer, etc.). Therefore, the potential areas of the sites, and several archaeological structures were identified, particularly earthen architecture (Fig. 5.2). This exercise of comparison between satellite images and the later field observations of the same structures allowed us to carry out a visual learning as the project developed, improving our detection capacity to rediscover the archaeological sites and new archaeological structures through satellite images.

After all this pre-field evaluation, the survey routes were designed, grouping the properties of each area in each county. Previous coordination with landowners also helped to improve the surveys. The visits to the properties were mostly conducted by an *ad-hoc* team of archaeologists and staff from each county, which greatly facilitated the explanation to the landowners of the activities to be carried out, while obtaining permission to access properties. In some counties, this linking task fell independently, or sometimes concurrently, to intermediate community organizations, such as school teachers, neighborhood associations, and environmental organizations.

In some properties, the archaeological sites were well known by the landowners, because the archaeological remains had been exposed during agriculture labor or other activities. This happens especially in Guarani sites, characterized by large amounts of pottery. The same happened for artificial mounds, "pit-houses", rock engravings or ring-enclosures. However, in many properties careful surveys had to be implemented in order to rediscover the sites. This also made it possible to detect other sites that had not yet been identified. Therefore, to develop the program, it was necessary to carried out surveys in agricultural lands, hills, large extensions of cattle fields, preserved areas with native vegetation, closed forests, caves, suburban and

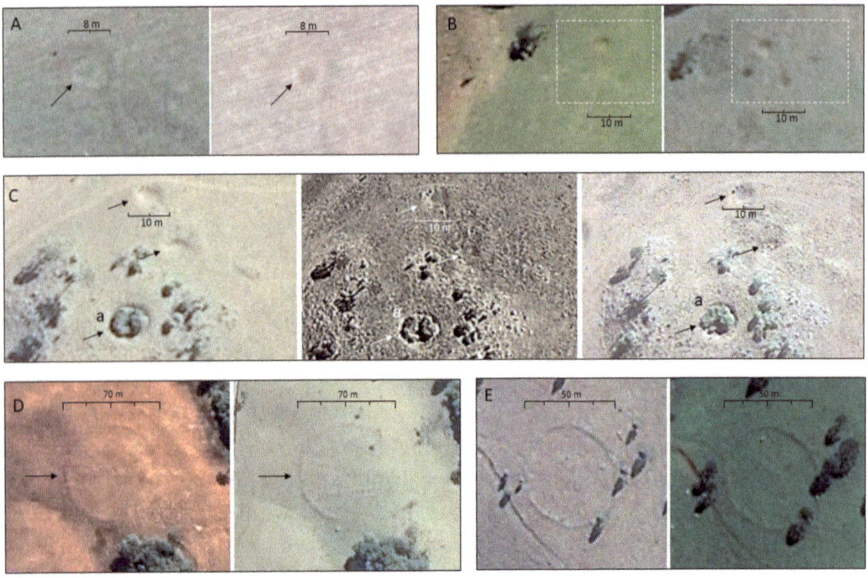

Fig. 5.2 Detection through images of archaeological structures (**a**) "pit-house" with disposition of the sediments excavated around, in the middle of a flat agricultural field. (**b**) several "pit-houses" in a flat cattle-field with natural grasses. (**c**) Three "pit-houses" in a hill used as cattle-field, one of them (structure "a") filled by growing trees. (**d** and **e**) ring-enclosure structures viewed with different lights.

urban areas. Systematic surface collections were made at a large number of sites, the stratigraphy was verified when this was possible and appropriate, and in some sites, charcoal, sediments and bones were collected for dating purposes and other studies. The resources used to approach the sites were basically land vehicles, covering ~60,000 km of paved and dirt roads during the 18 effective months of the project, but also boats had to be used to reach sites on islands, long surveys on foot, hill climbing, and rappelling (Fig. 5.3).

Combined actions between local authorities, communities, and the researchers were generally very effective in locating those archaeological sites that had some degree of specific location (e.g. name of the owner and/or name of the property), allowing us a privileged approach to the archaeological resources of the region, to the communities, landowners and local authorities, most of whom contributed to the field activities.

5.3.2 *Recording Archaeological Sites*

The IPHAN determined the minimum information that must be recorded of each archaeological site, including the name and contact details of the landowners, site-type, conservation state, topographical features, general maps to access the sites,

Fig. 5.3 Some examples of surveys and recordings during the field works. (**a**) Documentation of rock art on the flank of the hill accessed by abseiling (Rio do Campo County). (**b**) Recording the mortuary cave "Lapa da Medianeira" in the closed forest of the Xavantina County. (**c**) Photographic aerial record (by drone) at the Antônio De Liz de Souza I site, in the highlands of eastern Santa Catarina (Painel County). This image also shows the site's signaling plate already installed. (**d**) Surface collection at the L. Meier site (Mondaí County, lowlands of western Santa Catarina)

specific maps of their locations, photographs of the sites, and geographic coordinates (UTM, Datum Sirgas 2000). The latter should include the center of the sites, and their maximum axes, determined by coordinates. Here arose one of the classic methodological problems of field surveys, related to how an archaeological site is defined, which we discuss in the next section.

5.3.3 Identifying Archaeological Sites

Most of thosc sites with earthen structures (ring enclosures, pit-houses) and rock engravings were easily identified, as well as their midpoints and extensions. For open sites without architectural evidence, their recognition was based mainly by surface findings, and eventually, through test pits. For these last two cases, and given that the archaeological record is continuous in space, we looked for density peaks of findings to identify their potential "central areas" and the associated dispersion (Dunnell & Dancey, 1983; Ebert, 1992; Foley, 1980). Establishing a finding density threshold to identify archaeological sites is a matter of debate (Gallant, 1986; Hayes, 1991; Schiffer, 1987). Some projects used thresholds of five artefacts/ m^2 (Ebert, 1992: 48), or ten artifacts in areas smaller than 100 m^2 to establish the existence of an archaeological site (Pettigrew & Lebow, 1989). Besides, it is quite common in literature the use of an *ad-hoc* criteria, generally not explicit, and which are in the most closely related with the objectives of each research. This affects many studies of distributional archeology, including our region of study, where some archaeological maps indicate many "sites" which have been defined by highly debatable criteria. And certainly, this problem constituted an issue for our program, where we had to relocate sites defined by other researchers under very different standards. Since the definition of an archaeological site is mostly linked to the objectives of each survey, we established a conservative strategy of low-density thresholds of 5 artifacts in 400 m^2 as a minimum threshold (5 findings in 20 x 20 m quadrats). Besides, this conservative strategy allows these areas to be protected by Brazilian legislation, especially the Federal Law 3.924/196. Those spaces that during the surveys showed densities lower than this threshold, but that the original researchers or the owners indicated as areas of regular archaeological findings, should also be under legal protection, since more detailed investigations might find more artefacts or structures. Therefore, they were classified as "inconclusive sites" (see below).

A similar methodological problem arose when determining the size of the sites. This information is mandatory to complete the IPHAN electronic registration files, which requires, in addition to the geographic coordinates of the "central area", the coordinates of the "perimeter" of each site. The delimitation of sites with earthen and stone architecture (mounds, ring-enclosures, platforms, pit-houses, etc.) or rock engravings, is a relatively simple task. Here each axis was considered by taking a larger space than the visible structures to create buffer areas and to protect potential buried archaeological findings. In the same way, this was carried out in those sites with surface artefacts, where the axis of each site was calculated up to the perimeter of the distribution of the last finding, adding a buffer area, which in general depended on the geomorphology. The landowners´ observations, which often pointed out how far the artefacts were found during agricultural labors, were also taken into account concurrently. For deeply buried sites, test pits and geomorphological criterions were also considered. Thus, the polygonals consigned in the new database are preservation areas rather than precise extensions of each site, which require further investigations.

Another methodological problem showed up when comparing some rediscovered sites whose structure or composition did not exactly match the previous descriptions made by the original researchers. This was often the case with the number of pit-houses at some sites. The origin of these differences could partially be attributed to the greater forest coverage in the sites when the first investigations were carried out, which limited the archaeological visibility. Inverse cases have also been observed, where the number of structures is less, probably due to some of them being filled in. Other minor discrepancies are differences in the size of some structures, or site locations that are not entirely accurate, which in general is attributable to the current availability of better location tools. In a few cases we found greater discrepancies in the locations of sites. This was particularly verified when sites correspond adequately to the original descriptions and to the originally identified owners and properties, but erroneously located, sometimes dozens of kilometers away from the correct location. These incongruencies were especially verified in one particular research carried out several decades ago, involving the location of pit-houses and burial mounds in the Highlands of Santa Catarina.

5.3.4 Preservation States

Six categories were established according to the preservation state or status of each site: conserved, impacted, inconclusive, destroyed, flooded and without data. The first category includes those sites with a high degree of integrity. This is the case of ring-enclosures or "pit-houses" that have been preserved almost in their entirety, that is, when the rings and their associated mounds were complete or close to, or when the pit-houses were not artificially filled by sediments, their walls were well-preserved, and there was little disturbance due to the growth of vegetation or anthropic alteration. Regarding sites in stratigraphy, this definition includes those sites in which their layers seem to be "intact", where archaeological materials do not appear to have been redeposited or impacted by natural or anthropic activities. This category, of course, is a first sight appreciation, and a detailed analysis of the conservation status will require extensive excavations.

The "impacted" category includes those sites with structures or artifacts distribution with a substantive modification due to natural or anthropic disturbances. This case includes, for example, pit-houses filled with rocks or sediments, growing trees or partially excavated sites; ring-enclosures with some degree of destruction of their mounds or their peripheral rings, or rock engravings with modern stripes or graffities. For buried sites, the concept of "impacted" was applied to those where the plow or fluvial erosion has disturbed the archaeological layers. Those sites with mixed conservation (some layers or areas are impacted but others are preserved) were also typified as impacted.

The "inconclusive" category grouped those sites with a lower density of artefacts than the minimum threshold (see previous section), or where it was not possible to detect any findings, but considered previous researches, and/or oral information, making them potential areas for future and deeper research, protecting them with a legal status. The "destroyed" category refers to sites completely removed, generally due to mechanized agriculture, mining, afforestation or infrastructure construction. The "flooded" category are those sites submerged by dams or natural causes, where it is not possible to verify their degree of conservation nor their exact position. Finally, we incorporate the category of "no data", for those sites that have not been located due to generic or imprecise location, incomplete or erroneous data, or those where the landowners have not allowed surveys, which was the case for a few sites.

5.3.5 Signaling Archaeological Sites

During the implementation of the program, thirty archaeological sites were selected to be signaled by plaques. The selection criteria include some characteristics, such as a good degree of preservation, visibility or archaeological features, an outstanding record in relation to the average observed in the region, and easily accessible for the local community and public. The permits to signal the sites of the respective landowners were also a *sine qua non* condition. The plaque designs were defined by the IPHAN, using steel of 3.75 mm of thickness, with a size of 0.70 m x 1.15 m with electrostatic paint, which confers greater durability to the material. Two protecting and transparent films cover the surface. The plaques were fixed on 1.50 m high, chemically treated wood levers, placed in immediately adjacent areas to the sites (Fig. 5.4b). The selection of these sites, the information and layout of each plate, were designed by the authors.

5.3.6 Creating a Framework for Dialogue and Awareness for Conservation

Throughout the development of the program, a dissemination work was carried out on the importance of preserving archaeological sites with local authorities, and with the owners of the lands where they are located. This was carried out through personal interaction, supported by the distribution of educational brochures (Carbonera & Cecchin, 2015, Carbonera et al., 2016), one for adults, and another for children. During the execution of the program, the researchers also received multiple inquiries and questions from the landowners regarding what to do in the face of the fortuitous discovery of archaeological materials, what activities they should not carry out to encourage the conservation of the sites, and queries related to the legal framework on ownership of the findings and even of the land where the archaeological sites are located. Worth mention that in some cases, a few landowners had already spontaneously developed passive conservation measures, such as limiting livestock access, or avoiding using the area of the sites for afforestation, generally in those sites with earthen architecture (mounds, funerary structures, platforms and pit-houses). The research team also conducted many radio, television and graphic interviews with different local and state media, aimed at the general public, seeking to make the program and cultural heritage visible, and to involve the community to protect it (Fig. 5.4).

Fig. 5.4 (**a**) Interaction with local communities. A: presentation of the program in the Itapiranga County to the farmers of the MAB social movement (Movement of people affected by dams). All of them landowners, and mostly with archaeological sites in their properties. (**b**) Example of one of the plaques signaling an archaeological site (Contestado War Cemetery - historic site - Irani County). (**c**) Delivery of educational brochures to landowners, carried out in most of the sites recorded in the program. (**d**) one of many interviews conducted for state and local media, in this case conducted at one of the discovered sites

5.3.7 Constructing a New Archaeological Database

The actions described above allowed the identification and location of a large number of sites, as well as the discovery of new ones, all of them with a precise location, including nearly 400 pit houses identified at different sites. This new database include additional information such as new radiocarbon dates, digital topographies of sites with monumental structures, stratigraphic profiles, the identification of archaeological sequences and extensive photographic documentation. All this information was also added within the ongoing research programs in the region (Loponte & Carbonera, 2013, 2021). On the other hand, fieldwork confirmed the destruction of many sites.

5.4 Results

5.4.1 Overview

From 533 sites to be verified, 72 of them were duplicated because they were recorded twice in the IPHAN database. For example, the site originally registered in the Anita Garibaldi County "SC-UC-423: Silvino de Morais" (File SC00083), was duplicated with the "Silvino de Morais" site in the same county (File SC01054). These duplications were verified by concurrent events. Among them, all duplicate sites have the same information (same County, same locality, same landowner, same description of findings), although sometimes the names are slightly modified, or an alphanumeric code was added. This duplication was produced by the recording of the same sites into the electronic files of the IPHAN in two different administrative offices of this federal agency, carried out by different technicians in different times. This single act of analysis immediately allowed the database to be improved.

A second analysis of the original files was carried out with the authorities of each county. Among other results, 68 sites were identified in flooded properties by different dams of the Uruguay River, which was also confirmed during the surveys in the neighboring areas. Subtracting both the duplicated and flooded sites, 393 sites remained to be located, of which 277 were positively identified during the field surveys (70.5%) and 116 (29.5%) that could not be found, for the reasons stated in Sect. 5.2. Among the 277 located sites, 34 had been destroyed by different agents (see next section) before this program started, leaving 243 extant sites with different degrees of conservation. The field work also made it possible to identify 76 new sites not previously recorded in the IPHAN files. Thus, between the relocated sites that are still preserved and the new records, a new database of 319 sites was set up, integrated by different site-types (see examples in Fig. 5.5), and distributed according to their different features (Fig. 5.6).

The largest number of sites identified during the field work corresponds to Guarani sites, which are remarkably clustered on the margin of the Uruguay River in Areas 1 and 2 (see Fig. 5.1). Surface collections, test-pits or quadrats in different Guarani sites made it possible to obtain a large quantity of pottery, and to a lesser extent, faunal and human bones, lithic and ornamental artifacts. The Guarani record seems to distribute continuously along the Uruguay River from Itapiranga to the confluence of the Canoas and Pelotas rivers, with different density peaks. The second most abundant record are pit-houses, which also present a differential distribution, being concentrated in the counties of Areas 3 and 4 (see Fig. 5.1). All of them seem to belong to the Itararé-Taquara archaeological unit. Some of these sites are composed of a single structure, while others have up to 30. In all cases, all the structures of each site were measured and located with geographic coordinates, totaling 415 structures precisely located (average 5.6 pit-houses/site).

Most of the multicomponent sites correspond to Guarani and Itararé-Taquara occupations, and a few to pre-pottery occupations which are mainly surface findings, and to a lesser extent, artifacts in stratigraphic position. These contexts are

Fig. 5.5 Some site-types identified during the surveys. (**a**) ring-enclosure with central mounds, João Machado site (Highlands, Vargem County, new record). (**b**) rock engravings from the Morro Pelado II site (Highlands, Urubici County, originally recorded in the 1960s). (**c**) Exploration quadrat at L. Meier site (Lowlands, Mondaí County, new record). (**d**) Burial mound from site SC-CL-26 I (Highlands, Lages County, recorded in the 1970s, but incorrectly located in the original research)

quite variable, and cannot be easily unified under a single criterion, except for the lack of pottery. Some of them are middle or early Holocene sites. The category "Mounds, enclosures and platforms" can include associations between different types of structures, including pit-houses. Radiocarbon dates obtained during the execution of the program of some of these complex sites demonstrate that they may include structures with different ages (see below). This type of record is also clustered in Areas 3 and 4 (see Fig. 5.1). The less abundant sites are caves used for mortuary purposes, and rock engravings, also located in areas 3 and 4. Finally, historical sites (18th and 20th centuries) and caves corresponding to paleontological sites were also recorded.

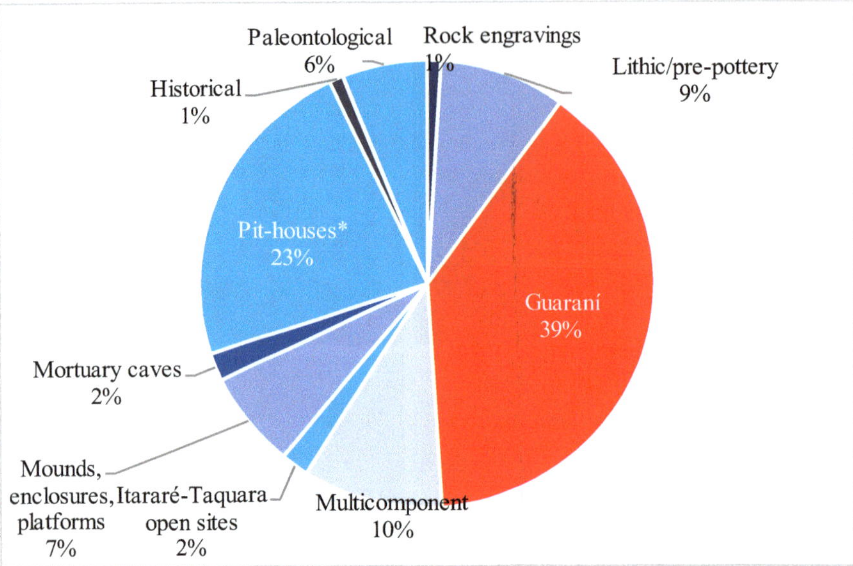

Fig. 5.6 Distribution of types of sites located during the survey (old and new records; n = 319 sites). *Pit-houses is a generic name for pit structures including residential sites and other functionalities

5.4.2 *Integrating Additional Information to the Database*

During the surveys, additional information was added to the database based on materials obtained in surface collections, test-pits, and from archaeological pieces or collections of great value donated by the landowners. Less frequently, rescue actions were carried out given the high possibility of loss or permanent damage to archaeological goods. Among them, the most common were the excavation of Guarani funerary urns, due to agricultural labor or water well excavations (or similar actions) that leave them exposed (Fig. 5.7). The surveys also provided the possibility of accessing some sites that remain under the water of the artificial lakes created by the dams but emerging few months during each year. Regarding this, we located and excavated the site U-318, whose record is preserved under a 30-cm layer of sand, clay, and silt deposited by the Itá dam lake. The findings included a large quantity of pottery, faunal remains, lithic artifacts, human bones, and structured hearts, showing a well-preserved archaelogical record (Fig. 5.8).

Fig. 5.7 Rescue actions in Itapiranga County. Left images, urn exposed and excavated at the Tarcisio Ternus site (new record), with the broken lid in image A, and without the lid in image A1. Images B and B1: extraction of funeral urn at SC-U-22 site. The pot lid was completely broken by the landowner, and recovered as sherds. The scales have 10 cm painted sections

Fig. 5.8 Excavations at U-318 site, subject to periodic flooding by the operation of the Itá dam. Above, the site between both arrows, emerged 1 m above the lake-level of the dam. Below, different views of quadrat 1

5.4.3 *The Destruction of the Archaeological Record*

The most frequent destructive agents of the archaeological sites observed in the region are dams. While these sites may be preserved underwater (see previous Sect. 5.4.1 and 5.4.2), they are mostly completely inaccessible as a heritage resource. Although the great incidence of the dams in the destruction of the archaeological heritage seems to be increased because the intensive surveys prior to dam construction, the large concentration of sites in the riverbanks of the Uruguay River do not seem to be a random archaeological pattern. On the contrary, it reflects the redundant use by pre-colonial societies of the banks of this great river. Nevertheless, in regional terms, the most destructive agent is agriculture. Although this process does not have as detailed a destruction record as that available for dams, agriculture slowly but consistently destroys the shallower archaeological record. This destruction is no longer limited to the river banks, but to the entire state. The topsoil modifications have occurred regularly almost every year for a century, although direct seeding has decreased the rate of plowing in regional terms. Notwithstanding this, it is notable how much of the later record has been modified by plowing, especially the record of the Itararé-Taquara and Guaraní archaeological units in the bottoms of the valleys and adjacent areas. In the highlands above 600 masl, earthen monumental architecture has been destroyed by extensive and mechanized agriculture of direct seeding, which is accompanied by mechanical land leveling. Thus, many mounds and platforms disappeared from the landscape in a few years (Fig. 5.9).

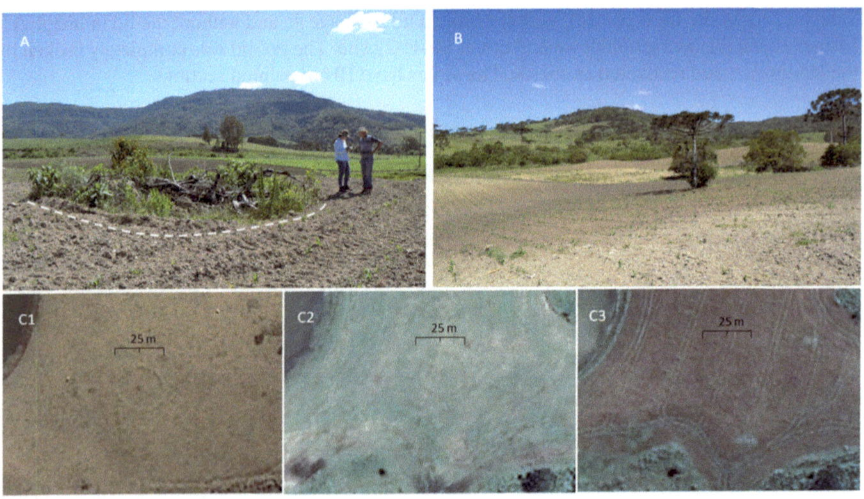

Fig. 5.9 Various phases of destruction of the archaeological record in the region. (**a**) Pit-house in the process of being filled with remains of plants, trees and sediments. (**b**) Area with filled pit-houses without any sign of them on the surface. (**c1**) satellite image of the ring-enclosure with a central mound at Ludovico Poleze I site in 2008, when the area was used for livestock. (**c2**) the same structure in 2018, after 3 years of mechanized agriculture. (**c3**) The same structure in 2019, after 4 years of mechanized agriculture

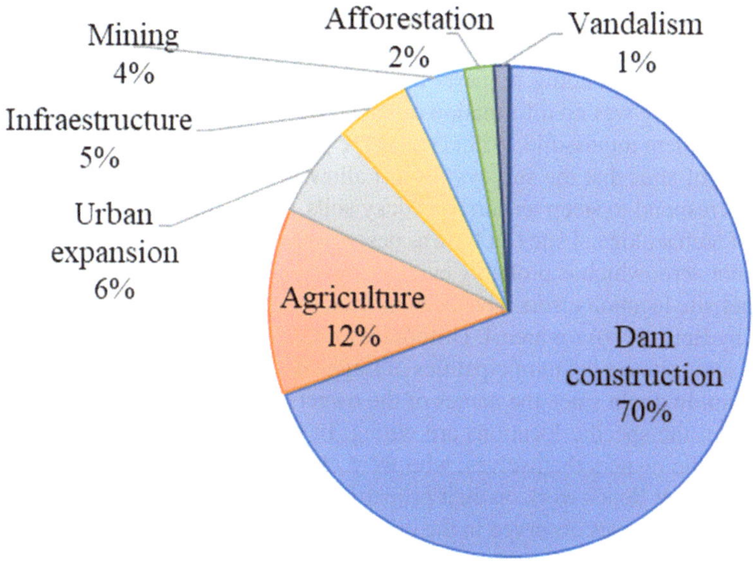

Fig. 5.10 Percentage of site destruction according to the agents involved

This leveling process also includes the filling of pit-houses (Fig. 5.9). Although in the latter cases the archaeological record could be preserved under the filler (mostly topsoil, but sometimes garbage of any kind), they cannot be detected from the surface nor by satellite imagery either, being invisible to detection by standard archaeological techniques (see Fig. 5.9). When an archaeological record is undetectable by standard archaeological techniques, it is practically unavailable as a heritage resource.

Other agents of destruction of the archaeological record that we have documented are urban expansion, construction of infrastructure in general terms (roads, bridges, etc.), mining and, to a lesser extent, afforestation. The latter involves the removal of roots with heavy machinery that destroys the structure of the archaeological record between the surface and one to one and a half meters deep. Figure 5.10 shows the incidence of each cause of the destruction of the archaeological record as it was documented during the execution of the PRESASC.

5.4.4 *Incomplete or Erroneous Records and Unidentified Sites*

A small fraction of sites included in the PRESASC were discovered during the implementation of CRM projects from the 1990s and 2000. Fortunately, most of these sites were located by geographic coordinates. However, and surprisingly, some sites have incomplete coordinates in the original reports. This is the case of 15 sites surveyed during studies related to the construction of the Foz de Chapecó dam.

Other sites of the same project are located with wrong coordinates, because they fall outside the county, sometimes outside the state, and even outside Brazil. Despite making the corresponding requests to the researcher who originally recorded all those sites, there was no information of the proper coordinates, making the location of some of them impossible. Within that same project, some GPS points indicate the existence of sites that the surveys did not allow to identify. Some of these coordinates correspond to steep terrain and rocky soils, where the probability of the existence of archaeological sites, at least as described in the original report, is extremely low if not zero, which is probably evidence of a wrong location, or nonexistent.

Other site location errors have been noticed when comparing some records published by Reis (1980) for area 4. Differences were particularly observed in the location of some sites in the municipalities of Lages and Painel. The identification of the sites is not in doubt since the names of the owners and the type of record described match, but the specific locations are wrong. In some cases, we have been able to interview the owners themselves, who were children or teenagers at the time, and who remember Reis's work on their properties. Other significant differences are the number of pit-houses observed in the original reports and those documented during the PRESASC. In most cases, our records include a greater number of structures than those recorded by Reis. These differences are probably due to greater clearing at present which increases archaeological visibility. In other cases, with respect to Reis's work, some pit-houses have been destroyed, or the location given by the author is uncertain, so these records could not be verified.

Another unexpected problem was the finding of sites in properties that previous preventive archeology studies had not recognized. Perhaps the most paradigmatic case corresponds to the Passo Fundo 1 site (Lages County). Previous environmental licensing studies, carried out in 2013, did not detect any archaeological sites, freeing up the land for the construction of condos. In the original report, the execution of numerous test-pits with negative results is observed, but none of them were carried out on monumental structures, which, in addition, were not identified. After that study, a few years later, three archaeological structures were identified in the same property by other researchers. Although these findings were not included in the IPHAN files at the time of the PRESASC, some results had already been published (Corteletti et al., 2016). During our surveys we recognized a fourth earthen structure close to the others (Fig. 5.11), and the former landowner recognized a fifth in the west area of this property, which was destroyed a few years before by afforestation.

5.5 Discussion

Distributional archaeology, as well as the various variants that are included under the generic term of landscape archaeology, require a precise database so that any model that is subsequently built is as rigorous as possible, regardless of the theoretical or interpretive framework (*e.g.* Anschuetz et al., 2001; Deetz, 1990; Dewar,

Fig. 5.11 Passo Fundo 1 site (Lages County). (**a**) circular platform. (**b**) artificial mound. (**c**) small mound with an excavated peripherical ring. (**d**) artificial mound. Radiocarbon dates yielded for structures A, B and D between 610 ± 20 (UCIAMS 238979) to 980 ± 20 (UCIAMS 238980) ^{14}C years BP. (work in progress)

1992; Ebert, 1992; Fowless, 2010; Last, 2020; Turner, 2018; several contributions in Bruno & Thomas, 2008, among others). A good database for distributional modeling can be laborious and expensive to obtain, and each resampling can distort or call into question previous conclusions obtained with smaller samplings. Nevertheless, generating models with the available sampling and testing them by increasing the sample sizes is an inseparable part of archaeological research and scientific progress.

Usually, the first step to model the distribution of the archaeological record is to look for the data. Ideally, this information should be included in the official records. Unfortunately, in Santa Catarina, not all archaeologists register the sites discovered during research, transforming the location of the sites on a map into a daunting task, because dozens of works must be consulted to form a unified database. On the other hand, the official records are not always updated with the information they already have, since, as we have seen, there are duplicate records, and not all new sites are quickly incorporated into the official databases. Moreover, the results obtained during the surveys have demonstrated that the capacity of archeology to develop an accurate database is reduced each year due to the destruction of the archaeological heritage, especially those of Guarani and Itararé-Taquara archaeological units, including pit-houses, and monumental earthen structures (Sect. 5.4.3).

A second issue that negatively impacts the construction of a solid database in the State is incorrect location or the record of non-existing sites (Sect. 5.4.4). The latter could be due to the identification of a few sherds or lithic artifacts, whose concentrations do not meet the minimum aggregation parameters to be considered sites, but rather as an expression of the continuous use of space by past societies. Including the site sizes into the databases, as required by the IPAHN, helps to address this problem, beyond methodological issues related to this topic (Sect. 5.3.3). We do not want to fail to mention here some aggregations of objects that doubtfully correspond to man-made lithic artifacts, but that have been used to identify pre-ceramic archaeological sites in some CRM reports. These doubtful or outright misidentifications of artifacts also affect our perception of the true distribution of the archaeological record along the landscape.

A third problem related to the biases identified during the program corresponds to those sites that were not recognized during preventative archeological studies. This bias is often due to the short time archaeologists must carry out the surveys before the infrastructure works start. Two paradigmatic cases can be mentioned here such as Passo Fundo I and Adão Sasanoviz sites. The first case was described in Sect. 5.4.4. The second corresponds to a site located in Caxambú do Sul County, which was not recorded during the surveys prior to the construction of the Foz de Chapecó dam (De Masi, 2012). Due to heavy rains, it was identified by the local municipality shortly before the dam lake was filled (Carbonera et al., 2021). This site has a Guarani occupation and a pre-ceramic level, but now it is underwater without having been properly sampled, and without having been integrated into the database at that time.

A fourth issue that should be considered as biasing the local databases are the differences derived from the archaeological visibility of the structures, especially the pit-houses and other earthen structures. Their distribution in the landscape has notable consequences for the models about settlement patterns, space use and other archaeological interpretations. Although today a large part of the original forest has been destroyed, and archaeological visibility has increased remarkably, large forested areas are still preserved. Lidar and similar tools can help to complete the archaeological information from these forested areas, but nothing for now can replace closely spaced parallel transects. Therefore, any distributional model cannot be considered accurate until an intensive and adequate survey is carried out in each specific area.

5.6 Conclusions and Perspectives

The results obtained during the execution of the PRESASC program show that the local archaeological databases are biased to different degrees, being necessary to analyze them before deriving archaeological models from them. While some of these biases may be limited, others, unfortunately, seem to have an unknown margin of error,

The destruction of the archeological record is probably the biggest bias of all, and the most worrisome one, because it not only destroys the cultural heritage, but also the archeological knowledge base, even if the sites have been sampled, since we cannot improve our knowledge of the past by adding new research, and previous recovered information cannot be independently tested by new investigations, which is one of the key processes of scientific advance.

Accurate and reliable databases are also crucial for cultural managers to define environmental licensing studies, research permits and preservation policies. The latter is a critical issue to heritage management for all countries, generally with scarce resources to cover extensive territories such many South American countries possess. In this sense, the PRESASC program established a new dynamic relationship between researchers, local cultural managers, and landowners. Most of the latter have become necessary partners for the conservation of the archaeological heritage, also becoming permanent informants of new findings in their properties or nearby lands. Indeed, until the execution of this program the preservation of the surveyed sites, with a few exceptions, has generally occurred fortuitously, but from now on, its conservation has moved, to a greater or lesser degree, into active actions by the local communities and landowners. This improvement in the quality of site preservation has not been achieved by coercion or through the imposition of legal instruments, but through the exchange of opinions and the search for consensus. However, beyond the legal terms of heritage protection, the compromises and achievements must be maintained over time through joint action by state and local authorities. The agreements reached with the owners do not extend to their heirs or to the new owners who acquire these properties. Also, eventually, the conditions of the properties and the economic situation of the owners may change over time. For this reason, long-term preservation programs require permanent monitoring by governments. In this sense, during the execution of the program we have seen signaling plaques of archaeological sites placed less than 10 years ago, which had fallen to the ground or unreadable. Other problems have also been highlighted. Among them, the persistent distrust in part of the population in relation to the possibility of expropriation of land if they notice the existence of archaeological sites, which even extends to several public officials of different counties, a situation that requires the dissemination of proper information to the citizens, and adequate training of the local authorities.

Finally, the preservation of the archaeological record, and the identification of record biases are two continuous tasks that as archaeologists we must face on a daily basis. The first is an inescapable social responsibility that we have as social scientists, while the second is part of the much-needed cautionary tales to interpret the past.

Acknowledgments A project like the one developed could only have been carried out due to the wide support of numerous people and institutions. Unfortunately, it is impossible to mention here all the people who helped us during the 18 months of field and laboratory work. To all of them, we deeply appreciate their collaboration. In terms of institutions, we want to thank IPHAN for the conception of the program and the support given to the researchers for the execution of the project, and to the Universidade Comunitária da Região de Chapecó (UNOCHAPECÓ-CEOM) for the logistical support without which the results would not have been achieved.

References

Anschuetz, K. F., Wilshusen, R. H., & Scheick, C. L. (2001). An archaeology of landscapes: Perspectives and directions. *Journal of Archaeological Research, 9*(2), 152–197.

Bruno, D., & Thomas, J. (2008). *Handbook of landscape archaeology*. Routledge Handbooks Online.

Campos, J. B. (Coord.). (2018). *Relatório Final do Programa de Resgate, Cercamento, Sinalização, Monitoramento Arqueológico e Educação Patrimonial para a Implantação da Linha de Transmissão 230 KV Foz do Chapecó – Pinhalzinho 2*. Criciúma. Processo IPHAN 01450.005649/2014-95.

Carbonera, M., & Cecchin, C. (2015). Como era antes? *O patrimônio arqueológico pré-colonial do oeste catarinense*.

Carbonera, M. & Loponte, D. (2021). *Archaeology of Southeastern South America*. www.assaprogram.com

Carbonera, M., Witte, G., & Vieira, S. H. (2016). *As pistas que revelam o passado: uma aventura arqueológica*. Êxito.

Carbonera, M., Mohr, M., Lino, J. T., & Loponte, D. (2021). Aspectos tecnomorfológicos e estílisticos da cerâmica pintada guarani do sítio Adão Sasanoviz (Alto Rio Uruguai). *Cadernos do Lepaarq XVIII, 35*, 263–291.

Corteletti, R., et al. (2016). News from the field ou como um projeto internacional começa a sair do papel. *Revista do Museu de Arqueologia e Etnologia, 27*, 197–212.

De Masi, M. A. N. (2005). *Relatório Final do Projeto de Salvamento Arqueológico Usina Hidrelétrica Campos Novos*. Florianópolis.

De Masi, M. A. N. (2012). *Relatório Projeto de Salvamento Arqueológico UHE Foz do Chapecó (Reservatório)*. Florianópolis.

Deetz, J. (1990). Landscapes as cultural statements. In W. M. Kelso & R. Most (Eds.), *Earth patterns: Essays in landscape archaeology* (pp. 2–4). University Press of Virginia.

Dewar, R. E. (1992). Incorporating variation in occupation span in settlement-pattern analysis. *American Antiquity, 56*, 604–620.

Dunnell, R. C., & Dancey, W. S. (1983). The Siteless Sulvey: A regional scale data collection strategy. *Advances in Archaeological Method and Theory, 6*, 267–287.

Ebert, J. I. (1992). *Distributional archaeology*. University of New Mexico Press.

Foley, R. A. (1980). *The Spatial Component of Archaeological Data*: Off-site Methods and Some Preliminary Results from the Amboseli Basin, Southern Kenya. In *Proceedings of the VIII Pan-African Congress in Prehistory and Quaternary Studies* (pp. 39–40).

Fowless, S. (2010). The southwest School of Landscape Archaeology. *Annual Review of Anthropology, 39*, 453–468.

Gallant, T. W. (1986). Background noise and site definition: A contribution to Sulvey methodology. *Journal of Field Archaeology, 13*, 403–418.

Goulart, M. (1995). *Projeto Salvamento Arqueológico Uruguai: Cadastro de Sítios Arqueológicos* (Vol. I). Univali.

Goulart, M. (Org.). (1985). *Levantamento de sítios arqueológicos Barragem de Itapiranga/SC*. Florianópolis.

Hayes, P. P. (1991). Models for the distribution of pottery around former agricultural settlements. In A. Schofield (Ed.), *Interpreting artefact scatters* (pp. 81–92). Contributions to Ploughzone Archaeology.

Last, J. (2020). *Landscape – Place – People: the values of spatial terminology in heritage and beyond*. Research Department Reports. Historic England (pp. 254–2020).

Loponte, D., & Carbonera, M. (2013). Arqueologia sem Fronteiras: projeto de cooperação binacional para o estudo arqueológico da província de Misiones (Argentina) e oeste de Santa Catarina (Brasil). *Revista Memorare, 1*(1), 43–50.

Milller, E. T. (1969). *Catálogo de sítios arqueológicos do Museu Arqueológico do Rio Grande do Sul*. Inédito.

Pettigrew, R. M., & Lebow, C. G. (1989). *An archaeological Sulvey of Trout Creek-Oregon canyon uplands, Harney and Malheur counties*. INFOTEC Research Inc.

Piazza, W. (1966). As grutas de São Joaquim e Urubici. *Série Antropologia, 1*, 1–53.

Piazza, W. (1969). Notícia arqueológica do vale do Uruguai. In *Museu Paraense Emílio Goeldi. Programa Nacional de Pesquisas Arqueológicas: resultados preliminares do segundo ano, 1966–67* (pp. 55–70). Museu Paraense Emílio Goeldi. (Publicações Avulsas, 10).

Piazza, W. (1971). Dados complementares à Arqueologia do Vale do Rio Uruguai. In *Museu Paraense Emílio Goeldi*. Programa Nacional de Pesquisas Arqueológicas: resultados preliminares do quarto ano, 1968–69. Belém: Museu Paraense Emílio Goeldi, 71–86. (Publicações Avulsas, 15).

Pontifícia Universidade Católica do Rio Grande do Sul. (1989). *Projeto Campos Novos*. Levantamento Arqueológico na região do alto Uruguai-Canoas, barragem de Campos Novos. Museu de Ciência e Tecnologia.

Pontifícia Universidade Católica do Rio Grande do Sul. (2001). *Salvamento Arqueológico em áreas do futuro reservatório da UHE Machadinho (Fase III): Relatório Técnico*. Museu de Ciência e Tecnologia.

Reis, M. J. (1980 [2007]). *A problemática arqueológica das estruturas subterrâneas no planalto catarinense*. Habilis.

Rohr, J. A. (1966). Os sítios arqueológicos do município de Itapiranga às margens do rio Uruguai, fronteira com a Argentina. *Pesquisas, Antropologia, 15*, 21–60.

Rohr, J. A. (1971). Os sítios arqueológicos do planalto catarinense, Brasil. *Pesquisas, Antropologia, 24*, 1–70.

Rohr, J. A. (1984). Os sítios arqueológicos de Santa Catarina. *Anais do Museu de Antropologia da UFSC, 17*, 77–168.

Schiffer, M. B. (1987). *Formation process of the archaeological record*. University of New Mexico.

Schmitz, P. I. (1957). Um paradeiro Guarani no Alto Uruguai. *Revista Pesquisas, 1*, 122–142.

Turner, S. (2018). Historic landscape characterisation: An archaeological approach to landscape heritage. In G. Fairclough, I. Sarlov-Herlin, & C. Swanwick (Eds.), *Routledge handbook of landscape character assessment* (pp. 37–50). Routledge.

Chapter 6
Traditional Erva-mate Production Systems: How Historical Ecology and Environmental History Can Inform Local and Global Approaches to Ecosystem Restoration

João Francisco Miró Medeiros Nogueira, Alessandra Izabel de Carvalho, and Evelyn Roberta Nimmo

Abstract Family farmers who continue to use traditional erva-mate agroforestry production systems in the Iguaçu River valley in Southern Brazil have a unique relationship with the forest environment in which they live and work. By focusing on the historical and contemporary trajectory of the connection between humans and erva-mate, this chapter examines how perspectives from Historical Ecology, Environmental History, and Landscape Archaeology can support the development of conservation strategies that are grounded in the perceptions, histories, and memories of local and traditional communities. Such strategies recognize the need to support sustainable livelihoods as well as sustainable ecological interactions that favor the preservation of biocultural diversity, local ecological knowledge, and ways of knowing and managing the environment. We examine the inter-generational

J. F. M. M. Nogueira (✉)
CEDErva (Centro de Desenvolvimento e Educação dos Sistemas Tradicionais de Erva-mate), Curitiba, PR, Brazil
e-mail: jfmmnogueira@gmail.com

A. I. de Carvalho
CEDErva (Centro de Desenvolvimento e Educação dos Sistemas Tradicionais de Erva-mate), Curitiba, PR, Brazil

Universidade Estadual de Ponta Grossa, Ponta Grossa, PR, Brazil
e-mail: alessandra@uepg.br

E. R. Nimmo
CEDErva (Centro de Desenvolvimento e Educação dos Sistemas Tradicionais de Erva-mate), Curitiba, PR, Brazil

Universidade Estadual de Ponta Grossa, Ponta Grossa, PR, Brazil

Laurier Centre for Sustainable Food Systems, Wilfrid Laurier University, Waterloo, ON, Canada
e-mail: enimmo@wlu.ca

A. C. Colonese, R. G. Milheira (eds.), *Historical Ecology and Landscape Archaeology in Lowland South America*, Interdisciplinary Contributions to Archaeology, https://doi.org/10.1007/978-3-031-32284-6_6

practices and management within the Araucaria Forest, the affect and sensorial assemblages involved in this human-environment system, and how these subjectivities can be leveraged to address current and future challenges related to changing climates and ecosystem degradation.

Keywords South Brazil · Long-term erva-mate agroforestry · Community identity · Biological diversity

6.1 Introduction

The consumption of erva-mate (yerba mate; *Ilex paraguariensis*) has been a strong cultural symbol and practice within the Rio de la Plata basin for thousands of years. With the arrival of the Spanish in the early sixteenth century, the practice of consuming the toasted leaves as an infusion in a gourd was quickly adopted from the Guarani and other indigenous groups of the region and became an important driver of the colonial Paraguayan economy. By the end of the eighteenth century, the consumption of erva-mate across the Spanish colonies in the Southern cone was widespread and continues to this day. While much discussion of the history of erva-mate has focused on its significant role in the colonial and republican economies of Paraguay, Argentina, and Brazil, little consideration is given to the ecological history of the tree and the impact its exploitation and cultivation had, and continues to have, on the Araucaria Forest landscape. Focus has been given to the iconic *Araucaria angustifolia* (pinheiro) tree in terms of how humans have interacted with the environment for thousands of years (Corteletti & Iriarte, 2018; Iriarte et al., 2017), with significant implications for forest cover in the contemporary landscape. However, other species have been just as important ecologically, culturally, and economically in Southern Brazil, and their role in historical processes of land use, forest conservation, and resource exploitation need further attention.

Integrated landscape approaches to addressing sustainability require a long-term understanding of landscape change and interaction, the cultural and ecological processes that shape not only the landscape but also how people interact with it and perceive it, as well as interdisciplinary collaborations that address current issues from a variety of perspectives. One of the key concepts that must be considered is biocultural diversity, which is the "diversity of life in all manifestations – biological, cultural, and linguistic – which are interrelated (and likely co-evolved) within a complex socio-ecological adaptive system" (Maffi & Woodley, 2010, p. 5). Importantly, this definition assumes not only that humans are integral parts of their environment, but also that "the past, present, and future of these human-environment interactions are conceptualized into a functioning whole" (Lennartsson et al., 2018, p. 188). By integrating a range of disciplines and methods, we can begin to understand the long-term cultural, social, and ecological processes that have helped to maintain forest cover across the landscape of Southern Brazil and from this develop policies that are consistent with forest use and conservation for those living in rural communities. However, this requires an understanding of how farmers perceive the forest, the environment in general, and their role as stewards of a highly threatened ecosystem.

By focusing on the historical trajectory of the connection between humans and erva-mate, this chapter aims to examine how perspectives from Historical Ecology, Environmental History, and Landscape Archaeology can support the development of conservation strategies that are grounded in the perceptions, histories, and memories of local and traditional communities. Such strategies recognize the need to support sustainable livelihoods as well as sustainable ecological interactions that favor the preservation of biodiversity, local cultures, and their ways of knowing and managing the environment. Our analysis is based on a series of Environmental Oral History interviews and site visits to a range of erva-mate producing farms conducted between 2017 and 2020, in which we discussed with small-scale farming families their histories, memories, knowledge about the forest and erva-mate management practices. These interviews were part of two larger interdisciplinary projects: one focusing on traditional forest management practices in the Araucaria Forest, conducted in collaboration with Embrapa Foresty (Forest division of the Brazilian Agricultural Research Corporation); and the other, an international partnership with the Laurier Centre for Sustainable Food Systems in Canada, examining the role of traditional knowledge and agroecology in supporting sustainable livelihoods and food security.

Therefore, the objective of this study is to examine the natural, historical, and socioenvironmental elements that shape and are shaped by the environmental identities and subjectivities of family farmers and communities who continue to use traditional erva-mate agroforestry production systems in the Iguaçu River valley in Southern Brazil (Fig. 6.1). To do so, we examine the historical trajectory of this human-environment system and use environmental oral histories and farm visits to consider the affect and sensorial assemblages involved in the inter-generational practices and management within the Araucaria Forest, and how this is manifest in the materiality and landscape of the property. Finally, we consider how these subjectivities can be leveraged to address current and future challenges related to changing climates, food security, and ecosystem degradation.

6.2 Methodology

Research and fieldwork were organized based on methodological approaches that enable us to interpret how the materiality of traditional and agroecological erva-mate systems harbors elements related to the physical space – constituted by its relationship with the ecosystem – and, at the same time, present elements that reflect the interactions with and subjective experiences of such a complex system. Environmental subjectivities can "address the formation process of how individuals construct and reconstruct a set of discursive relationships with 'nature'" (Zhang, 2019, p. 489) and recognize that "the boundaries between the 'I' and the

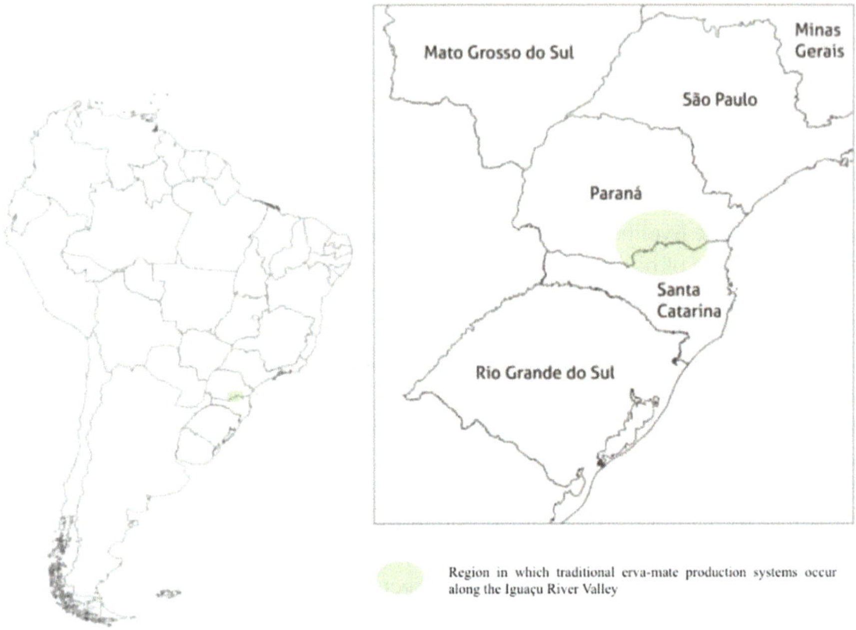

Fig. 6.1 Map of study region. (After CEDErva, 2017)

environment are porous, and subjectivity is shaped by a human being's engagement with their total environment, not just their social environment" (Singh, 2013, p. 191).

Thus, according to Berque (2013), it is necessary to consider the fact that the construction of narratives is a result of how a subject's meaning in space is permeated by both spatial and temporal layers. Such layers imply meanings with an impact on the subject's own spatial performance. Therefore, a sensitive reading of a landscape is one that considers an interpretation of the layers of meaning in their entirety. As such, three aspects of space that potentially permeate phenomenological subjectivity must be taken into account which are implicated in how the subject acts in space: the "natural (geological, evolutionary, cyclical); that of society (the history of human relationships); that of the person (of the individual's experience) contemplating the landscape, both in reality and in representation" (Berque, 2013, p. 7).

With this in mind, our methodology aimed to recognize, identify, and establish criteria that resemble the visited landscapes: the ecosystem of the Araucaria Forest; historicity of small-scale rural properties and traditional communities; and the life trajectories experienced by members of traditional and agroecological erva-mate systems. Thus, a thorough analysis of the historicity of the Araucaria Forest was undertaken to examine how the forest has shaped and been shaped by current forest management practices related to erva-mate production. This historical context is a key element of the analysis as it helps to orient our understanding of how current practices and subjectivities are intimately linked with the forest environment.

Oral history interviews were also undertaken as part of the methodology as a means of documenting and co-creating knowledge together with our community partners in order to better understand the "ecosystem at multiple scales in time and space" and begin to deconstruct the dominant view of a landscape and the interactions of people with it, by enabling the development of a dialogue between people and the landscape (Arce-Nazario, 2007, p. 119).

The field research took place between 2017 and 2020 and included interviews with small-scale rural farmers and other stakeholders. A total of 45 interviews were recorded, including men and women, young people, and elders. All interviewees were from the same socio-environmental region across a range of municipalities that have common cultural and historical elements. During the majority of the field visits the research team walked with the interviewee and other family members through the '*ervais*', or erva-mate agroforestry systems, that have been managed for generations by the family (in some interviews this was not possible due to reasons including weather conditions and logistical considerations). These walks, which were led by the agents actively managing the space, were an important stage of the interpretative methodology, enabling us to verify the subjectivities present in the materiality of the landscape of these systems.

By adopting this methodological approach, we sought to conduct interviews focused on environmental oral history, which attempts to move away from the traditional interviewer-interviewee dynamics, enabling the construction of dialogues that are multidimensional and incorporate numerous perspectives (Nimmo et al., 2020). Interviews were guided by a list of semi-structured questions, with the goal of activating memories, bringing the lived experiences of the environment to the fore and enabling a dialogue between the farmers, the small-scale rural property, and the management system.

Thus, the field visits were divided into two instants that, together, allowed us to collect data and experiences for subsequent analysis:

- **semi-structured interviews** with the farmers and, in some cases, with the whole family (i.e., adult children, spouses). Interviews were recorded with a digital voice recorder, transcribed, and coded for analysis. Interviews provide evidence of sentiments and feelings as a category of historical production and affectivity as a spatial relationship (Laverdi & Mastrângelo, 2013);
- **a tour through the property**, stopping at specific structures or places related to historical agricultural practices and the management and processing of erva-mate developed in the properties. The main purpose of this stage was to document and analytically comprehend spatial dynamics within the landscape of the rural property.

Using both a narrative interview and an experiential setting involving walking and speaking with the farmers, we could intertwine the narratives with both the expressions of how each farmer related and interacted with the forest environment, and the historicity of management techniques that organize the space as practiced by the rural society (Fournier, 2010, p. 473).

After each farm visit, detailed field notes, ideas, and impressions were documented, and sketches of each property were developed to represent the layers of use, management, and meaning. To code the interviews, a hierarchical tree of themes and subthemes related to the interview questions were developed, with the possibility of adding themes that emerged during the visits. The interviews, drawings, photographs, and field experiences were then used to consider the ways in which memory, history, and practice intersect in traditional erva-mate production systems.

6.3 Historical and Territorial Context: The Araucaria Forest and the Long-Term Exploitation of Erva-mate

Brazil is known internationally for the vast diversity of its forest landscapes. With an estimated biota between 1.2 and 1.4 million species, it is one of the most biologically rich countries in the world (Toledo & Barrera-Bassols, 2015). The country is divided into six large biomes, each with unique characteristics in terms of climate, vegetation, and fauna (Brasil, 2021). The Brazilian Atlantic Forest, which extends latitudinally along the coast of Brazil, advancing inland at various points due to the humidity and climatic influence of the Atlantic Ocean, covers about 15% of the country and is made up of nine different types of vegetation cover (Brasil Law 11.428/2006; IBGE, 2012). It coincides with the most densely populated region in the country and has suffered extensive deforestation and intense land conversion (Dean, 1996) (Fig. 6.2).

At the time of European contact, a large portion of the three states that make up southern Brazil (Paraná, Santa Catarina and Rio Grande do Sul), was covered by the forest ecosystem known as Araucaria Forest (Mixed Ombrophilous Forest (IBGE, 2012) or Araucaria moist forest (FAO, 2001; WWF, 2000)), a biome within the Brazilian Atlantic Forest. The dense, humid Araucaria Forests are characterized by low luminosity in the understory due to the long-living pinheiro (*A. angustifolia*) that dominates the forest canopy. When the European colonial project began its expansion into the Rio de la Plata basin, this ecosystem occupied an area of about 182,000 km^2 (Maack, 2012). With the historical process of unsustainable exploitation of its ecological resources and extensive landscape conversion, the area of occurrence of vegetation cover of the Araucaria Forest has been vastly reduced. Research suggests that of the 16.8 million hectares (ha) of Araucaria Forest that occurred across Paraná at the end of the nineteenth century only 880,000 ha remained at the end of the 1980s (Gubert Filho, 1988).

Today, remnants of this ecosystem rarely exceed 50 ha, and only 0.62% of the biome occurs in protected areas. Specifically in the state of Paraná, only 0.3% of the total remaining forests is classified as advanced secondary forest or altered primary forest (Castella & Britez, 2004). The ongoing modification of the regional landscape means that the extensive and biodiverse Araucaria Forest has been converted into a fragmented landscape of monoculture crops interspersed with small forest remnants that are generally isolated and not interconnected (Lacerda, 2016).

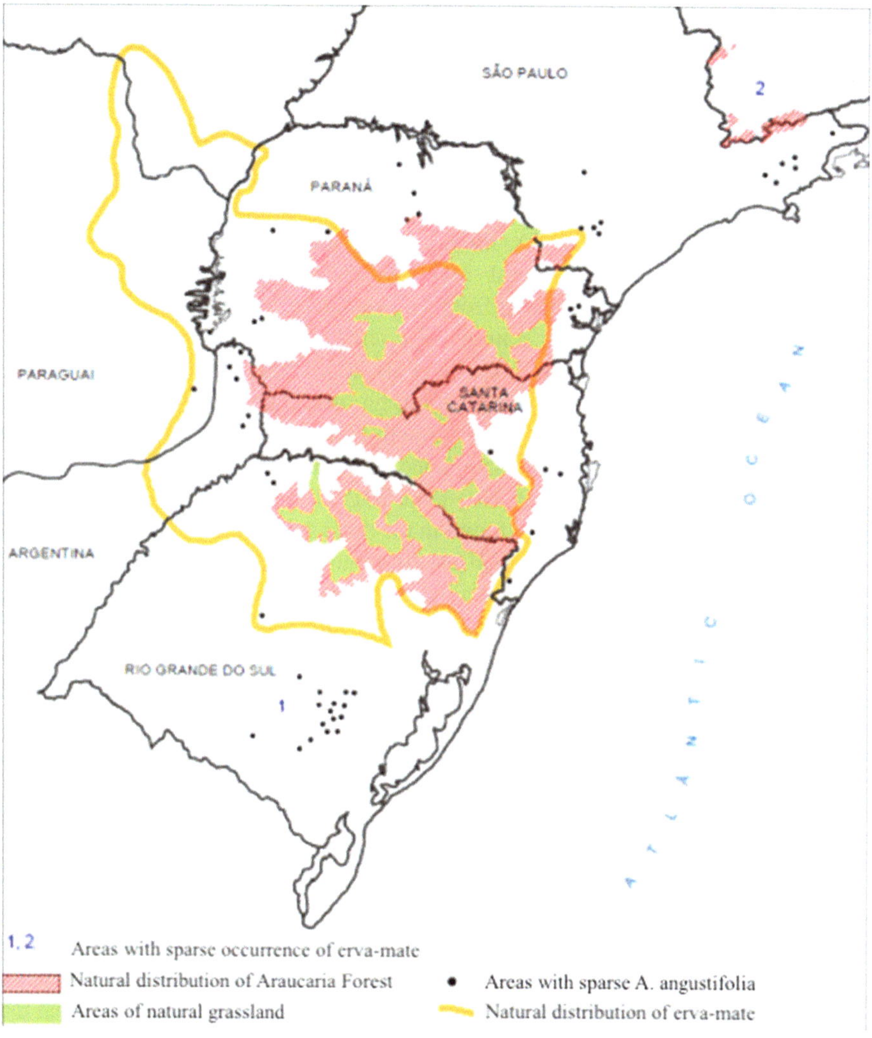

Fig. 6.2 Distribution of the Araucaria Forest and erva-mate. (After Oliveira & Rotta, 1985; Zampier, 2001)

Outside of the large Conservation Areas focused on preserving the remaining old-growth Araucaria Forests, in Southern Paraná a significant proportion of the remaining Araucaria Forest fragments are linked with traditional agroforestry practices related to the management of erva-mate (Chaimsohn & Souza, 2013; Lacerda et al., 2020; Marques, 2014). Endemic to the ecosystem, erva-mate production is characteristic of traditional small-scale agriculture in the Iguaçu River valley and according to Radomski et al. (2014), its management represents an opportunity to develop forest conservation strategies in association with agricultural production and income generation for family farmers. These conditions offer an opportunity to maintain the forest ecosystem alongside traditional cultural practices, which have resisted constant pressure from the agroindustrial complex in the region, which is linked to the conversion of forest cover into monocultures of soy, corn, tobacco, pine, and eucalyptus.

Traditional systems of erva-mate production are imbued with ecological knowledge and sustainable environmental practices that can be leveraged to develop conservation and restoration strategies for the last remnants of the Araucaria Forest in the southern states of Brazil. Traditional erva-mate systems are comprised of a series of sustainable agricultural and agroecological practices (Caporal & Costabeber, 2004) that are derived from knowledge of forest and ecological management transmitted across generations, also referred to as local ecological knowledge (LEK; Berkes et al., 2000; Lacerda et al., 2020). The basis of this knowledge stems from the Guarani and other indigenous groups living in the region at the time of colonial expansion, that developed through constant interaction with the forest ecosystem over generations and maintained by traditional and settler populations over the last 500 years (Nimmo & Nogueira, 2019).

Currently, erva-mate is mainly found in the Araucaria Forests of southern Brazil and north-eastern Argentina, as well as the mountainous regions in the tropical moist deciduous forest of eastern Paraguay (Gerhardt, 2013; Oliveira & Rotta, 1985). The species grows well in the shaded understory and trees can grow to up to 30 m in height in the forest (Carvalho, 2003). However, because of consistent management of trees over generations, most trees are maintained at a maximum height of 10 m, and between 20 and 40 cm in diameter at breast height (DBH). The species is commonly associated with *A. angustifolia* and *Ocotea porosa* (imbuia; Inoue et al., 1984). Because of the close affinity and symbiosis between these species, one of the initial names used to classify the forest, as given by the geologist and naturalist Reinhard Maack, was Yerba Mate Forest[1] (Maack, 1931).

The leaves of erva-mate contain a range of compounds such as antioxidants, caffeine, theobromine, vitamins, among others, making the infusions from its leaves beneficial to human health while also offering significant opportunities for use in a range of industries, including pharmaceutical, beverage, and cosmetics (Alkhatib, 2018; Gan et al., 2018; Horn et al., 2018). Previous studies have suggested that the stimulant effects of erva-mate was one of the main reasons for the management, cultivation, and consumption of its leaves as an infusion by pre-Columbian populations in South America (López, 1974). In analyzing the historiographic sources on the consumption of erva-mate among the Guarani throughout the sixteenth century, Garavaglia (2008) states that the use of erva-mate for medicinal purposes had spread among the diverse indigenous peoples of colonial Paraguay before the arrival of European colonizers. People of the Kaingang indigenous groups – descendants of the Jê linguistic family, some of whom were horticulturalists that occupied the highland forests and grasslands throughout what is now Southern Brazil (Da Luz, 2011; Koch & Corrêa, 2010) – also collected and consumed the leaves, which they referred

[1] Maack uses the phrase *Herva Mate Wald* to describe the forest.

to as *kògünh* or congonha (Haverroth, 1997). The habit of consuming mate, however, was quickly adopted by the colonial population and as early as the 1520s the processed leaves of the tree were used as a form of currency in the burgeoning colonial economy (Folch, 2010).

Similar to processes of colonization throughout the Americas, interaction with European colonizers in the Rio de la Plata region had drastic and devastating consequences for the indigenous peoples who suffered constant abuse, exploitation, and massive cultural extermination (Monteiro, 1994; Ribeiro, 1985; Cunha, 2013). However, the ecological knowledge developed and held by indigenous peoples about their environments and the species within (Elisabetsky, 1986) were essential for the technological and social innovations necessary for the colonial project to succeed (Cañizares-Esguerra, 2005; Harris, 2005; Norton, 2017). The production of erva-mate offers an example of one such resource as it was an important economic driver that helped to support the implementation of the social, political, religious, and economic structures imposed by colonization.

With the arrival of the Spanish in the sixteenth century in the Rio de la Plata region, the conquistadores rapidly adopted the cultural practice of drinking mate and appropriated the techniques of collecting and processing the leaves for commercial purposes, with the tree becoming one of the most important economic resources in colonial Paraguay and Argentina, and later in Southern Brazil (Folch, 2010). As with other natural resources in colonial Paraguay, erva-mate was part of a system of forced indigenous labor, known as the *encomienda* system (Austin, 2015; Lockhart, 1969; Service, 1951, 1971), in which *encomenderos* extracted personal service, or other forms of tribute, from their indigenous tributaries and in return the *encomendero* was required to provide religious instruction and protection (Haubert, 1990; Saeger, 1981). Indigenous labor was used on the haciendas owned by the *encomenderos* to raise cattle and grow crops, but, because of the abundance of native erva-mate growing in the mountains near Maracayú (also referred to as Maracaju or Mbaracayu), *encomenderos* forced their Indigenous laborers to travel the long distance to harvest and process the leaves. Due to the abundance of the resource, the region of Maracayú, located in northern Paraguay, remained the main source of erva-mate extraction for much of the colonial period.

The arrival of the Society of Jesus at the end of the sixteenth century in colonial Paraguay coincided with the moment in which erva-mate had been established as one of the main commercial goods. Although the Jesuit fathers initially were opposed to the consumption of mate – because of the exploitation of the indigenous laborers through the *encomienda* system, but also because they saw it as a vice – they eventually came to monopolize the market. Two types of mate were produced for consumption: *caàmini*, the highest quality product and a specialty of the Jesuit missions, in which the erva-mate was processed into a fine powder, including only the leaves; and *yerba de palos*, produced by the Jesuits as well as the Spanish *encomenderos*, which offered a courser grind, often mixed with the leaves of other trees and inclusions such as twigs and small stones (Furlong, 1962). By 1704, the Jesuits and the Guaranis in the missions developed the technology to germinate the seeds of the tree and cultivate erva-mate in plantations, a technology that had evaded the

Spanish. The development of *yerbales*, or orchards, of erva-mate was a key turning point in the success of the Jesuit missions in colonial Paraguay (Nimmo & Nogueira, 2019).

After the expulsion of the Jesuits from Paraguay in 1767, the technology and knowledge related to the planting of erva-mate seems to have disappeared. Although the exploitation of the erva-mate orchards near the abandoned mission complexes continued until at least the middle of the nineteenth century (Gallardo, 1898), the actual cultivation of new orchards fell out of use even through the peak of production from the late nineteenth to early twentieth centuries. The boom in production again relied exclusively on natural populations. It was only at the beginning of the twentieth century in Argentina (Linhares, 1969; Martins 1926), and in the 1970s in Brazil (Chaimsohn & Souza, 2013) that the cultivation of seedlings from seeds gained widespread use.

In Paraná, the process of cultural interaction and colonization throughout the seventeenth to nineteenth centuries involved various movements and migrations of indigenous and African descendent groups, caboclos[2], missionaries, *tropeiros* (horse, cattle and mule drivers), and settlers. At the end of the eighteenth century and throughout the nineteenth century, the occupation of the forested land throughout Paraná was related to the transportation of cattle and mules, with outposts for resupply and rest emerging along the route from Rio Grande do Sul to Sorocaba in São Paulo State. Nevertheless, the extraction of erva-mate persisted as part of the local economy for indigenous and caboclo communities.

Although the occupation of Paraná by European colonists was intermittent until the second half of the nineteenth century, the region continued to be inhabited by indigenous groups and caboclo populations (Gerhardt, 2013). However, in the mid nineteenth century the Brazilian Imperial government began to promote the settlement of the 'vacant lands' (*terras devolutas*) in the interior hinterlands of the southern states through the establishment of settler colonies with immigrants from Central Europe, especially the Ukraine and Poland (Oliveira, 2001; Wachowicz, 1981). In parallel with the project of European occupation of the region, the exploitation and speculation of native plant species in the region exploded, bringing about drastic changes to the forest landscape and displacing the majority of the indigenous and caboclo communities.

With the recent occupation and settlement of the forest hinterlands of Paraná, the extraction of erva-mate grew significantly in the early 1900s, accompanied by the harvesting of its vast forest resources. According to official export statistics analyzed by Martins (2005), the logging of Paraná's forests between 1911 and 1941 consumed an average of 106,432 tons of pinheiro wood per year. Considering that one pinheiro produces, on average, 1500 kg of wood (Martins, 2005), we can

[2] Caboclos are an autochthonous population formed of Kaingang and Guarani indigenous groups that remained in the forest, escaped slaves, *tropeiros*, and other descendants of mixed European heritage who dwelled in the forest at the end of the eighteenth century who played an important role in the absorption and transmission of a large body of knowledge related to the forest (Sahr & Cunha, 2005).

estimate that approximately 71,000 pinheiro trees were sold abroad per year over that 30-year period. In other words, during this period alone, more than 2 million pinheiro trees were cut, processed, and sold on the foreign market. Yet, these estimates do not consider the numbers related to internal trade, the use of pinheiro in rural properties, as well as wood that was burned, discarded, or used as fuel for steamboats and trains that transported the wood to the ports.

In 1928, the botanist Frederico Carlos Hoehne (1930) travelled through the Araucaria Forest and described the transformation of the forest landscape in Paraná and its conversion into "destruction and desert". In his travel chronicle, entitled *Araucarilândia*, Hoehne warned of the imminent disappearance of the natural vegetation cover in the state of Paraná. At the time, the author described the vast forest exploitation in the region of the Rio Negro valley, the largest tributary of the Iguaçu River:

> All the banks of the Rio Negro and its different tributaries are bordered by mixed forests in which the 'Imbuia' dominates or has already dominated. These forests are also characterized by the presence of 'Mate' and 'Pinheiro'. However, in a few decades, perhaps only the memory of these forests will remain. *Ilex paraguariensis*, St. Hilaire will certainly survive, which, in addition to being the most resistant of the three, has the advantage of the fact that it is only exploitable while alive. (Hoehne, 1930, p. 51)

Hoehne's description highlights the intense destruction of the forest biome in the state of Paraná and at the same time underscores that the management of erva-mate helped, both in the past and in the present, to avoid the complete decimation of the forest. One of the key qualities of erva-mate is the fact that the tree has significant potential for regeneration and regrowth. For this reason, exploitation is based on its continuity as a living species, a situation very different to the harvesting of timber trees, i.e., pinheiro or imbuia. Because of their ability to resprout, erva-mate trees were and continue to be pruned in cycles of 3–4 years, depending on their regenerative capacity (Linhares, 1969; Garavaglia, 2008), a cycle that has changed little since the process was first described by the Jesuit Antonio Ruiz de Montoya in 1610 (Montoya, 1639).

According to Inoue et al. (1984), a productive erva-mate agroforestry system requires the natural vegetation cover of the Araucaria Forest with a concentration of erva-mate trees in the understory. Thus, despite the extensive processes of land conversion of the forest landscape in the region over the last 150 years, the existence of erva-mate within the Araucaria Forest ecosystem ensured that some of forest environments in which this tree occurs still exist today (Fig. 6.3; Nogueira, 2021). Currently, many of the remaining patches of Araucaria Forest in the states of Paraná and Santa Catarina are associated with the presence of traditional and agroecological erva-mate management systems (Marques et al., 2019; Lacerda et al., 2020). Understanding the connection and history between farming families and traditional and indigenous communities that maintain these practices and the forest are essential to develop strategies to protect not only the ecosystem but also the ecological and cultural knowledge that underpins this unique agroforestry system.

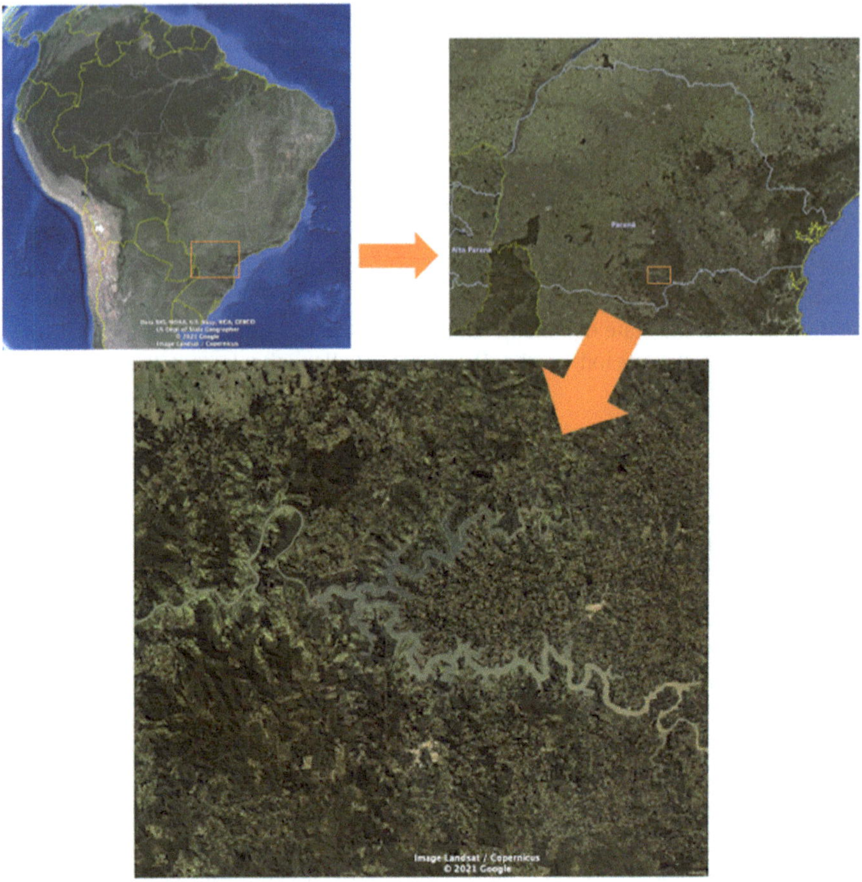

Fig. 6.3 Landscape of the region demonstrating the impact of erva-mate on the continuation of forest cover in the region

6.4 Results and Discussion

Based on the methods used and the historical context examined above, it is possible to begin to interpret the layers of temporality that imbue meaning into the management and construction of the forest landscape within traditional erva-mate systems. The results presented below demonstrate how meaning and affect are generated through a complex interaction with the environment. Materiality, which is the physical outcome of the organization of space in terms of functionality, resilience, and historicity, enables us to understand how meanings are attributed to the landscape by farming families as a result of the experience and historical interactions with the Araucaria Forest.

6.4.1 *Affective and Sensorial Assemblages in the Materiality of Traditional Erva-mate Production Systems*

The accumulation of inter-generational management practices applied to erva-mate on small-scale farms along the Iguaçu River valley show important traces of the ecological, environmental, and social history and materiality of the Araucaria Forest in the region. Elements dispersed within these traditional systems reflect activities of the past that are echoed in the present, demonstrating the biocultural materiality of the system, as well as the sensorial and sentimental affect *erveiro*[3] families have with the erva-mate and the forest environment. According to Naquin and Holton (2002), affect is a cognitive bias based on emotions through which an individual relates to and understands his or her experiences. While affect can be positive or negative, what interests here is positive affect, in which positive emotional states are experienced in a situation (Naquin & Holton, 2002), or even a place, based on experience, attitudes and values (Tuan, 2012). Yet affect does not act alone – it is made up of sensorial assemblages in which material and immaterial elements are interlaced with memory, feelings, and time. As Hamilakis (2017, p. 170) explains, "a fundamental property of all assemblages is their sensorial and affective import; that assemblages are arrangements of material and immaterial entities; and that they are also about material and sensorial memory, as well as about the engendering of diverse temporalities; and finally, that assemblages necessitate the deliberate agency and intervention of social actors." Thus, we can read traditional erva-mate systems as socioecological sensorial assemblages in which farming families have a high degree of affinity and affect in relation to place, time, and memory of the environment, and these are written and felt through daily practice, affective labor, and material symbols across the landscape.

At the property scale, traces of decades of erva-mate harvesting, species and shade management through felling and pruning of the forest, and forest fires conducted in the environment where erva-mate is produced attest to the physical traces of management practices over time (Fig. 6.4). Many properties managed for several generations are home to centenary orchards (Fig. 6.5) which are imbued with memories of working with grandparents and parents in the forest during the winter, as harvesting is done when the plants are not sprouting new leaves. During interviews, many families relate the affective memories associated with collecting erva-mate, drying the leaves on an open fire (*sapecando*), and tending to the carijo[4] throughout the night in the forest. These memories are entangled with other species such as

[3] Small-scale traditional erva-mate producers in the study refer to themselves as *erveiros*, whereas those who run businesses and produce the finished product are referred to as *ervateiros*, with an implicit class and ideological distinction (Nimmo et al., 2020).

[4] A *carijo* is the traditional structure used to roast erva-mate leaves in the forest. The structure is often temporary and consists of a hole dug into the ground for a fire from which the heat was directed toward a bamboo or other wood scaffold to roast the leaves (Da Luz, 2011; Chaimsohn & Souza, 2013).

Fig. 6.4 Erva-mate tree with traces of recent and historic pruning, creating trees embedded with history and meaning. (Source: the authors)

Fig. 6.5 Centenary erva-mate trees managed for several generations. (Source: the authors)

pinheiro, as its *pinhão* seeds are collected at the same time of the year, and are often invoked in memories of childhood, as well as affective and shared labor, and community festivities.

Within traditional erva-mate agroforestry systems, sustainable agriculture practices and LEK have been maintained alongside the biological diversity of the forest, and one of the most important elements that binds families to these systems is affect, feelings of belonging, and identities associated with the Araucaria Forest environment. Affect permeates interactions with the forest and connects memories with the

Fig. 6.6 Esequiel Hupalo demonstrating the pinheiro tree planted by his father – a geosymbolic and sensorial landmark on his property. (Source: the authors)

environment, and is manifest in socioecological relationships with and biocultural diversity of the environment. An interview with the erva-mate producer Esequiel Hupalo in the municipality of Antônio Olinto, demonstrates clearly how memories and affect are imprinted on the forest (Fig. 6.6):

> It was my father who gave this pinheiro tree to me. He planted the pinhão seeds, which according to him were not even germinating. He said that I should use this wood as needed. But now that there is this law that doesn't allow us to cut trees,[5] so the pinheiro ended up staying where it is, and every time I see it, I remember my deceased father. It ended up being a memory, right? (Nogueira, 2021)

The various manifestations of affectivity existing within erva-mate systems are intertwined with the physical memories and sensations of erva-mate, which can be classified as intra-forest geosymbolic landmarks (Bonnemaison, 1981). These landmarks are focal points that give meaning to the landscape within rural properties where the traditional erva-mate systems are maintained. They intertwine identity and belonging of *erveiro* families with the forest. This phenomenon is intensified through socioecological materiality, which takes into account the places and sensorial feelings of memory within properties and situates farming families as an integral part of the process of managing the forest landscape.

Beyond the organizational elements of the landscape that directly intersect with the forest ecosystem, other phenomenological factors accentuate the feelings of belonging of these communities with their location and experiences. Affectivity for

[5] The Brazilian Forest Code (Brasil, 2012; Machado, 2016) and the Atlantic Forest Law (Brasíl, 2006) place significant restrictions on the amount and type of management that can occur in forests on private property. The felling of threatened tree species, such as pinheiro and imbuia, are in most cases prohibited.

Fig. 6.7 *Cuia* of erva-mate with a metal straw to sip the infusion. Sharing mate is an important element of conducting interviews with traditional erva-mate producers. (Source: the authors)

the environment and with erva-mate cultivation is manifest in the act of sharing a *cuia* of mate, the gourd in which *mate* is traditionally consumed. Erva-mate leaves are placed in the *cuia* and hot water is poured over to create the infusion that is consumed through a metal straw (Fig. 6.7). The act of sharing and drinking *chimarrão* has a strong cultural significance for traditional erva-mate producers. When a visitor arrives at a property that uses these systems, they are always received with a warm kettle and a *cuia* of mate to share. The mate used is often grown and processed on the property, in a small artisanal *barbaqua* for the family's own consumption. The ritual of preparing and sharing mate is an important element to start the conversation, a symbol of hospitality, friendship, and shared struggles and histories. Thus, each *cuia* of mate has its own specific flavor that is instilled not only with specific characteristics of the erva-mate orchard from which the leaves are harvested, but also the years of use, shared memories, and conversations.

The culinary traditions of traditional erva-mate families based on generations of accumulated experience and the use of native heirloom species or landraces on the property are also important affective elements that tie families to the forest. The cakes made of pinhão collected during the cold winter months, the empanadas filled with ricotta made from fresh cow's milk, the *torresmo* (pork scratching) fried in lard from heirloom pigs bred free-range on the farm, drinking toasted mate tea, all bring together feelings and sensations of pleasure, and a sense of belonging to the forest and the place. Because it is a system that is embedded with sensorial assemblages of good food, difficult but satisfying work, and time spent with family and friends from the community, the continuation of these systems is intimately tied not only to understanding and knowledge of the forest, but also the feelings of belonging and affect that *erveiros* have for the environment in which they live (Nogueira, 2021; Marques et al., 2019).

Fig. 6.8 Abandoned *barbaquá* constructed in the early twentieth century on the property of Luis Szymanek, in the community of Rio do Banho, Cruz Machado, Paraná. (Source: the authors)

The scope of socioecological materiality and sensorial assemblages can also be found in the built structures within properties that demonstrate the historicity of the relationship between traditional farmers and their environment. For example, the structure within which erva-mate is processed, called a *barbaquá*, is an important structure and testament to local ecological history (Fig. 6.8). The *barbaquá* is an evolution from the *carijó* structure as it is permanent and includes a range of processing activities. However, the technology used in the *barbaquá* developed directly from the knowledge of indigenous peoples on how to process the plant, which to this day is still divided into three basic stages: flash drying (*sapecar*), roasting, and macerating the leaves. Further, the building itself is constructed of elements from the forest, for example imbuia or pinheiro wood felled and processed on the property, along with the animal traction system once used to power the macerators (*malhador* or *maiadô*). Many of the *barbaquás* that still stand on traditional erva-mate producing properties throughout the region were used for generations during the twentieth century but were abandoned and fell into disrepair in the 1980s, due to legal and sanitary restrictions imposed as a result of modernizing agro-industrial policies (Chaimsohn & Souza, 2013). Nevertheless, they are still a strong symbol of belonging based on memory and experiences of the individuals involved in the cultivation of traditional erva-mate (Miranda & Urban, 1998; Nogueira, 2021).

Traditional erva-mate production systems are the confluence of a multitude of agroecological and social practices, particularly linked to the idea of belonging and identity of *erveiro* families. From an ecological perspective, traditional erva-mate production is also associated with agroecological cultivation and the management of genetic resources of range of native South American plants, such as heirloom varieties of corn (*Zea mays*), cassava (*Manihot esculenta*) and sweet potato (*Ipomea potatoes*), as well as plants with cultural and nutritional value, such as beans

(*Phaseolus vulgaris*). The maintenance of these plant varieties is associated with knowledge sharing and seed fairs that occur throughout the region. Therefore, the continuity of the traditional erva-mate systems constitutes a relevant possibility for preserving agrobiodiversity in the Araucaria Forest (Marques, 2014).

The relationships between farmers and these traditional systems invoke profound knowledge and subjectivities that make the social technologies used in forest management much more significant than mere gestures of daily labor. These sensorial assemblages create a strong bond between how these farmers see themselves, their histories, and their future as stewards of the forest environment. It is these particular subjectivities that are essential to understand how those using traditional agroecological practices, such as the erva-mate systems described herein, can be important actors in locally- and culturally-relevant strategies for ecosystem conservation and restoration.

6.5 Conclusions and Considerations for Conservation

The shared elements that occur across traditional erva-mate production systems create common connections of environmental subjectivities and feelings of stewardship and belonging in the Araucaria Forest that are key elements of sustainable development and ecological conservation and restoration in the region. However, throughout the production chain, from the cultural and ecological knowledge maintained by the families to the final consumer, there is a need to recognize the added value arising from the socioecological realities and agroecological practices within the system. Such a recognition can offer greater autonomy and self-determination for farming families both economically and in terms of management strategies within their system, while also supporting the continuation of these systems in the face of pressures from the agroindustrial complex and the movement toward productivity-based monoculture systems, which are currently promoted by the large erva-mate producers in the region.

The assemblage of knowledge, practice, affect, and memory that inform traditional erva-mate farming families and indigenous communities in the region represent possibilities to develop and implement regenerative agriculture and agroforestry technologies in the Araucaria Forest (Lacerda et al., 2020). Strengthening the local network of *erveiros* and documenting the sound ecological practices in their respective orchards can lead to innovations that encourage forest restoration associated with agricultural production and sustainable rural livelihoods, including legal frameworks for forest protection that are better-adapted to, and supportive of, agroforestry practices, the dissemination and promotion of agroforestry management techniques, and the implementation of public policies aimed at improving the quality of life of the populations involved. Research and outreach collaborations between communities, research institutions, and civil society organizations are key aspects of developing and disseminating these innovations. For example, participatory studies with *erveiros* and Embrapa Forestry, in partnership with a range of other state and civil society organizations, have replicated, documented, and leveraged this knowledge to develop models that can be used to restore degraded areas. That is, based on practices maintained in traditional erva-mate production systems, degraded landscapes can be reforested using agroforestry systems that not only provide improved socioecological services to the property and the region, but also offer consistent economic return for farming families (Radomski et al., 2014; Lacerda, 2019a, b).

In Paraná, where erva-mate is economically, ecologically, and socially important, over the last 30 years, institutions and communities have come together to strengthen the sustainable practices used in these agroforestry systems, while also improving dialogue across environmental and social agencies. One outcome of this long-term engagement is the creation of the 'Observatory of Traditional and Agroecological Erva-mate Systems' (*Observatório dos Sistemas Tradicionais e Agroecológicos de Erva-mate*), a strategic council made up of representatives from 32 non-governmental and civil society organizations, universities, research and rural extension institutions, farmers unions, municipal governments and communities (Lacerda et al., 2020; Nimmo et al., 2020). The purpose of this council is to protect these systems and support actions to ensure a more resilient and respectful future based on the knowledge inherent in traditional erva-mate production. The Observatory acts as a mediator in the dialogue across a range of organizations and institutions that develop public policies, and brings to these discussions several priorities as defined by the community, including prioritizing regenerative agricultural practices that support diversification, engaging and supporting the active participation of women and youth, preserving agrobiodiversity across the landscape, mitigating the effects of climate and food insecurity, and incorporating the knowledge of indigenous and traditional populations as guiding principles within public policies related to sustainable development and restorative agriculture.

Recently, the Food and Agriculture Organization of the United Nations (FAO) and the UN Environment Program (UNEP) have named 2021 to 2030 the UN Decade of Ecosystem Restoration (UN, 2021a, b). The goals of this program are to inspire governments and civil society to develop and catalyze initiatives to restore terrestrial and marine ecosystems across the planet through the engagement of traditional populations, indigenous peoples, and research institutions. The UN Decade of Ecosystem Restoration complements efforts to achieve the 17 Sustainable Development Goals (SDGs), but our research demonstrates that restoration and conservation strategies must consider the environmental subjectivities and sensorial assemblages inherent in traditional agricultural systems to develop solutions that are relevant to the communities in which they will be implemented, and therefore have the greatest impact. As such, collaborative approaches that explore concepts from Historical Ecology, Environmental History, and Landscape Archaeology can help to integrate and value different perceptions of agroecosystems, and ensure that communities, their sensorial assemblages of LEK, history, memory, and affect, are central to strategies and frameworks that can bring about transformational change at the landscape scale.

References

Alkhatib, A. (2018). Yerba maté and exercise performance. *Medicine & Science in Sports & Exercise, 50*(7), 1525. https://doi.org/10.1249/MSS.0000000000001583

Arce-Nazario, J. A. (2007). Landscape images in Amazonian narrative: The role of oral history in environmental research. *Conservation and Society, 5*(1), 115–133.

Austin, S. M. (2015). Guaraní kinship and the encomienda community in colonial Paraguay, sixteenth and early seventeenth centuries. *Colonial Latin American Review, 24*(4), 545–571. https://doi.org/10.1080/10609164.2016.1150039

Berkes, F., Colding, J., & Folke, C. (2000). Rediscovery of traditional ecological knowledge as adaptive management. *Ecological Applications, 10*, 1251–1262.

Berque, A. (2013). *Thinking Through Landscape*. Routledge.

Bonnemaison, J. (1981). Voyage autour du territoire. *Espace géographique, 4*(10), 249–262.

Brasil. (2006). *Lei no 11.428 de 22 de dezembro de 2006. Dispõe sobre a utilização e proteção da vegetação nativa do Bioma Mata Atlântica, e dá outras providências*. Diário Oficial da União: Brasília, 22 de dezembro de 2006. Retrieved on June 26, 2021, from http://www.planalto.gov.br/ccivil_03/_Ato2004-2006/2006/Lei/L11428.htm.

Brasil. (2012). *Lei N° 12.651, de 25 de Maio de 2012; Dispõe Sobre a Proteção da Vegetação Nativa*. Retrieved June 26, 2021, from https://www2.camara.leg.br/legin/fed/lei/2012/lei-12651-25-maio-2012-613076-normaatualizada-pl.pdf

Brasil. Ministerio do Meio Ambiente. (2021). *Biomas do Brasil*. Retrieved June 26, 2021, from https://antigo.mma.gov.br/biomas.html

Cañizares-Esguerra, J. (2005). Iberian colonial science. *Isis, 96*(1), 64–70. https://doi.org/10.1086/430679

Caporal, F. R., & Costabeber, J. A. (2004). *Agroecologia: alguns conceitos e princípios*. MDA/SAF.

Carvalho, P. E. R. (2003). *Espécies arbóreas brasileiras*. Embrapa.

Castella, P. R., & Britez, R. M. (2004). *A Floresta com Araucária no estado do Paraná*. MMA.

Centre for Development and Education of Yerba Mate Traditional Systems (CEDErva) (2017). *Sobre os sistemas tradicionais de erva-mate*. Retrieved June 21, 2021, from http://www.cederva.com/sobre.html

Chaimsohn, F. P., & Souza, A. M. d. (Eds.). (2013). *Sistemas de produção tradicionais e agroflorestais de erva-mate no centro-sul do Paraná e norte catarinense: contribuições para a construção do processo de indicação geográfica*. IAPAR.

Corteletti, R., & Iriarte, J. (2018). Recent advances in the archaeology of the Southern Proto-Jê People. In C. Smith (Ed.), *Encyclopedia of global archaeology* (pp. 1–11). Springer. https://doi.org/10.1007/978-3-319-51726-1_3037-1

Cunha, M. C. (2013). *Índios no Brasil: história, direitos e cidadania* (p. 2013). Companhia das Letras.

Da Luz, M. (2011). *Carijos e Barbaquás no Rio Grande do Sul: Resistência camponesa e conservação ambiental no âmbito da fabricação artesanal de erva-mate*. Master's thesis, Universidade Federal do Rio Grande do Sul.

Dean, W. (1996). *A ferro e fogo: a história e a devastação da Mata Atlântica brasileira*. Companhia das Letras.

Elisabetsky, E. (1986). Etnofarmacologia de algumas tribos brasileiras. In D. Ribeiro (Ed.), *Suma Etnológica Brasileira* (1st ed., pp. 134–149). Vozes/FINEP.

Folch, C. (2010). Stimulating consumption: Yerba Mate myths, markets, and meanings from conquest to present. *Comparative Studies in Society and History, 52*(1), 6–36. https://doi.org/10.1017/S0010417509990314

Food and Agriculture Organization (FAO). (2001). *Global ecological zoning for the global forest resources assessment 2000* (Forest Resources Assessment Working Paper, 56). Retrieved June 22, 2021, from http://www.fao.org/3/ad652e/ad652e00.htm

Fournier, P. (2010). Pour une histoire de l'aménagement des espaces ruraux. *Annales du Midi: revue archéologique, historique et philologique de la France méridionale, 272*, 469–476.

Furlong, G. (1962). *Misiones y sus pueblo de guaraníes*. Imprenta Balmes.
Gallardo, C. R. (1898). *La indústria yerbatera en Misiones* (Oficina Nacional de Agricultura). San Jorge.
Gan, R.-Y., Zhang, D., Wang, M., & Corke, H. (2018). Health benefits of bioactive compounds from the genus *Ilex*, a source of traditional caffeinated beverages. *Nutrients, 10*(11), 1682. https://doi.org/10.3390/nu10111682
Garavaglia, J. C. (2008). *Mercado interno y economía colonial. Tres siglos de historia de la yerba mate* (2nd ed.). Prohistoria Ediciones.
Gerhardt, M. (2013). *História Ambiental da erva-mate*. PhD thesis. Universidade Federal de Santa Catarina.
Gubert Filho, F. A. (1988). *Levantamento de Áreas de Relevante Interesse Ecológico no Estado do Paraná*. Instituto Ambiental do Paraná.
Hamilakis, Y. (2017). Sensorial assemblages: Affect, memory and temporality in assemblage thinking. *Cambridge Archaeological Journal, 27*(1), 169–182. https://doi.org/10.1017/S0959774316000676
Harris, S. J. (2005). Jesuit scientific activity in the overseas missions, 1540–1773. *Isis, 96*(1), 71–79. https://doi.org/10.1086/430680
Haubert, M. (1990). *Índios e Jesuítas no tempo das missões*. Companhia das letras.
Haverroth, M. (1997). *Kaingang: um Estudo Etnobotânico do uso e a classificação das plantas na Área Indígena Xapecó (oeste de SC)*. Master's thesis. Universidade Federal de Santa Catarina.
Hoehne, F. C. (1930). *Araucarilandia. Observações Geraes e Contribuições ao Estudo da Flora e Phytophysionomia do Brasil*. Secretaria da Agricultura, Industria e Commercio do Estado de São Paulo; Directoria de Publicidade.
Horn, T. B., Nascimento, J. d., Vogel, P., & Faccin, C. (2018). Evolução histórica do cultivo e usos da erva-mate. In N. J. Ferla, G. L. da Silva, & L. Johann (Eds.), *A cultura da erva-mate e os ácaros: situação atual e perspectivas* (pp. 25–40). Editora Evangraf.
Inoue, M. T., Roderjan, C. V., & Kuniyoshi, Y. S. (1984). *Projeto Madeira do Paraná*. Fundação de Pesquisas Florestais do Paraná.
Instituto Brasileiro de Geografia e Estatística (IBGE). (2012). *Manual técnico da vegetação brasileira* (2nd ed.). IBGE.
Iriarte, J., Smith, R. J., De Souza, J. G., Mayle, F. E., Whitney, B. S., Cárdenas, M. L., Singarayer, J., Carson, J. F., Roy, S., & Valdes, P. S. (2017). Out of Amazonia: Late-holocene climate change and the Tupi–Guarani trans-continental expansion. *Holocene, 27*(7), 967–975. https://doi.org/10.1177/0959683616678461
Koch, Z., & Corrêa, M. C. (2010). *Araucária: a floresta do Brasil Meridional*. Olhar Brasileiro.
Lacerda, A. E. B. (2016). Conservation strategies for Araucaria Forests in Southern Brazil: Assessing current and alternative approaches. *Biotropica, 48*(4), 537–544. https://doi.org/10.1111/btp.12317
Lacerda, A. E. B. (2019a). Série erva-mate sombreada (1): Sistema de produção de erva-mate baseado no manejo tradicional: bracatingais dominados por bambus (taquarais). *Comunicado Técnico Embrapa Florestas, 439*.
Lacerda, A. E. B. (2019b). Série erva-mate sombreada (2): Sistema de restauração produtivo agroflorestal. *Comunicado Técnico Embrapa Florestas, 400*.
Lacerda, A. E. B., Hanisch, A. L., & Nimmo, E. R. (2020). Leveraging traditional agroforestry practices to support sustainable and agrobiodiverse landscapes in southern Brazil. *Land, 9*(6), 176. https://doi.org/10.3390/land9060176
Laverdi, R., & Mastrángelo, M. (2013). *Desde las profundidades de la Historia Oral*. Imago Mundi.
Lennartsson, T., Eriksson, O., Iuga, A., Larsson, J., Moen, J., Scholl, M. D., Westin, A., & Crumley, C. L. (2018). Diversity in ecological and social contexts. In C. L. Crumley, T. Lennartsson, & A. Westin (Eds.), *Issues and concepts in historical ecology: The past and future of landscapes and regions* (pp. 182–239). Cambridge University Press. https://doi.org/10.1017/9781108355780
Linhares, T. (1969). *História econômica do Mate*. Livraria José Olympio.

Lockhart, J. (1969). Encomienda and Hacienda: The evolution of the great estate in the Spanish Indies. *Hispanic American Historical Review, 49*(3), 25–49. https://doi.org/10.2307/2511778
López, A. (1974). The economics of yerba mate in seventeenth-century South America. *Agricultural History, 48*(4), 493–509.
Maack, R. (1931). Urwald und Savanne im Landschaftsbild des Staates Parana. *Zeitschrift der Gesellschaft für Erdkunde zu Berlin, 3*(4), 95–116.
Maack, R. (2012). *Geografia Física do Estado do Paraná*. Editora UEPG.
Machado, F. (2016). *Brazil's new forest code: A guide for decision-makers in supply chains and governments*. WWF Brazil.
Maffi, L., & Woodley, E. (2010). *Biocultural diversity conservation, a global sourcebook*. Earthscan.
Marques, A. C. (2014). *As paisagens do mate e a conservação socioambiental: um estudo junto aos agricultores familiares do planalto norte catarinense*. PhD thesis. Universidade Federal do Paraná.
Marques, A. C., Reis, M. S., & Denardin, V. F. (2019). Yerba mate landscapes: Forest use and socio-environmental conservation. *Ambiente e sociedade, 22*, e02822.
Martins, A. R. (2005). *Livro das árvores do Paraná*. Instituto Ambiental do Paraná, Imprensa Oficial do Estado do Paraná.
Miranda, N., & Urban, T. (1998). *Engenhos e barbaquás*. Posigraf.
Monteiro, J. M. (1994). *Negros da terra: índios e bandeirantes nas origens de São Paulo*. Companhia das Letras.
Montoya, A. R. d. (1639). *Conquista espiritual hecha por los religiosos de la Compañia de Iesus en las Provincias del Paraguay, Parana, Uruguay, y Tape*. En la imprenta del Reyno.
Naquin, S. S., & Holton, E. F. (2002). The effects of personality, affectivity, and work commitment on motivation to improve work through learning. *Human Resource Development Quarterly, 13*(4), 357–376.
Nimmo, E. R., & Nogueira, J. F. M. M. (2019). Creating hybrid scientific knowledge and practice: The Jesuit and Guaraní cultivation of yerba mate. *Canadian Journal of Latin American and Caribbean Studies/Revue canadienne des études latino-américaines et caraïbes, 3*(44), 347–367.
Nimmo, E., Carvalho, A. I., Laverdi, R., & Lacerda, A. E. B. (2020). *Oral history and traditional ecological knowledge in social innovation and smallholder sovereignty: a case study of erva-mate in Southern Brazil Ecology and Society, 25*(4), art17. https://doi.org/10.5751/ES-11942-250417
Nogueira, J. F. M. M. (2021). *Historicidade e significado nas paisagens dos sistemas tradicionais de produção de erva-mate na bacia do alto Iguaçu, sul do Paraná*. Master's thesis, Universidade Estadual de Ponta Grossa.
Norton, M. (2017). Subaltern technologies and early modernity in the Atlantic world. *Colonial Latin American Review, 26*(1), 18–38. https://doi.org/10.1080/10609164.2017.1287322
Oliveira, Y. M. M. d., & Rotta, E. (1985). Área de distribuição natural de erva-mate (Ilex paraguariensis). *Seminário sobre atualidades e perspectivas florestais*. EMBRAPA-CPNF.
Oliveira, R. C. (2001). O silêncio dos vencedores: genealogia, classe dominante e Estado do Paraná. Moinho do Verbo.
Radomski, M. I., Lacerda, A. E. B., & Kellermann, B. (2014). *Sistemas agroflorestais: restauração ambiental e produção no âmbito da floresta Ombrófila Mista*. Documentos EMBRAPA Florestas, 276.
Ribeiro, D. (1985). *O Povo Brasileiro: a formação e o sentido do Brasil*. Companhia das Letras.
Saeger, J. S. (1981). Survival and abolition: The eighteenth century Paraguayan Encomienda. *The Americas, 38*(1), 59–85. https://doi.org/10.2307/981082
Sahr, C. L. & Cunha, L. A. G. (2005). O significado social e ecológico dos Faxinais: Reflexões acerca de uma política agrária sustentável para a região da mata com araucária no Paraná. *Emancipação, 5*(1), 89–104.

Service, E. R. (1951). The Encomienda in Paraguay. *Hispanic American Historical Review, 31*, 230–252. https://doi.org/10.2307/2509030

Service, E. R. (1971). *Spanish-Guarani relations in early colonial Paraguay*. Greenwood Press.

Singh, N. M. (2013). The affective labor of growing forests and the becoming of environmental subjects: Rethinking environmentality in Odisha, India. *Geoforum, 47*, 189–198. https://doi.org/10.1016/j.geoforum.2013.01.010

Toledo, V. M., & Barrera-Bassols, N. (2015). *A memória biocultural: a importância ecológica das sabedorias tradicionais*. Expressão Popular.

Tuan, Y. (2012). *Topofilia: um estudo da percepção, atitudes e valores do meio ambiente*. Eduel.

United Nations (UN). (2021a). *The sustainable development goals*. Retrieved May 20, 2021, from https://sdgs.un.org/goals

United Nations (UN). (2021b). *United Nation decade on ecosystem restauration (2021–2030)*. Retrieved May 20, 2021, from https://www.decadeonrestoration.org/

Wachowicz, R. C. (1981). *O camponês polonês no Brasil*. Fundação Cultural Casa Romário Martins.

WWF. (2000). *Araucaria moist forests*. Retrieved June 22, 2021, from https://www.worldwildlife.org/ecoregions/nt0101

Zampier, A. C. (2001). *Avaliação dos níveis de nutrientes, cafeína e taninos após adubação mineral e orgânica, e sua relação com a produtividade na erva-mate (Ilex paraguariensis St. Hil.)*. Master's thesis. Universidade Federal do Paraná.

Zhang, J. (2019). Tourism and environmental subjectivities in the anthropocene: Observations from Niru Village, Southwest China. *Journal of Sustainable Tourism, 27*(4), 488–502.

Part II
Our Shared Past

Chapter 7
An Archaeology of Social Jê Landscapes at Urubici, Santa Catarina

Rafael Corteletti, Bruno Labrador, and Paulo Antônio Dantas DeBlasis

Abstract This paper summarizes recent archaeological research at the upper Canoas valley, Santa Catarina, also known as the Urubici region. This area constitutes a long-term focus of occupation by Southern Jê speaking societies, where archaeological sequences overlap with historical and ethnographic accounts, from circa 1800 BP to the beginning of the twentieth century. While mapping a wide range of settlements across the regional landscape, a basic chronological control for the Southern Jê long permanence in the area has been established. Furthermore, some referential sites have been excavated in greater detail, to refine settlement chronology and explore their internal configuration. In this sense, Jê landscapes are composed of a diversity of places and spaces of coexistence and circulation with different functionalities, which interact with one another through a socio-cosmological structure of their own. All along this period, Jê societies thrived and expanded, having created ingenious earthworks and a proper cultural landscape, full of symbolic references, which can be perceived to this day. It has been an history of expansion, but also of resistance and fight. It is our hope that archaeology might help original peoples to recover their dignity and their rights.

Keywords Southern Jê · Southern Brazilian highlands · Late Holocene · Jê speaking societies · Cultural landscape

---

R. Corteletti (✉)
Instituto de Ciências Humanas – UFPel, Pelotas, Brazil
e-mail: rcorteletti@ufpel.edu.br

B. Labrador
Museu de Arqueologia e Etnologia Oswaldo Rodrigues Cabral – UFSC, Florianópolis, Brazil
e-mail: bruno.labrador@ufsc.br

P. A. D. DeBlasis
Museu de Arqueologia e Etnologia – USP, São Paulo, Brazil
e-mail: deblasis@usp.br

A. C. Colonese, R. G. Milheira (eds.), *Historical Ecology and Landscape Archaeology in Lowland South America*, Interdisciplinary Contributions to Archaeology, https://doi.org/10.1007/978-3-031-32284-6_7

In this chapter, the Jê archaeology of the Urubici region, Santa Catarina, is discussed. The Urubici region is defined by the upper Canoas River basin, an open valley amidst the massive scarps of the Southern Brazilian plateau, one of the highest and coldest patches of the southern Brazilian highlands. This area constitutes a long-term focus of occupation by Southern Jê speaking societies, where archaeological sequences overlap with historical and ethnographic accounts, from circa 1800 BP to the beginning of the twentieth century.

This dense and characteristic Jê occupation has been studied by our research group in the past 10 years. While mapping a wide range of settlements across the regional landscape, a basic chronological control for the Southern Jê long permanence in the area has been established. Furthermore, some referential sites have been excavated in greater detail, to refine settlement chronology and explore their internal configuration. The results are summarized here, and an interpretive model integrating archaeological and ethnographic information is advanced.

7.1 The Southern Jê

The archaeology of Southern Jê groups dates to c. 2200 cal year BP and extends to the beginning of the twentieth century. It has been first defined by its diagnostic small and thin-walled globular ceramics, still its most typical and characteristic feature (Schmitz, 1988; Beber, 2005). Another principal characteristic of these societies is the construction of earthworks, including pit houses of different sizes and internal organization (DeSouza et al., 2016b; DeSouza, 2018), as well as elaborated mound and enclosure complexes (Iriarte et al., 2008, 2013; DeMasi, 2009; DeSouza et al., 2016a; Robinson et al., 2017). Additionally, collective burials in caves (Saldanha, 2008; Corteletti, 2012) and rock art at spectacularly placed locations have also been described (Riris & Corteletti, 2015; Corteletti et al., 2021).

A deep and long-term trajectory has been established for this Native American people by linguists, archaeologists, anthropologists, and historians, emphasizing the historical continuity between the archaeological Taquara-Itararé Tradition and the modern Southern Jê groups – the *Kaingang* and *Laklãnõ* (*Xokleng*) people (see details in Iriarte et al., 2017). Under the general label of Southern Jê and its shared material culture, there is an increasing recognition that this broadly defined tradition spreads over a vast and diverse ecological area, encompassing a remarkable range of local variability in social and ritual organization. The material evidence of their presence can be found throughout the southern Brazilian cone and beyond, from the Atlantic coast to the west of the Paraná River, encompassing at least four wide geomorphological and climatic domains and their ecotones (Corteletti & Iriarte, 2018) (Fig. 7.1a).

To the east, the coastal plain with lagoons and *restingas*; further inland, the eastern elongated mountain ranges and open valleys covered by Atlantic rain forest. Towards the west, beyond the steep *cuestas* of the Southern Brazilian plateau, the flat highland extensions covered with grasslands and, at the valleys and deep

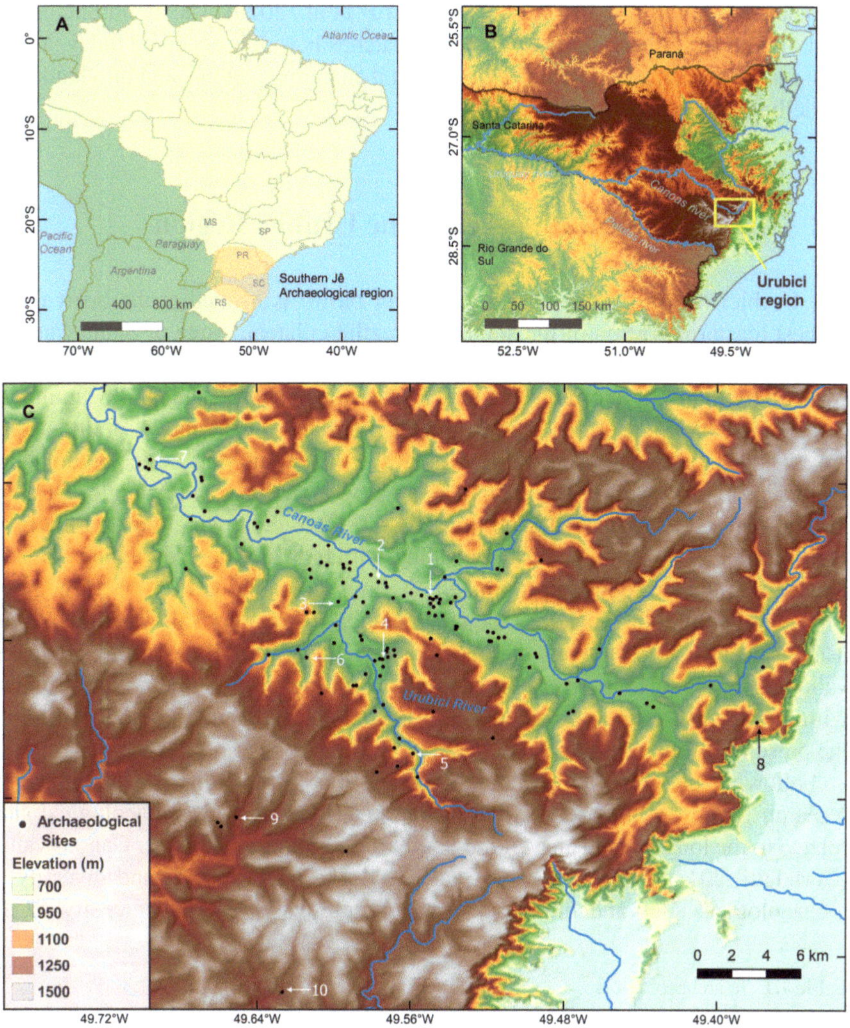

Fig. 7.1 (**a**) Southern Jê archaeological region. (**b**) Urubici archaeological region, in Santa Catarina State. (**c**) Upper Canoas valley, in Urubici, with a plenty of Jê archaeological sites, where: 1. Bonin; 2. Anderman; 3. Urubici 4; 4. A. Rodrigues; 5. Aripol; 6. Morro do Avencal; 7. Copetti; 8. Casa de Pedra; 9. Vacas Gordas 3; 10. Santa Bárbara

canyons, Paraná pine (*Araucaria angustifolia*) forest. These highlands softly descend westwards towards the open areas near the Paraná River and its tributaries, densely covered by subtropical forest (Fig. 7.1b).

The management of *Araucaria* Forest by the Southern Jê populations has long been postulated (Bitencourt & Krauspenhar, 2006; Iriarte & Behling, 2007). It has recently been supported by interdisciplinary studies of archaeology, paleoecology and palaeoclimatology (Robinson et al., 2018), phylogeographic studies of *Araucaria* genome (Lauterjung et al., 2018) and forest composition patterns associated to archaeological data (Pereira Cruz et al., 2020). The symbiotic expansion of both Jê settlements and pine forest patches is widely accepted today, a good case for the human presence on environmental changing patterns and into landscape modelling in southern Brazil.

7.2 Archaeological Research at the Urubici Region

The Urubici region (the upper Canoas River basin) is located at the heart of the highest (eastern) portion of the southern Brazilian plateau, State of Santa Catarina (Fig. 7.1b). The weather in these highlands is humid most of the year, with cold winters (with usual morning frost and occasional snow) and warm summers. *Araucaria* subtropical forest and highland grasslands, or *campos*, dominate the area. The Canoas headwaters are located at the top of this very desiccated plateau, around 1800 m above sea level (a.s.l.). Like all watercourses in this region, it flows through incised canyons and deep forested valleys to get to an elongated flat open plain, around 1000 m a.s.l. This open valley, with high and steep mountain walls by its sides, extends through approximately 35 km, widening up to 3 km (Fig. 7.1c). This configuration distinguishes the upper Canoas from other valleys at the region, usually deep incised canyons on the sandstone substrate. All along this elongated valley bottom is sided by river terraces and low rolling hills, making a sharp contrast with the steep surrounding walled scarps of the plateau. Here, material vestiges of the presence of the southern Jê are plenty.

Following previous publications (Piazza, 1966, 1969; Rohr, 1971, 1984), our research group has intensely surveyed the area, locating 124 archaeological sites related to the long-term history of the Southern Jê groups in the upper Canoas valley (Corteletti, 2012; Labrador, 2018; Table 7.1). The large number and diversity of archaeological sites and their long-lived chronology reveal the intensity of the

Table 7.1 Jê archaeological sites per categories and types at the Urubici region

Categories of sites	Types of sites
Earthworks	Pit houses villages
	Mounds
	Enclosures
	Mound and enclosure complexes
	Pit houses and mounds
Open-air	Litho-ceramic
Rock art/petroglyphs	In pit houses
	In rock shelters
	In palaeoburrows
Rocky substrate	Rock shelters with collective burials
	Rock shelters
	Lithic stations
	Palaeoburrows

Southern Jê occupation at the Urubici region (Piazza, 1969; DeMasi, 2001; Corteletti, 2012; Almeida, 2014; Labrador, 2018).

Settlement includes villages, crop locations and lithic stations, as well as ceremonial and mortuary places, some of them located at rock shelters with parietal representations, such as Morro do Avencal, studied in some detail (Corteletti, 2012; Riris & Corteletti, 2015; Corteletti & DeBlasis, 2018; Corteletti et al., 2021). Besides rock art and specific traits of the material culture, available chronology also connects the archaeological proto-Jê to the ethnographically known and contemporary Jê speaking societies, thus making a strong case for an extensive indigenous history of the southern Jê people.

Most of the valleys of southern Brazilian highlands are V-channelled canyons deep carved on basaltic and sandstone rock substrate. Thus, at most of them, the major Jê settlements are located at the higher part of the slopes and at the hilltops (Schmitz et al., 2002; Saldanha, 2005; Corteletti, 2008; Robinson et al., 2017). In the upper Canoas valley, Southern Jê settlement distribution is subtly different. In this open and large, bottom-flat, U-shaped fluvial corridor, with a predominance of sandstone and metasedimentary older rock formations (Scheibe, 1986), most sites have been found along the open plain at the bottom of the valley, amidst its older and higher terraces and surrounding slopes (Figs. 7.1c and 7.2).

Indeed, more than 75% of the sites are in the lower slopes and at the bottom of the valley (Corteletti, 2012; Labrador, 2018), in clayey, moderately drained, and humic cambisol areas (Fasolo et al., 2004; Corteletti & DeBlasis, 2018; Fig. 7.1c). Due to climatic conditions, these areas have been described as "restricted to summer crops" (Streck et al., 2002). Thus, at Urubici, Jê settlements have beneficiated from such open geomorphological scenario, creating a peculiar cultural landscape rather distinct from other surrounding highland settlements. But, as we shall see, social landscape organization here seems to share the essential ontological configuration and structure common to Jê societies all over the plateau, adapted to a different and peculiar landscape.

Fig. 7.2 View of Upper Canoas Valley with flat bottom and steep hills surrounding. Left: From top of Campestre Hill looking the wide view to east. Right: into the valley, near Bonito Creek, looking to south. (Photos: Rafael Corteletti)

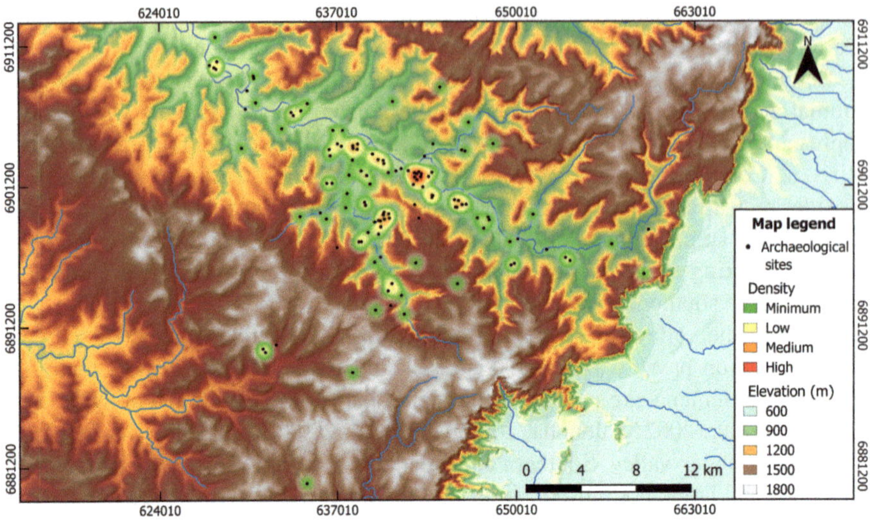

Fig. 7.3 Clusters of sites at upper Canoas valley

In addition to the preferential location of sites at the bottom of the valley, where the soils are more fertile, point density (Kernel density) analysis shows that, along the course of the upper Canoas and some of its tributaries, sites cluster into aggregates, usually closer to confluences between larger streams, on average 4 km away from each other (Fig. 7.3). Similar site clustering patterns have been described at the lower Canoas valley, around 160 km westward (DeMasi, 2005), Campo Belo do Sul (D'Ávila, 2017) and Pinhal da Serra (Saldanha, 2005). In order to get a better understanding of this settlement disposition and concomitance, test excavations have been performed in several pit house villages, one ceremonial site and a few rock art bearing rock shelters. Results are summarized in the following pages, and site distribution discussed.

7.3 Southern Jê Settlements in the Urubici Region

As we have seen, sites are organized in clusters of different sizes across the valley. In between them scattered occurrences might occur, but there are patches where no archaeological evidence has been found. If, by eight to six hundred years ago, these clusters represent a concomitant and dense Jê occupation throughout the valley, some locations have much longer history, revealing a gradual process of territorial expansion and, perhaps, earlier cyclical mobility, reaching up to at least around 1800 years ago.

Bonin is one of these enduring locations. It consists of a circular pit house village, with at least 30 distinct pit house units, some of them composed by two, occasionally three, geminated pit structures, settled in a large and low hill, not far from

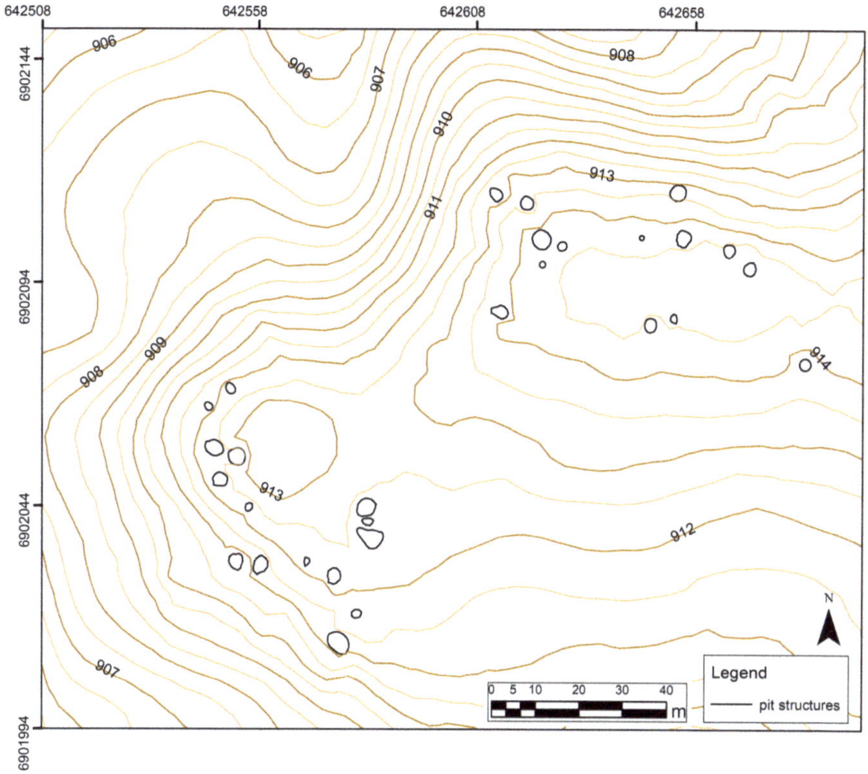

Fig. 7.4 Bonin pit house village topographic plan

the main river (Fig. 7.4). The flat central area, void of earthwork features, exhibits though fireplaces and disperse lithic and potsherd scatters. Excavations in some of these pit structures have revealed habitational vestiges, mostly kitchen refuse and lithic tools. Lots of charcoal and burnt potsherds into well delimited stone-made cook places make it clear quotidian culinary activities, using at least 30 different pots from well-known Taquara-Itararé (Jê) ceramic ware types (Corteletti, 2012; Souza Soares, 2019; Sprenger, 2020).

Starch grains and phytolith analysis of food remains incrusted on potsherd walls has revealed traces of several cultigens in Jê people's diet, including maize (*Zea mays*), pumpkin (*Cucurbita* sp.), beans (*Phaseolus* sp.), cassava (*Manihot esculenta*) and yams (*Dioscorea* sp.) (Corteletti et al., 2015). Although cultivation among the proto-Jê is widely acknowledged, this study was pioneer in providing hard evidence for the presence of these cultigens, documenting their variety as well. Additionally, flotation of sediments from pit fires excavated in 2016 and 2017 allowed to recover charred seeds of maize (*Zea mays*), *pinhão* (*Araucaria angustifolia*), beans (*Phaseolus* sp.), and probably *gabiroba* (*Campomanesia* sp.), plus some from species of Arecaceae, Euphorbiaceae and Fabaceae families, and many other unidentified. Poorly preserved charred bone remains from small fishes and birds and a fragment of *Mycetopodidae* sp. shell has also been recovered by flotation, strengthening the wide range of dietary possibilities available to the Je pre-colonial societies.

The distribution of the pit houses at the site reveals two clusters, or sides, pretty much like two horseshoes facing each other (Fig. 7.4). Radiocarbon dating shows that these "halves" are concomitant from around eight hundred years ago up to historical times, making up a typical Jê pit house village. Excavations at two of these pit structures, however, show that they have been installed over earlier, deep-carved oven pits,[1] dating from at least 1200 years ago (Table 7.2). Such evidence reveals the earlier presence of Jê people in the valley, and that this nice and appraisable spot has been visited, or even settled, long before the larger village installed there some centuries later. It might well be that the occupation of this location has been continuous but, at this point, available data so far amassed does not support that yet.

The divided, dual pit house distribution at Bonin evokes another very characteristic Jê social feature. Indeed, dual social organization of ethnographically documented Jê villages, whether at the central Brazil (e.g., Wüst & Barreto, 1999) or among the southern Jê (Veiga, 2006; Crépeau, 2005) seems expressed in its spatial organization in pre-colonial times. Such a dual pattern has been documented at a ceremonial funerary location elsewhere (Robinson et al., 2017), but at Bonin, a typical Jê pit house village in the very heart of the Urubici area, it reinforces the longevity of this social configuration among the southern Jê, establishing an additional connection between archaeology and ethnography of these people. Also, the chronology we are unveiling in the Urubici sites show that Bonin is contemporaneous of other villages and site clusters, integrating a wider, regional settlement system at the valley.

Ceremonial sites, thought of as places of convergence (Zedeño & Bowser, 2009), also integrate this regional system. They seem to have been built not before a thousand years ago approximately, as documented at Anderman, another excavated site in the area. This is a circular earthwork enclosure with a diameter of around 20 m and 30 cm high containing a 1 m high and 3 m wide mound, located on a plain older river terrace. Studies at other research areas (e.g., Robinson et al., 2017) have disclosed the funerary nature of these structures, sometimes containing dozens of cremated burials.

Former excavations at the center of the mound of this site have reported only burnt earthen strata and retrieved flakes and a small globular ceramic vessel (Rohr, 1971, 1984). Our own test excavations did not reveal much more besides some additional lithics and potsherds. A hearth at the basal area of the mound, where a bone fragment of wild boar (*Tayassu pecari*) has been found, shows that this ceremonial structure has been built by the end of the pre-colonial era, when Jê settlements at the area were densest. Another dating, from the encircling ring, reveals that ring and mound have been built approximately at the same time (Table 7.2).

[1] This is another typical Jê archaeological feature, reported in several other locations (La Salvia, 1968; Beber, 2005; DeSouza, 2018). This kind of pit oven also appears in ethnographical reports, as used for roasting (Paula, 1924; Lavina, 1994).

Table 7.2 Radiocarbon chronology for the Urubici region

Date ^{14}C BP	2 sigma cal. Date BC/AD (years)	13/12C ratio	Lab Code	R_Date	Site	Description	Depth (cm)	Material dated	Source
3650 + −30	2133–1882 BC	−27.15	UGAMS 30076	BON-Q11N3	Bonin	Pit house village	20–30	Charcoal	This study
2770 ± 41	989–806 BC	−22.04	UGAMS 30087	ADE-QEA2	Ademar Rodrigues	Pit house village	40	Charcoal	Labrador (2018)
2130 ± 25	179–51 BC	−29.76	UGAMS 30067	AVE1-PAN20	Morro do Avencal	Rock shelter with rock art	195–205	Charcoal	Corteletti et al. (2021)
1840 ± 40	122–340	−17.062	N/A	St. Bar.01	Santa Bárbara	Rock shelter with burials	N/A	Bone	DeMasi (2001)
1330 ± 36	660–855	−25.05	UGAMS 30082	BONQ4Z128	Bonin	Pit house village	128	Charcoal	This study
1220 ± 42	694–989	−30.40	UGAMS 30088	ARI-T1A12	Aripol	Rock shelter	5	Charcoal	Labrador (2018)
1210 ± 25	771–972	−26.20	UGAMS 25384	Bonin 12B	Bonin	Pit house village	180–190	Charcoal	This study
1180 ± 20	885–988	N/A	UGAMS	St. Bar.02	Santa Bárbara	Rock shelter with burials	N/A	Bone	Almeida (2014)
1090 ± 27	900–1031	−26.49	UGAMS 30073	COPE22N9	Copetti	Pit house village	80–90	Charcoal	This study
884 ± 200	771–1450	N/A	SI 227	Casa de Pedra	Casa de Pedra	Rock shelter with rock art	30–50	Charcoal	Piazza (1969)
690 ± 40	1286–1396	−27.38	UGAMS 30086	AND-T3A7	Urubici 4	Oversized pit house	20	Charcoal	Labrador (2018)
690 ± 25	1290–1393	−26.53	UGAMS 32425	URU4-QCA12	Anderman	Mound and enclosure complex	13	Charcoal	Labrador (2018)
640 ± 40	1296–1415	−26.90	Beta 298,216	BON 1-6070	Bonin	Pit house village	60–70	Charcoal	Corteletti et al. (2015)

(continued)

Table 7.2 (continued)

Date ^{14}C BP	2 sigma cal. Date BC/AD (years)	13/12C ratio	Lab Code	R_Date	Site	Description	Depth (cm)	Material dated	Source
640 ± 39	1296–1415	−26.93	UGAMS 30085	URU4-QFA4	Urubici 4	Oversized pit house	50	Charcoal	Labrador (2018)
630 ± 25	1314–1414	−27.29	UGAMS 32426	AND-T1A13	Anderman	Mound and enclosure complex	35	Charcoal	Labrador (2018)
610 ± 50	1299–1441	−26.5	Beta 298215	BON 1-3040	Bonin	Pit house village	30–40	Charcoal	Corteletti et al. (2015)
600 ± 33	1316–1441	−24.28	UGAMS 30079	URU4-Q12/11A3	Bonin	Pit house village	60–80	Charcoal	This study
600 ± 37	1318–1436	−26.49	UGAMS 30083	BONE22QBN8	Urubici 4	Oversized pit house	59	Charcoal	Labrador (2018)
560 ± 38	1325–1452	−28.04	UGAMS 30084	URU4-FEI1	Urubici 4	Oversized pit house	137	Charcoal	Labrador (2018)
470 ± 28	1428–1612	−27.44	UGAMS 30074	BON-Q6N1	Bonin	Pit house village	0–10	Charcoal	This study
360 ± 20	1487–1642	−26.40	UGAMS 25385	Bonin 15	Bonin	Pit house village	10–20	Charcoal	This study
350 ± 29	1465–1650	−26.45	UGAMS 30075	BON-Q4N2	Bonin	Pit house village	10–20	Charcoal	This study
350 ± 35	1489–1648	−25.76	UGAMS 30081	BON-Q7N2	Bonin	Pit house village	10–20	Charcoal	This study
340 ± 31	1490–1654	−26.51	UGAMS 30077	BONE22QCN7	Bonin	Pit house village	30–40	Charcoal	This study
340 ± 32	1496–1654	−24.41	UGAMS 30078	BONE21QAN4	Bonin	Pit house village	60–70	Charcoal	This study
340 ± 25	1501–1649	−26.00	UGAMS 25388	Bonin Ex2	Bonin	Pit house village	0–10	Charcoal	This study

330 ± 25	1504–1653	−25.40	UGAMS 25386	Bonin 22	Bonin	Pit house village	30–40	Charcoal	This study
320 ± 90	1446-present	N/A	SI 597	Vacas Gordas 3	Vacas Gordas 3	Open air site	60	Charcoal	Piazza (1969)
310 ± 34	1501–1796	−9.54	UGAMS 30080	BON-Q4N1	Bonin	Pit house village	0–10	Maize seed	This study
310 ± 25	1507–1665	−25.40	UGAMS 25387	Bonin EX1	Bonin	Pit house village	0–10	Charcoal	This study
310 ± 25	1507–1665	−26.30	UGAMS 25383	Bonin 12A	Bonin	Pit house village	10–20	Charcoal	This study
190 ± 20	1670-present	−26.84	UGAMS 30068	AVE1-PFN4	Morro do Avencal	Rock shelter with rock art	30–40	Charcoal	Corteletti et al. (2021)
180 ± 26	1671-present	−24.84	UGAMS 30072	COPE22N5	Copetti	Pit house village	40–50	Charcoal	This study
120 ± 25	1698-present	−24.05	UGAMS 30066	AVE1-PAN6	Morro do Avencal	Rock shelter with rock art	55–75	Charcoal	Corteletti et al. (2021)
Modern	Modern	−28.27	UGAMS 30069	AVE1-PA1A1	Morro do Avencal	Rock shelter with rock art	On the wall	Rock pigment	Corteletti et al. (2021)
Modern	Modern	−28.27	UGAMS 30070	AVE1-PA1A3	Morro do Avencal	Rock shelter with rock art	On the wall	Rock pigment	Corteletti et al. (2021)
Modern	Modern	−27.98	UGAMS 30071	AVE1-PA1A6	Morro do Avencal	Rock shelter with rock art	On the wall	Rock pigment	Corteletti et al. (2021)

The dates were calibrated using ShCal20 atmospheric curve (Hogg et al., 2020), in OxCal program (Bronk Ramsey, 2009), version v4.4 (Bronk Ramsey, 2021), issued in 14/05/2021

Another type of ceremonial site is Urubici 4, consisting of two large atypical neighboring pit houses excavated on the sandstone bedrock and a small mound nearby, located on a low hill at the side of the valley bottom, about 60 m higher from it. The site has been first excavated by Rohr (1972, 1974), who reports the very straight vertical walls of the smaller structure, fireplaces and quite a few lithic tools and potsherds. He noticed a peculiar feature on the vertical wall, like a chimney conduct, "full of thick charcoal shanks" (Rohr, 1972: 34–35), and also several inscriptions engraved on the rock wall.

Revisiting the site (Labrador, 2018), we have found several small fireplaces around and near both structures, as well as scatters of lithics and potsherds, a context quite similar to other reported large pit house sites (e.g., Schmitz et al., 2002). A trench cut into the biggest pit house has revealed, alongside lithic tools and potsherds, a large pit oven right in the middle of the house floor. Four radiocarbon dates situate these structures around eight to six hundred years ago (Table 7.2 and Fig. 7.5; see chronological discussion below).

These oversized pit houses (after DeSouza et al., 2016b) are still poorly understood. They do not seem to be simple residential places, but many and diverse activities have been performed therein and around. They seem to be a place of social convergence, whether "ceremonial" or not, and have been interpreted as special locations where people come together, possibly at different occasions and for different purposes. Anyway, at this site it has been possible to establish a full connection between earthwork structures, Taquara-Itararé ceramic style and some specific rock art graphic patterns as an archaeological trace complex referring to the Jê peoples at the end of pre-colonial times.

7.4 Southern Jê Chronology and Territoriality at the Urubici Region

At a larger, macro-regional scale, demographic analysis of archaeological data shows that Jê populations, in southern Brazil and surrounding areas, have experienced at least two moments of expansion, around 1300 and 500 cal year BP approximately (DeSouza & Riris, 2021). In some areas, such as Urubici, the occupation seems to be continuous throughout the period, as depicted by chronological data amassed in the last years (Table 7.2). One of these sites, Bonin, a pit house village arranged around a central open area, seems to have been occupied for more than one thousand years, although it might not have been continuous. The dates for the Urubici region reveal two very clear occupation horizons for the proto-Jê (Fig. 7.5).

For the most ancient dates, at Ademar Rodrigues and Morro do Avencal rock shelters, and even from the central plaza at Bonin, cultural affiliation is elusive since sampling contexts are not clearly connected to the proto-Jê. At the rock shelter sites, samples have been collected into deeper anthropic coal-rich layers without clearly associated diagnostic material culture, such as pottery. Interestingly, they match

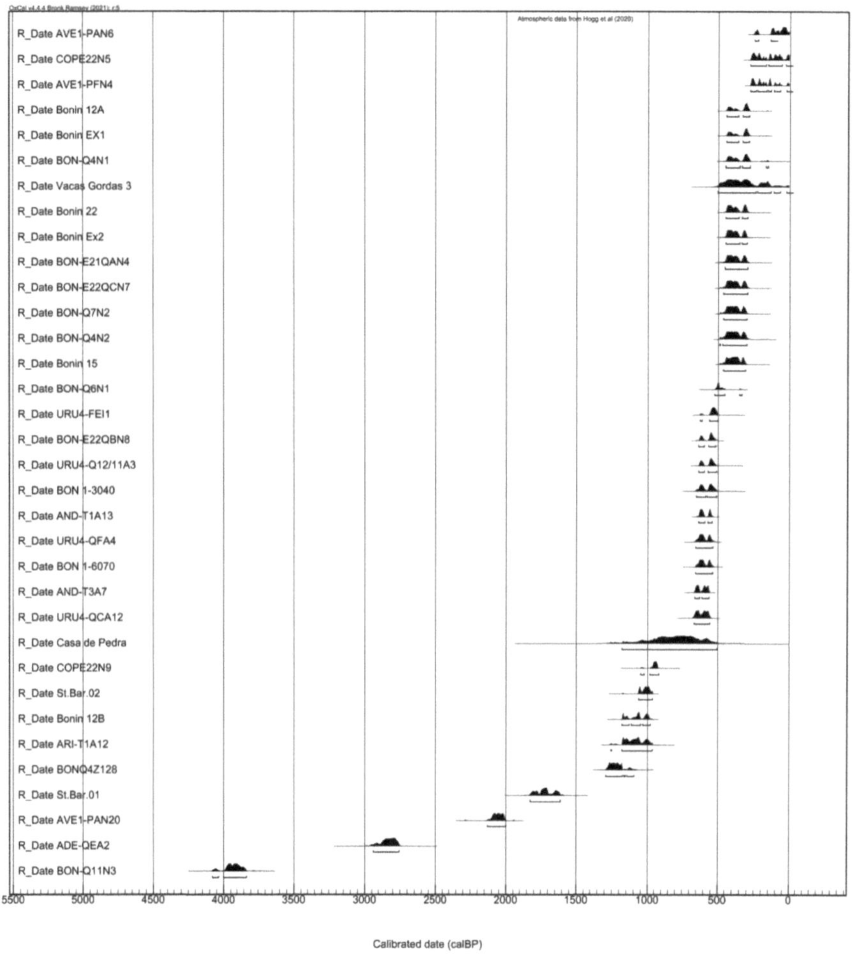

Fig. 7.5 Calibrated radiocarbon dates for the Urubici region

pretty well with Santa Bárbara, a dated shelter not too far from the valley, around 1800 years ago, where several burials have been found (Rohr, 1971; DeMasi, 2001).

Lithic vestiges of older Umbu Tradition (Schmitz, 1987; Dias, 2012), attributed to hunter-gatherer societies that lived in this same area long before the arrival of Jê speaking societies, are frequently found in superficial small scatters throughout the area, even nearby (or even into) Jê settlements – what might be the case of the flat open central plaza at Bonin, an excellent location that might have been reoccupied throughout the Holocene. Thus, although the existence of non-ceramist Jê speaking groups has been postulated (Schmitz et al., 1999, 2009), these dates might well refer to a moment before the arrival of Jê people in the area.

The first clear Southern Jê horizon in this area occurs around 1300–1000 years ago, at Aripol rock shelter and at Bonin, an open-air intensely reoccupied pit house village on a later period. This horizon also occurs into the floor of one pit house at Copetti, another village located on a hilltop near the banks of a bend of the Canoas River, 16 km downstream. This village contain mounds, platforms and at least 48 clustered pit houses similar in size and shape to the Bonin ones, but with fewer archaeological remains (Corteletti, 2012). At Bonin, datings of 1330 BP and 1210 BP have been obtained from charcoal collected at the bottom of deep-buried pit ovens encircled by blocks of sandstone and basalt configuring well-designed combustion structures, both re-occupied some 700 years later as habitational areas.

In addition to these, two sites not far from Urubici, dating from this period, have also been reported, with no indicative of pottery. At Santa Bárbara, a rock shelter with burials on the headwaters of the upper Canoas region, 24 km to the south from the valley, human bone samples have been dated around 1200 BP (Almeida, 2014). At Casa de Pedra, another huge rock shelter with rock art carved on sandstone, located some 30 km upstream, a few lithics have been recovered in an anthropic layer dating from around 900 BP (Piazza, 1969; see Table 7.2).

At this period, it seems that Jê occupation is scattered throughout the valley, but already occupying localities that have remained as long-standing living places towards the following centuries, although it is not clear yet if all archaeologically documented clusters were occupied by that time. Available chronology does not allow to confirm that occupation has been continuous, or permanent, throughout the horizons detected by radiocarbon datings. It is quite possible, even probable, that more intensive dating might confirm the long-term permanent presence of the Jê people at the Urubici area.

Lithic vestiges of older Umbu Tradition (Schmitz, 1987; Dias, 2012), attributed to hunter-gatherer societies that lived in this same area long before the arrival of Jê speaking societies, are frequently found in superficial small scatters throughout the area, even nearby (or even into) Jê settlements – what might be the case of the flat open central plaza at Bonin, an excellent location that might have been reoccupied throughout the Holocene. Thus, although the existence of non-ceramist Jê speaking groups has been postulated (Schmitz et al., 1999, 2009), these dates might well refer to a moment before the arrival of Jê people in the area.

The first clear Southern Jê horizon in this area occurs around 1300–1000 years ago, at Aripol rock shelter and at Bonin, an open-air intensely reoccupied pit house village on a later period. This horizon also occurs into the floor of one pit house at Copetti, another village located on a hilltop near the banks of a bend of the Canoas River, 16 km downstream. This village contain mounds, platforms and at least 48 clustered pit houses similar in size and shape to the Bonin ones, but with fewer archaeological remains (Corteletti, 2012). At Bonin, datings of 1330 BP and 1210 BP have been obtained from charcoal collected at the bottom of deep-buried pit ovens encircled by blocks of sandstone and basalt configuring well-designed combustion structures, both re-occupied some 700 years later as habitational areas.

7.5 Modelling a Settlement System and Territoriality from a Jê Perspective

Together with other studies (e.g., Saldanha, 2005; DeMasi, 2005; Copé, 2006; D'Ávila, 2017; DeSouza, 2018), available data for the Urubici area allows to infer Southern Jê territoriality based on stable settlements, sustained by hunting, gathering, and fishing associated with food production (Corteletti et al., 2015). Short distance movements, related to resource management or social network connections like ceremonies, exchange, etc., are sustained by archaeological (Corteletti, 2012; DeSouza et al., 2016a; Labrador, 2018) as well as ethnoarchaeological (Machado, 2016) perspectives, and reinforced by the distribution of clustered communities dispersed throughout the territory (Fig. 7.3). In this analysis, we will explore Jê meanings for spatial and landscape configurations.

Societies have places for promoting and celebrating interactions between people, as well as with nature and the supernatural (Zedeño & Bowser, 2009). Living territory is full of meaningful places, imbued within peoples' memories and landscape perception. In this sense, the diversity of sites described in Urubici indicates the occurrence occupational features and spaces of circulation interacting in a scenario in which it is possible to ascertain to them specific functions such as housing, crop production, and ritual locations structuring the landscape. At Urubici, different places with culturally defined functions seem to follow social and cosmological Jê principles.

Figure 7.6 shows a site cluster on the terraces and lower slopes in the bottom of the Canoas valley. Several pit houses and open-air litho-ceramic locations indicate a living area, with local daily household activities. At a distance of about 2 km to the east of this living area, three ceremonial mound and enclosure locations and, further upstream, a rock shelter with collective burials, can be found. This configuration corroborates ethnographical information from *Kaingang* and *Laklãnõ* (*Xokleng*) peoples that burial places were located outside, far from the villages (Henry, [1941] 1964; Mabilde, [1897] 1988; Veiga, 2006), usually to the east (Crépeau, 2002). Such locational disposition refers to the spatial organization of the *Kaingang's* shamanic territories (Rosa, 2005).

Shamanic territory of *Kaingang* people is composed of three levels, which might interconnect in some circumstances. The underground level is formed by the domain of the "world of the dead" (*nügme*). The surface level is constituted by "house", "clean space" and "forest" domains, where people live and act. The "world above" level is formed by the domain "sky" (*fãg kavã* or *kaikã*). Each domain contains different spaces, where activities take place, and also specific border zones (Rosa, 2005). To approach the landscape as experience, we will make an exercise of ethnographic analogy examining the surface level, where the archaeological sites can be found, in the domains of "clean space" and "virgin forest", exploring its interaction as well with other domains.

The domain of "clean space" is where villages are built. One of its borders is the *sítio*, cultivation locations usually integrating different plant species (crops).

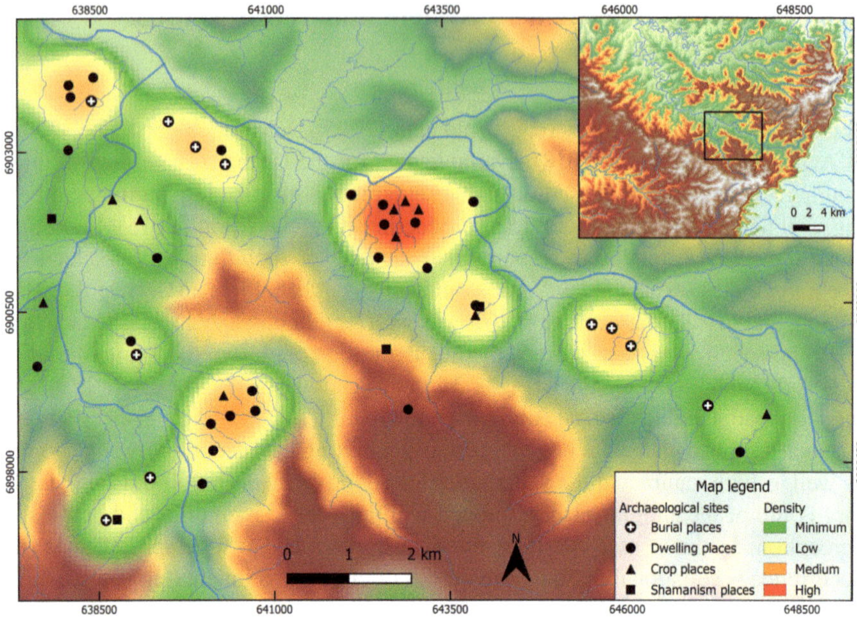

Fig. 7.6 A cluster and its site disposition in the Urubici area

Distance between villages and *sítios* are variable, and many of them lay along routes of displacement between villages, thus producing larger and socially integrated "clean spaces". In turn, the domain of the "virgin forest" is the place where all the things that were not transformed by the human hand can be found. This is the domain of spirits and wild animals (like felines and raptors). The *Kaingang* hunt on the borders of the sociological plan of this domain: mountains, the forest, the bushy canyon, the river and the waterfall. The cosmological plan of "virgin forest" also connects to other borders, such as the house of the spirits, the house of the owner of the animals and the house of *Migtán*, who controls the ferocious animals and is invisible to a common *Kaingang*, appearing only to shamans (*kujà*) and dreamers. Former cemeteries are another border of this domain (as in Fig. 7.6) but, today, due to the circumscription that land demarcation for indigenous reservation imposes, they were transferred to the domain "clean space" (Rosa, 2005).

The archaeological site clusters (Fig. 7.3) are interpreted as "clean space" domains along the upper Canoas valley. Within and around each of these clusters, pit house villages and related *sítios* (crop areas) can be found. Rock art locations and burial sites are usually distant from pit house villages, away from clustering areas and outside the "clean space" domain. Thus, they are interpreted as borders of the "virgin forest" domain. The parietal art at Morro do Avencal rock shelter (Riris & Corteletti, 2015; Corteletti et al., 2021) portrays, among other representations, animals such as birds and felines, that belong in the "virgin forest" domain. Such representations suggest that, at this place, shamans (*kujà*) could access spiritual entities like *Migtán*, surrounded by densely forested mountains and with a view towards a splendid 100 m high waterfall (a sociological border of "virgin forest" domain) (Fig. 7.7).

As most settlements lie in relatively flat areas at, or close to, the bottom of the valley, a least-cost pathways model (Weathley & Gillings, 2002; Conolly & Lake, 2010), connecting archaeological sites of the region, converge to the lower portions of the valley, tracing their way across the landscape practically parallel to the streams. Some paths go from a pit house site to an open-air litho-ceramic location and, after that, to the lower portion of the valley (see details in Corteletti & DeBlasis, 2018). This recurrent displacement pattern (place of dwelling – place of cropping – floodplain) shows that the social sphere of the southern Jê people concentrates at the bottom of the valley, and can be compared to *Kaingang's* shamanic territorial organization, as most *sítios* (border of "clean space" domain) are located along the paths connecting villages (focal instances of "clean space").

"House" and "clean space" domains are located mostly into the lower portions of the valley, open areas with flat or soft topography, fertile soils, and full access to a wide range of ecotones and resources. This domesticated territory is surrounded by the "virgin forest" domain, areas with higher slopes (and, eventually, peripherical, less occupied zones of the bottom of the valley), undomesticated territory, wilderness where hunting takes place, but also where the dead "live", and the spirits can be contacted (Fig. 7.2).

Fig. 7.7 The iconic landscape view from the Morro do Avencal rock art site, a border of the "virgin forest" domain. (Photo: Rafael Corteletti)

As we have seen, Jê landscapes in the Urubici region are composed of a diversity of places and spaces of coexistence and circulation with different functionalities, which interact with one another through a socio-cosmological structure of their own, a long-lasting cultural and adaptive-transformative process carried out by Jê populations. Places of dwelling, places of burial, places of shamanism, mountains, rivers, forests, and grasslands become sensible landmarks that culturally define (build) the landscape. This landscape, integrating place and people, creating memories and identity, thus constitutes a patched and complex pattern of territoriality (Zedeño, 2010; Corteletti & DeBlasis, 2018).

7.6 Final Words

As we have seen in this paper, archaeological research at the upper Canoas valley, the heart of the Urubici region, reinforces the long-term history of the Jê presence in those highland territories, for more than a thousand years. All along this period, Jê societies thrived and expanded, having created ingenious earthworks and a proper cultural landscape, full of symbolic references, which can be perceived to this day. It has been an history of expansion, but also of resistance and fight. It is our hope that archaeology might help original peoples to recover their dignity and their rights.

Acknowledgements The archaeological research at the Urubici region has been part of the *Jê Landscapes of Southern Brazil: Ecology, History and Power in a Transitional Landscape during the Late Holocene*, a research project coordinated by the University of São Paulo and the University of Exeter, with participation of several other Brazilian and British institutions. We thank the Fundação de Auxílio à Pesquisa do Estado de São Paulo (FAPESP), Brasil, and the Arts and Heritage Research Council (AHRC) from United Kingdom, for their support (grants FAPESP 2012/51328-3 and 2014/07754-3 and AHRC AH/K004212/1). We also thank Manoella Souza Soares for Fig. 7.4.

References

Almeida, L. C. (2014). *Ocupação pré-colonial no planalto catarinense: os sítios arqueológicos do município de Urubici (SC) sob a perspectiva da geoarqueologia*. Unpublished Thesis – Universidade Federal de Santa Catarina, Florianópolis.

Beber, M. V. (2005). O sistema do assentamento dos grupos ceramistas do planalto sul-brasileiro: o caso da Tradição Taquara/Itararé. *Documentos, 10*, 5–125. IAP – UNISINOS.

Bitencourt, A. L. V., & Krauspenhar, P. M. (2006). Possible prehistoric anthropogenic effect on Araucaria angustifolia (Bert.) O. Kuntze expansion during the Late Holocene. *Revista Brasileira de Paleontologia, 9*, 109–116.

Bronk Ramsey, C. (2009). Bayesian analysis of radiocarbon dates. *Radiocarbon, 51*(1), 337–360.

Bronk Ramsey, C. (2021). *OxCal* version 4.4.4, available on OxCal Project.

Conolly, J., & Lake, M. (2010). *Geographical information system in archaeology* (338p). Cambridge University Press.

Copé, S. M. (2006). Narrativas espaciais das ações humanas. História e aplicação da arqueologia espacial como teoria de médio alcance: o caso das estruturas semi-subterrâneas do planalto Sul-brasileiro. *Revista de Arqueologia., 19*, 111–123.

Corteletti, R. (2008). *Patrimônio Arqueológico de Caxias do Sul* (200p). Ed. Nova Prova.

Corteletti, R. (2012). *Projeto Arqueológico Alto Canoas- PARACA: Um estudo da Presença Jê no Planalto Catarinense*. PhD thesis, Universidade de São Paulo, 323p.

Corteletti, R., & DeBlasis, P. (2018). Arqueologia Jê do Sul do Brasil: ambiente, sistema, poder e experiência na paisagem de Urubici, Santa Catarina. *Revista Memorare, 5*(2), 132–164.

Corteletti, R., & Iriarte, J. (2018). Recent advances in the archaeology of Southern proto-Jê people. In C. Smith (Ed.), *Encyclopedia of global archaeology* (2nd ed.). Springer Publishing.

Corteletti, R., Dickau, R., DeBlasis, P., & Iriarte, J. (2015). Revisiting the economy and mobility of southern proto-Jê (Taquara-Itararé) groups in the southern Brazilian highlands: Starch grain and phytoliths analyses from the Bonin site, Urubici, Brazil. *Journal of Archaeological Science, 58*, 46–61.

Corteletti, R., Souza Soares, M., Labrador, B., & DeBlasis, P. (2021). Southern Jê engravings at Morro do Avencal: Preliminary archaeometrical analysis and interpretation of a rock shelter in Southern Brazil. *Journal of Archaeological Science: Reports, 35*(2021), 102721.

Crépeau, R. (2002). A prática do xamanismo entre os Kaingang do Brasil meridional: uma breve comparação com o xamanismo Bororo. *Horizontes Antropológicos, 8*, 113–129.

Crépeau, R. (2005). *"Os Kamé vão sempre primeiro" Dualismo social e reciprocidade entre os Kaingang* (pp. 9–33). Anuário Antropológico.

D'Ávila, L. N. (2017). *Caminhos e Lugares: modelo de mobilidade regional e sistema de assentamento entre os proto-Jê Meridionais na região de Campo Belo do Sul, SC*. Master thesis, Universidade Federal de Pelotas, 295p.

DeMasi, M. A. N. (2001). Pescadores e Coletores da Costa Sul do Brasil. *Pesquisas Antropologia* 57, IAP – UNISINOS.

DeMasi, M. A. N. (2005). *Relatório Final – Projeto de Salvamento Arqueológico Usina Hidrelétrica Campos Novos*. Unpublished version, 277p.

DeMasi, M. A. N. (2009). As Terras Altas do Sul do Brasil e o Litoral de Santa Catarina, a arqueologia dos mortos e evidências de hierarquia social. In *Anais do XV Congresso SAB* (20p). Belém.

DeSouza, J. G. (2018). Rethinking households, communities and status in the southern Brazilian highlands. *Journal of Anthropological Archaeology, 52*, 44–58.

DeSouza, J. G., & Riris, P. (2021). Delayed demographic transition following the adoption of cultivated plants in the eastern La Plata Basin and Atlantic coast, South America. *Journal of Archaeological Science, 125*(2021), 105293.

DeSouza, J. G., Corteletti, R., Robinson, M., & Iriarte, J. (2016a). The genesis of monuments: Resisting outsiders in the contested landscapes of southern Brazil. *Journal of Anthropological Archaeology, 41*, 196–212.

DeSouza, J. G., Robinson, M., Corteletti, R., Cárdenas, M. L., Wolf, S., Iriarte, J., Mayle, F., & DeBlasis, P. (2016b). Understanding the chronology and occupation dynamics of oversized pit houses in the southern Brazilian highlands. *PLoS One, 11*, e0158127.

Dias, A. S. (2012). Hunter-gatherer occupation of south Brazilian Atlantic Forest: Paleoenvironment and archaeology. *Quaternary International, 256*, 12–18.

Fasolo, P. J., Potter, R. O., Carvalho, A. P., et al. (2004). *Levantamento de Reconhecimento de Solos do Estado de Santa Catarina – Solos – SHP*. EMBRAPA Solos/EMBRAPA Floresta. Disponível em. http://mapoteca.cnps.embrapa.br/geoacervo/

Henry, J. [1941] (1964). *Jungle people. A Kaingáng tribe of the highlands of Brazil*. Vintage Books, 215p.

Hogg, A., Heaton, T., Hua, Q., Palmer, J., Turney, C., Southon, J., Bayliss, A., Blackwell, P., Boswijk, G., Bronk Ramsey, C., Petchey, F., Reimer, P., Reimer, R., & Wacker, L. (2020). SHCal20 Southern Hemisphere calibration, 0–55,000 years cal BP. *Radiocarbon, 62*, 759–778.

Iriarte, J., & Behling, H. (2007). The expansion of Araucaria Forest in the southern Brazilian highlands during the last 4000 years and its implications of the Taquara/Itararé Tradition. *Environmental Archaeology, 12*(2), 115–127.

Iriarte, J., Gillam, J. C., & Marozzi, O. (2008). Monumental burials and memorial feasting: An example from the southern Brazilian highlands. *Antiquity, 82*, 947–961.
Iriarte, J., Copé, S., Fradley, M., Lockhart, J. J., & Gillam, J. C. (2013). Sacred landscapes of the southern Brazilian highlands: Understanding southern proto-Jê mound and enclosure complexes. *Journal of Anthropological Archaeology, 32*, 74–96.
Iriarte, J., DeBlasis, P., DeSouza, J. G., & Corteletti, R. (2017). Emergent complexity, changing landscapes, and spheres of interaction in Southeastern South America during the Middle and Late Holocene. *Journal of Archaeological Research, 25*, 251–313.
La Salvia, F. (1968). Resumos das Pesquisas Arqueológicas no Planalto – Rio Grande do Sul. *Pesquisas, Antropologia, 18*, 101–113. IAP – UNISINOS.
Labrador, B. (2018). *Sistema de Assentamento Proto-Jê Meridional no Alto Rio Canoas*. Master thesis, Museu de Arqueologia e Etnologia, Universidade de São Paulo.
Lauterjung, M. B., Bernardi, A. P., Montagna, T., Candido-Ribeiro, R., Costa, N. C. F., Mantovani, A., & Reis, M. S. (2018). Phylogeography of Brazilian pine (*Araucaria angustifolia*): Integrative evidence for pre-Columbian anthropogenic dispersal. *Tree Genetics & Genomes, 14*, 36.
Lavina, R. (1994). *Os Xokleng de Santa Catarina: uma etnohistória e sugestões para os arqueólogos*. Master thesis, IAP-UNISINOS.
Mabilde, A. P. T. [1897] (1988). O índio Kaingáng no Século XIX. Arqueologia do Rio Grande do Sul, Brasil. *Documentos, 2*, 141–172. IAP – UNISINOS.
Machado, J. S. (2016). Caminhos e Paradas. Perspectivas sobre o território Laklãnõ (Xokleng). *Revista do Museu de Arqueologia e Etnologia, 27*, 179–196, USP.
Paula, J. M. (1924). Memoria sobre os botocodos do Paraná e Santa Catharina organisada pelo serviço de protecção aos selvicolas sob a inspeccao do Dr. José Maria de Paula. *Annaes do XX Congresso Internacional de Americanistas, Rio de Janeiro, 1922*(1), 119–137.
Pereira, C. A., Giehl, E. L. H., Levis, C., Machado, J. S., Bueno, L., & Peroni, N. (2020). Precolonial Amerindian legacies in forest composition of southern Brazil. *PLoS One, 15*(7), e0235819.
Riris, P., & Corteletti, R. (2015). A new record of pre-Columbian engravings in Urubici (SC), Brazil using polynomial texture mapping. *Internet Archaeology, 38*. https://doi.org/10.11141/ia.38.7
Piazza, W. F. (1966). *As Grutas de São Joaquim e Urubici*. Florianópolis, UFSC.
Piazza, W. F. (1969). A área arqueológica dos Campos de Lages. In *PRONAPA – Resultados Preliminares do Terceiro Ano (1967-1968)*. Belém, Museu Paraense Emílio Goeldi, Publicações Avulsas, n. 13, p. 63–74.
Robinson, M., Iriarte, J., Souza, J. G., Corteletti, R., Ulguim, P. F., Fradley, M., Cárdenas, M. L., DeBlasis, P., Mayle, F., & Farias, D. S. E. (2017). Moieties and mortuary mounds: Dualism at a mound and enclosure complex in the southern Brazilian Highlands. *Latin American Antiquity, 28*(2), 232–251.
Robinson, M., Souza, J. G., Maezumi, Y. S., Cárdenas, M. L., Pessenda, L., Prufer, K., Corteletti, R., Farias, D. S. E., Mayle, F., DeBlasis, P., & Iriarte, J. (2018). Uncoupling human and climate drivers of late Holocene vegetation change in southern Brazil. *Scientific Reports, 8*, 7800.
Rohr, J. A. (1971). Os sítios arqueológicos do Planalto Catarinense, Brasil. *Pesquisas Antropologia, 24*, 72p. IAP-UNISINOS.
Rohr, J. A. (1972). *Cadernetas de campo originais (1971–1972). Notas sobre o livro Tombo da Igreja de Urubici*. Acervo do Instituto Anchietano de Pesquisas, IAP-UNISINOS.
Rohr, J. A. (1974). Die vorgeschichtlichen grubenwohnungen in Brasilien. In *Jahrbuch der Familie* (pp. 178–183). Sociedade Cultural e Beneficente Padre Reus.
Rohr, J. A. (1984). Sítios arqueológicos de Santa Catarina. *Anais do Museu de Antropologia, 17*, 77–168, Ed. UFSC.
Rosa, R. R. G. (2005). O Território Xamânico Kaingang vinculado às Bacias Hidrográficas e à Floresta de Araucária. *Cadernos do LEPAARQ, 2*(4), 99–115, Ed. UFPel.
Saldanha, J. D. M. (2005). *Paisagem, Lugares e Cultura Material: Uma Arqueologia Espacial nas Terras Altas do Sul do Brasil*. Master thesis, PUCRS, 177p.
Saldanha, J. D. M. (2008). Paisagens e sepultamentos nas terras altas do sul do Brasil. *Revista de Arqueologia, SAB, 21*, 85–95.

Scheibe, L. F. (1986). A Geologia de Santa Catarina – Sinopse Provisória. *Geosul, 1*, 7–38.

Schmitz, P. I. (1987). Prehistoric hunters and gatherers of Brazil. *Journal of World Prehistory, 1*, 53–126.

Schmitz, P. I. (1988). As Tradições Ceramistas do Planalto Sul-Brasileiro. Arqueologia do Rio Grande do Sul, Brasil. *Documentos, 2*, 75–130. IAP – UNISINOS.

Schmitz, P. I., Rosa, A., Izidro, J. M., Haubert, F., Krever, M. L. B., Bitencourt, A. L. V., Rogge, J. H., & Beber, M. V. (1999). Içara: Um Jazigo Mortuário no Litoral de Santa Catarina. *Pesquisas Antropologia, 55*, 164p. IAP – UNISINOS.

Schmitz, P. I., Rogge, J. H., Rosa, A., Beber, M. V., Mauhs, J., & Arnt, F. V. (2002). O Projeto Vacaria: Casas Subterrâneas no Planalto Rio-Grandense. *Pesquisas Antropologia, 58*, 11–105. IAP – UNISINOS.

Schmitz, P. I., Arnt, F. V., Beber, M. V., Rosa, A., & Rogge, J. H. (2009). Taió, no Vale do Rio Itajaí, SC – O encontro de antigos caçadores com as casas subterrâneas. *Pesquisas Antropologia, 67*, 185–320. IAP – UNISINOS.

Souza Soares, M. (2019). *Geoquímica de Solos Arqueológicos na Identificação de Áreas de Atividades: um quadro geográfico para o Sítio Bonin/SC*. PhD thesis, 211p. Universidade Federal do Paraná.

Sprenger, A. C. (2020). Entre os meandros do rio e as profundezas do vale: análises e reflexões sobre a cerâmica Jê do Sul, sítio Bonin/SC. In *Monography of graduation in archaeology* (94p). Universidade Federal de Pelotas.

Streck, E. V., Kämpf, N., Dalmolin, R. S. D., Klamt, E., Nascimento, P. C., & Schneider, P. (2002). *Solos do Rio Grande do Sul* (107p). EMATER/RS, Ed. UFRGS.

Veiga, J. (2006). *Aspectos Fundamentais da Cultura Kaingang* (256p). Ed. Curt Nimuendajú.

Weathley, D., & Gillings, M. (2002). *Spatial technology and archaeology: The archaeological applications of GIS* (250p). Taylor & Francis.

Wüst, I., & Barreto, C. (1999). The ring villages of Central Brazil: A challenge for Amazonian archaeology. *Latin American Antiquity, 10*, 1–23.

Zedeño, M. N. (2010). The archaeology of territory and territoriality. In B. David & J. Thomas (Org.), *Handbook of landscape archaeology* (pp. 210–217). Left Coast Press.

Zedeño, M. N., & Bowser, B. J. (2009). The archaeology of meaningful places. In: Bowser and Zedeño (Org.), *The archaeology of meaningful places* (pp. 1–14). University Utah Press.

Chapter 8
Archaeology of Fishing of the Earthen and Shell Moundbuilders (Cerritos and Sambaquis) of the Patos Lagoon, Southern Brazil, 3200–200 Years BP

Rafael Guedes Milheira, Flávio Rizzi Calippo, and Manuel Haimovici

Abstract Faunal remains deposited in archaeological sites are important sources for understanding the eco-historical processes and dialectic relations between human populations and the environment since Pre-Columbian times. They make it possible to present archaeologically framed models of biodiversity and an ecological past which, when compared to contemporary concerns, can drive political and technical decisions meant to solve current environmental problems. In this chapter, we analysed the data regarding long-term fisheries in the Patos Lagoon (located on the southern coast of Brazil) from 3200 years BP, based on the archaeological record of earthen (Cerritos) and shell (Sambaquis) moundbuilders of the Atlantic coast and Pampa biomes. By considering the data on captured fishes, radiocarbon dating, the position of the archaeological sites and the environmental and climate changes that occurred since the mid-Holocene, we inferred the role of fisheries in the societies and the climatic factors that drove the past practices. It was found that throughout the last three millennium fish were an important element in the diet of indigenous populations around Patos Lagoon year-round with emphasis on the spring and summer months, when adult specimens of some species enter the interior of the lagoon. Otoliths of small specimens of the sciaenid fishes *M. furnieri* and *Pogonias courbina* and catfish of the genus Genidens were dominant, suggesting fishing occurred

R. G. Milheira (✉)
Department of Anthropology and Archaeology, Federal University of Pelotas, Pelotas, Brazil
e-mail: milheirarafael@gmail.com

F. R. Calippo
Centro de Ciências da Natureza, Coordenador do Laboratório de Arqueologia Marítima e Subaquática – LARQSUB/CCN/UFPI, Universidade Federal do Piauí, Teresina, Brazil
e-mail: calippo@ufpi.edu.br

M. Haimovici
Pesquisador Sênior, CNPQ, Instituto de Oceanografia, Universidade Federal do Rio Grande – FURG, Rio Grande, Rio Grande do Sul, Brasil
e-mail: manuelhaimovici@gmail.com

A. C. Colonese, R. G. Milheira (eds.), *Historical Ecology and Landscape Archaeology in Lowland South America*, Interdisciplinary Contributions to Archaeology, https://doi.org/10.1007/978-3-031-32284-6_8

mostly in shallow shoreside waters with fishing gear of limited efficiency. Although fishing was associated with the exploitation of the most abundant species, the data shows that the sizes of fish caught were smaller than those available in the environment due to the use of fishing gear that allowed the capture of mostly smaller specimens.

Keywords Patos Lagoon · Late Holocene · Earthen mounds (Cerritos) · Shell mounds (Sambaquis) · Coastal fisheries

8.1 Introduction

From a paleoenvironmental perspective, the remains of plants and animals in archaeological sites are an important source of information regarding the long-term dialectical relationship between human populations and the environment. By applying archaeology it is possible to access the record of human behaviour materialised in artefacts, bones of exploited and consumed animals, phytoliths, pollen grains and starches from managed and processed plants, coal and ash from bonfires and buried human bones, all arranged in a sedimentary package also composed of physical and chemical signatures of cultural activities and natural processes. This matrix of human-ecological information can be analysed to describe processes of environmental transformation, as well as to perceive the most dramatic moments of these changes over time. This information can be used not only to compose historical scenarios of interaction between humans and the environment, but also to understand long-term human-oriented ecological disturbances and serve as a point of departure to drive actions aimed at environmental and heritage conservation in the present (Balée, 2006).

The archaeological-ecological record is generally more complex, evident and significant among populations that lived in the same places for long periods. Behavioural recurrence in places with specific functions tends to make the record of human activities more archaeologically identifiable (Heckenberger, 2001). Archaeological sites intentionally built as mounds, for example, acquire a double importance as they are matrices of behavioural actions "sealed" in stratigraphic packages. This is the case of sites such as that of the earthen mounds (Cerritos) and shell mounds (Sambaquis). These two families of identifiable architectural forms were used and built by indigenous populations that inhabited the Pampa biome and the southern Brazilian Atlantic coast during the recent Holocene. Thus far, more than 3000 "*Cerritos de indios*" have been recorded in southern Brazil, Uruguay and northeastern Argentina between approximately 5000 and 200 years BP (López Mazz, 2001; Bracco et al., 2008; Milheira & Gianotti, 2018). They are earthen mounds ranging from 30 cm to 7 m in height, which occur singly or in groups of up to a hundred, usually arranged in grassland and wetland environments, such as fields and swamps. The sambaquis, in turn, are shell mounds reaching up to 6 m in height and appear on almost the entire Brazilian Atlantic coast. Similar to the cerritos, they are found in both isolation and in groups of dozens of mounds. They generally are associated with marine and lagoon environments, and have been dated between approximately 8500 and 1000 years BP (Gaspar, 2000; Calippo, 2010; De Blasis et al., 2007). In this chapter, we did not include information on the Guarani indigenous groups due to the lack of quantitative zooarchaeological data regarded to diet and economy in the research area.

The Patos-Mirim complex that surrounds the watershed of Patos and Mirim Lagoons, between the southern coast of Brazil and Uruguay, is a one-of-a-kind environment to trace the ecological history of fishing populations from 3200 years BP, when the first recorded occupations took place (Schmitz, 1976; Milheira et al., 2017). In the Patos-Mirim complex, more than 400 cerritos and 11 sambaquis have been recorded, with zooarchaeological studies pointing to a wide significance of animal species in the composition of the sites, reflecting different pre-colonial indigenous strategies of capture, consumption and economic exploitation of the regional fauna. Through Abundance Index methods used in Zooarchaeology, such as the NISP (Number of Identified Specimens), the MNI (Minimum Number of Individuals) and the sizes of otoliths of fish in archaeological sites, it is possible to estimate the spectrum of ichthyological fauna, its relative abundance in the history of the sites and the sizes of fish captured.

In order to support such a historical-ecological approach to fishing in the Patos Lagoon, this chapter presents the main characteristics of the environment in which fishing was carried out and the information available on the ecology, diversity and abundance of species caught by the cerritos and sambaquis populations, between 3200 and 200 years BP. A synthesis of knowledge on pre-colonial indigenous occupation and use of food resources (including aquatic ones) is presented. Published and unpublished data on fish remains (notably otoliths) and radiocarbon dating from several archaeological sites in the region were considered, which supported the inference about the main species captured and the sizes of the specimens consumed over the more than 3000 year period under consideration. Finally, based on the interpretation of this data, discussions about the types of fishing practised, the artefacts used and the role of fish and fishing locations for these ancient indigenous populations are also presented.

8.2 The Estuary of Patos Lagoon and Its Ichthyological Biodiversity

The area under analysis in this project (Fig. 8.1) is located in the southern Coastal Plain of Brazil, and was formed from 400,000 years BP by a process linked to discontinuous sedimentary deposition events known as the Laguna-Barreira System (Tomazelli et al., 2017). This process is evidenced by the presence of at least four waves of sediment transported during the last marine transgressions and regressions that occurred between the Tertiary and Quaternary which isolated and formed

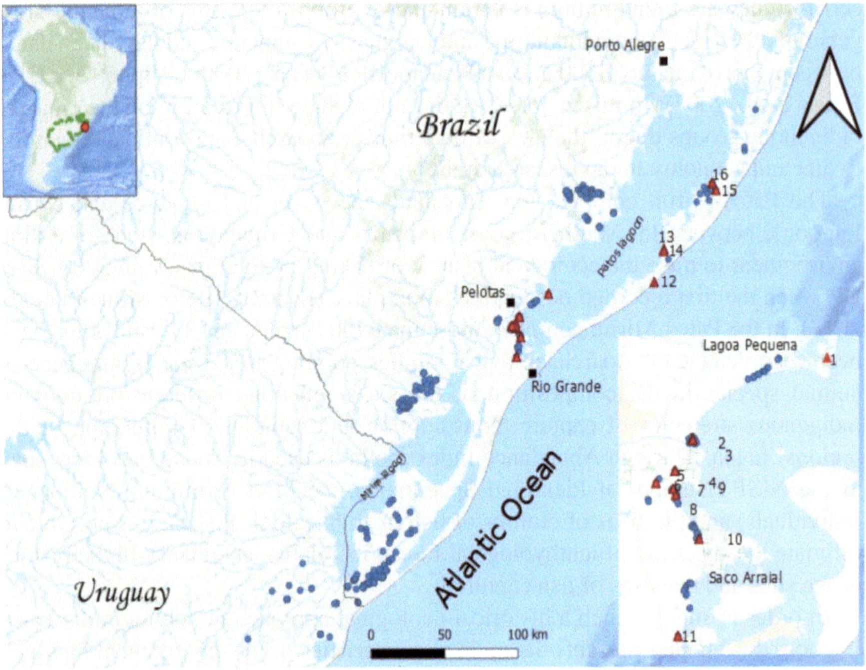

Fig. 8.1 Location of the research area between the Pampa biome and the Atlantic coast with the position of the cerritos in the Patos-Mirim complex (circles) and the studied sites (triangles) with emphasis on the area of the Patos Lagoon estuary. Studied archaeological sites: (1) PT-02-cerrito da Sotéia, (2, 3) Cerrito PSG-02 and Cerrito PSG-07, in Pontal da Barra, (4) RS-LS-11 Cerrito Ariano de Souza, (5) RS-RG-20, (6) RS-RG-21, (7) RS-RG-48, (8) RS-RG-49, (9) RG-8, (10) CI, (11) C-5 -II, (12) RS-LC-59 Sambaqui Capão D'areia, (13) RS-LC-16 Sambaqui Farol do Capão da Marca, (14) RS-LC-14 Sambaqui Capão da Marca, (15) RS- LC-42 Cerrito Nilton Dutra, (16) RS-LC-21 Cerrito Ondina Brazil

lagoon bodies (including the Patos-Mirim complex which appeared approximately 5500 years BP). The current geological structure of the Coastal Plain is composed of deposits of both marine and continental natures, modelled by weather, alongside alluvial and fluvial sedimentation. It results in a complex and diversified topography, which involves floodplains, terraces and deposits of fluvial network trough, flat terrain, dunes and paleodunes (Tomazelli et al., 2017).

As a result of these transgressive-regressive processes several lakes, supplied by river drainage from the mountains in the West, by the Atlantic Ocean in the East and by precipitation, were formed in the Coastal Plain. Laguna dos Patos (30°S and 32°S) is the largest obstructed lagoon in the world and one of the most voluminous bodies of water in southern Brazil, with an area of approximately 10,227 km^2. It is located between the state of Rio Grande do Sul and the east coast of Uruguay and is connected to Mirim Lagoon through the 75 km long São Gonçalo Channel (Simon & da Silva, 2015; Burns et al., 2019). The southern portion of the Patos Lagoon is home to an estuary zone, considered one of the greatest natural fisheries in Brazil.

This zone comprises approximately 963.8 km^2 (10% of the total lagoon area), between the Barra breakwater of the Rio Grande and an imaginary line between the island of Feitoria and Ponta dos Lençóis (Oliveira & Bemvenuti, 2006).

8.3 Fauna and Fishing in the Patos Lagoon and Adjacent Region

The estuarine portion of the Patos Lagoon is a large area for reproduction, spawning and breeding of a variety of organisms, such as fish and crustaceans that utilize the estuary throughout their life cycle, as well as more intermittent species, which inhabit the area for only a certain period, such as spawning and breeding seasons. The shallow (<2 m) and more protected areas (such as *Saco do Arraial* and *Lagoa Pequena*) (Fig. 8.1), stand out for the number of smaller fish. Meanwhile, in channels with a depth greater than 2 m, larger fish occur in more abundance, serving as movement corridors for the species (Oliveira & Bemvenuti, 2006).

The ichthyological fauna of the Patos-Mirim complex ranges includes over 300 marine, estuarine and freshwater species with different behaviours in habitat use (Chao et al., 1985). Among the most frequent species in the estuarine portion of the Patos Lagoon recorded in the post-colonial period are the whitemouth croaker (*Micropogonias furnieri*), the marine catfishes (*G. barbus* and *G. planifrons*), the black drum (*Pogonias courbina*), the silverside (*Odontesthes bonariensis*), the mullet (*Mugil liza*), the red flounder (*Paralichthys orbignyanus*), pink shrimp (*Farfantepenaeus paulensis*), blue crab (*Callinectes sapidus*), the menhaden (*Brevoortia pectinata*), and the small clupeid (*Lycengraulis grossidens*). In addition, there are common species in the environs of the São Gonçalo Channel, such as the trahira (*Hoplias malabaricus*) and the freshwater silver catfish (*Rhamdia* sp.). More frequent marine species in the coastal strip occasionally penetrate the lagoon by means of the wedge of salt water that flows through the channel and spreads to the deepest areas. In periods of low flow, this saltwater influx can reach as far as the shore of *São Lourenço beach*, nearly 100 km from the mouth. The presence of marine mammals such as the porpoise (*Tursiops truncatus gephyreus*) and the sea lion (*Otaria flavescens*) are also recorded in the Patos Lagoon (Castello & Pinedo, 1977; Pinedo, 1990).

The weights recorded by species in the available fisheries statistics indicate the most abundant species in the Patos Lagoon and adjacent coastal region. The percentages of the main species in the total landings of artisanal fisheries in the period of 1945–1975 were: croaker 35.3%, catfish 23.3%, menhaden 8.5%, mullet 8.4%, yellow hake 7.2%, black drum 4.7% and anchovies 1.5%. This period was chosen because it represents a time before the sharp decline and even the collapse of the main fisheries in the last four decades (D'Incao et al., 2002; Haimovici & Cardoso, 2017; Thykjaer et al., 2020). We observed that, together, croaker and catfish, which were also the most common taxa in zooarchaeological studies, represented 63.4%

of landings from contemporary fisheries in the Patos Lagoon, of which three species have their life cycle linked to the estuarine region.

The whitemouth croaker reaches relatively large lengths at up to 70 cm in total length and over 35 years of age in Rio Grande do Sul populations. It is considered mature between 2–4 years old at 25–40 cm and spawns between November and April in coastal waters close to the mouth of the Patos Lagoon (Haimovici et al., 2021). Adults enter the lagoon in spring, and young specimens appear throughout the year in the channel and coves of the estuarine region, where they find the ideal conditions for growth and feed preferentially on benthic invertebrates (Costa et al., 2014).

Marine catfish are fished both in the lagoon and on the adjacent coast, while juveniles are commonly found in the estuary and near ocean beaches (Reis, 1986a; Chao et al., 1985). Catfish are considered long-lived and slow growing species. *G. barbus* reaches sexual maturity at 40–43 cm in total length and between 8 and 9 years of age (Reis, 1986b). According to Reis (1986c), sexually mature *G. barbus* and *G planifrons* enter the Patos Lagoon from the sea in August and September before spawning, which takes place in November and December. Later the females return to the ocean, while the males remain orally incubating the eggs until their release between January and February. Recent studies indicate that adults of *G. barbus* return to spawn preferentially in the environments where they were born (Avigliano et al., 2017a). However, Avigliano et al. (2017b) suggest that the species shows great variety in relation to the environments where it completes its life cycle, with individuals that migrate annually between the estuary and the sea or that reside permanently in their habitat of birth, either in the sea or in sweet water. The "guri" catfish *G. genidens* is a freshwater and estuarine species that completes its life cycle in Patos Lagoon as evidenced in a recent study of stable isotopes by Pereyra et al. (2016).

The black drum is a species that can reach up to 140 cm in length, weigh up to 40 kg, and can live beyond 55 years of age, occurring both in the marine and estuarine environments. In the Patos Lagoon and adjacent region, spawning takes place between November and March, amongst individuals with 2–5 years of life, and with sizes between 40 and 70 cm. Food, both in the estuary and in adjacent waters, consists predominantly of bivalve molluscs, small gastropods, and crabs (Haimovici, 1997; Santos et al., 2019).

8.4 Earthen Mounds (Cerritos) and Shell Mounds (Sambaquis), Fishing Peoples of the Pampas and Atlantic Coast

The potential for catching fish in the lagoon environment has certainly attracted fishing populations since pre-colonial times, as attested by more than 400 cerritos (Milheira et al., 2019a, b), at least 11 sambaquis (Ribeiro & Calippo, 2000) and dozens of Guarani occupations (Milheira & Wagner, 2014), almost all showing evidence of the exploitation of aquatic resources. Both cerritos and sambaquis, while sharing some characteristics, also have differences in terms of architectural structure and material culture.

In terms of economy, the earthen moundbuilders (cerriteiros), at the end of the Mid-Holocene, exploited a wide spectrum of animal and plant resources through hunting, fishing, gathering and also planting botanicals such as corn (*Zea mays*), beans (*Phaseolus* sp.) and pumpkin (*Cucurbita* sp.) (Iriarte, 2006; del Puerto, 2015). From 2500 BP onwards, a certain specialisation in the exploitation of some animals (deer and cavy) is known, which is accompanied by incipient experiences in animal management, associated with changes in the ownership of the territory and resources in general (Moreno, 2016). From 2500 years BP, there is also an economic diversification evidenced by the significance of fishing, especially in the environment of the Patos and Mirim Lagoons, where the cerritos demonstrate a meaningful exploitation of marine and estuarine-dependent species (Schorr, 1975; Ulguim, 2010, 2018; Sens, 2020; Chanca et al., 2021).

In this sense, if in the interior of Uruguay the cerriteiros seem to have been sustained by the fauna of mammals typical of wetlands, in the estuary of the Patos Lagoon, fishing seems to have played the central role (Milheira et al., 2019a). Pottery, invented around 3000 BP, was also an important technological improvement, reflecting new ways of processing, cooking, consuming and storing food. Animals and plants, therefore, were structuring resources of the economy of the earthen moundbuilders – a society that could be classified as a mixed economy of hunter-gatherers specialised in flooded environments (Iriarte, 2006) and oriented to fishing in some contexts (Chanca et al., 2021; Milheira et al., 2019a).

In the estuary of Patos Lagoon, approximately one hundred cerritos were recorded between the municipalities of Rio Grande, Pelotas, Capão do Leão, Mostardas and Tavares, since the first half of the twentieth century (Figs. 8.1 and 8.2). Between the 1960s and 1980s, a narrative (currently contested) was constructed regarding the Cerriteiro groups, defining them as populations with a low level of cultural complexity, a system of wide mobility determined by the abundance of seasonal species, existing between approximately 2500 and 200 years BP. An exhaustive identification work of species allowed Schorr (1975) and Schmitz (1976) to understand fishing and the exploitation of aquatic resources as the central factor in the social organisation of the cerriteiros, in which the estuary of the Patos Lagoon acted as a central area of seasonal encampments occupied during the hottest seasons of the year (spring and summer), given the abundance of estuarine-dependent species such as croakers, catfish and black drums. On the other hand, the rarity of mullet in the archaeological record, a species most present in the autumn and winter periods in the estuary, was suggested as an indicator that the indigenous occupations did not occur in the winter.

However, more recently, in the municipalities of Pelotas and Capão do Leão, a survey of 43 cerritos was carried out in the wetlands of Pontal da Barra, Lagoa do Fragata, and Lagoa Pequena (Fig. 8.2). Excavations at six of these sites have provided the most robust chronological information for cerritos in southern Brazil to date, with radiocarbon dates ranging from approximately 2400 to 800 years BP

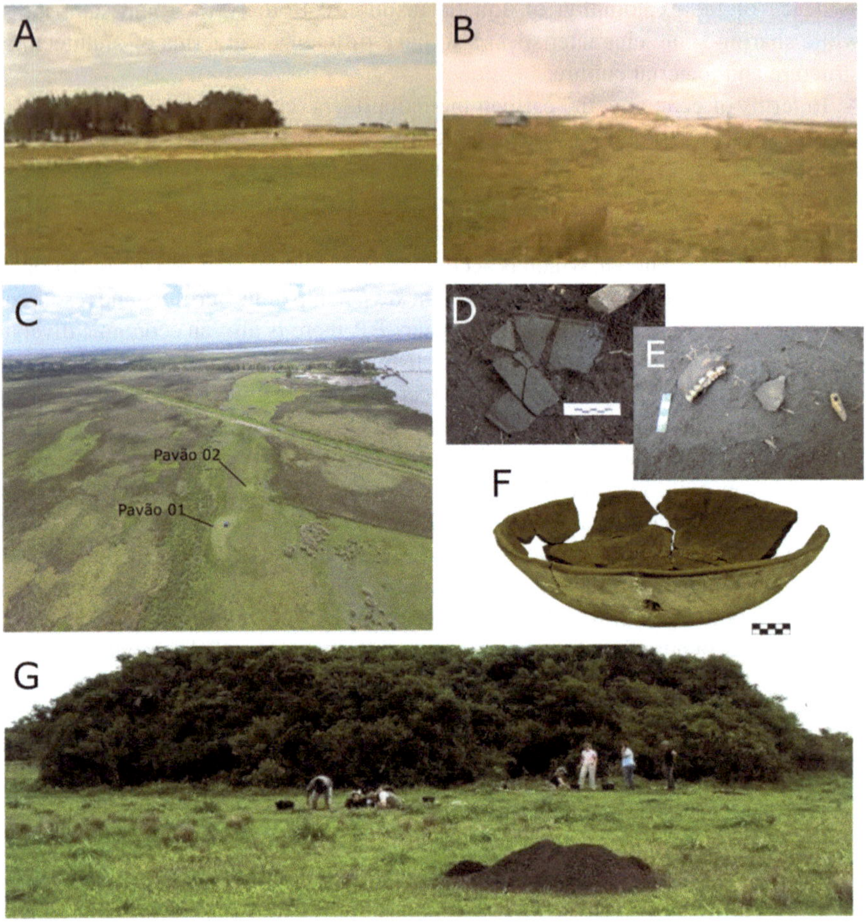

Fig. 8.2 Archaeological sites and material culture of cerritos and sambaquis in the Atlantic coast and Pampa region: (**a**) Sambaqui RS-LS-16, located in Mostardas; (**b**) sambaqui RS-LS-59, located in Tavares; (**c**) Cerritos Pavão 01 and 02, located on the banks of the São Gonçalo channel; (**d**); Ceramic sherd with suspension hole located in the cerrito PSG-07, Pontal da Barra; (**e**) Human jaw associated with a pendant made from a dolphin tooth and ceramics, located at cerrito PSG-02, Pontal da Barra; (**f**) Cerrito's ceramic vessel located at the CH2D01 site, Uruguay; (**g**) Cerrito PSG-06, Pontal da Barra in Pelotas

(Milheira et al., 2017, 2019a). The combination of radiocarbon dating, analysis of material culture and faunal remains suggests long-term occupation may have been compatible with the development of semi-sedentary villages, in which the estuarine environment would be stably and permanently occupied throughout the entire periods of the year and not just in the summer, as previously suggested by Schorr (1975) and Schmitz (1976). Oliveira (2006), Ulguim (2010, 2018), and Sens (2020), in turn, identified traces of mullet at the RG-48, PT-02, PSG-02, and PSG-07 sites respectively, which also suggests that fishing would have taken place permanently throughout all seasons of the year.

The cerritos of the estuary of the Lagoon of Patos are mounds with circular and elliptical shape, dimensions that vary from 20 to 120 m in the major axis (average of 26 m of radius), between 0.45 and 1.20 m in height (average of 0.85 m in height), and are generally located in riparian forests along rivers and lakes. A complex stratigraphic record is a distinguishing feature of these sites. Archaeological deposits are dark, rich in charred plant and fauna remains and artefacts, including a large number of ceramic, lithic and sometimes human bone fragments. Human remains normally occur disjointed, highly fragmented and spatially dispersed, following the same distribution pattern as the rest of material culture. The excavations were carried out from the top to the base of the mound structures, at artificial levels of 5 cm, 10 cm or 20 cm, with individual spatial recording of areas of activities and materials. The excavations at the sites covered between 1 and 4.5 m^2 of excavated area (average of 3.5 m^2), corresponding to 0.9 m^3 and 5.4 m^3 (average of 2.94 m^3) of excavated volume. Dating was performed in 13 cerritos with dates ranging from approximately 2500 to 200 years BP.

Among the sambaquis, for many years the inferred economic model was based on the premise of a heavy mollusc exploitation. However, this idea was revised in the 1990s, when it was realised that the marine, lacustrine and mangrove fauna, derived from fishing, had a much more important role in the economy than simple shell collecting (Figuti, 1992). Fishing would have been a foundational practice in the life of the shell moundbuilders (sambaquieiros) since the beginning of the occupations, around 8500 years BP. It was certainly a factor that allowed for significant demographic densification, especially during the Mid-Holocene, when there was a process of expansion of the sambaquieiro's territory denoted by the construction of gigantic mound structures and their functional diversification. The variation in the types of structures is evidence of important social transformations and practices. For example, it is now known that the species that were fished not only served to feed the body at the dietetic level, but also had a symbolic dimension, since fished animals became commonly associated with funerary deposits (Gaspar, 2000; Klokler, 2010).

As fishing societies, the sambaquieiros' centre of daily life were the lagoon, sandbanks, and the mangroves environments. The groupings of sites, the visibility between them and the notion of territory amongst these ecosystem would offer greater resource predictability (DeBlasis et al., 2007; Colonese et al., 2014; Scheel-Ybert & Boyadjian, 2020). Domesticated and managed plants were also important sources of food and certainly played a major role in the diet and economy of sambaquieiros societies. Species such as sweet potato (*Ipomoea batatas*), squash (*Cucurbita* sp.), yam (*Dioscorea* sp.) and corn (*Zea mays*) have been variably consumed since approximately 4000–3000 years BP, and this consumption of plants is also indicated by the stone milling tools and by dental marks that denote a carbohydrate-based diet, suggesting dietary diversity and complementarity between exploited fauna and flora; a typical case of mixed economy (Wesolowski et al., 2007; Scheel-Ybert & Boyadjian, 2020).

8.5 Spectrum of Zooarchaeological Fauna

The Patos Lagoon is a privileged region for zooarchaeology studies, with extant indexes of several animal and, specially, fish species identified from bones and otoliths. Regarding the cerritos studied in the estuary, located in the municipality of Rio Grande (RG-21, RG-49, RG-20, C-1, RG-08, C-5-II and RG-48), dated between approximately 2500 and 200 years BP (Table 8.1), Schorr (1975) and Oliveira (2006) identified and tallied a series of mammal species, such as cavy (*Cavia aperea*), raccoon (*Myocastor coypus*), armadillo (*Dasypus* sp.), armadillo (*Novemcinctus novemcinctus*), opossum (*Didelphis* sp.), deer (*Mazama* sp.), birds, reptiles, crustaceans, molluscs and fish such as *Cynoscion leiarchus, Hoplias malabaricus, Rhamdia* sp., *Mugil* sp., *Pimelodella* sp., *Pimelodus maculatus, Loricariichthys anus, Cichlidae, Sinbranchus marmoratus, P. courbina, M. furnieri, Genidens* sp., demonstrating that the ichthyological fauna is the most abundant and corresponds to the highest index of consumed and exploited biomass. In the RS-LS-11-Ariano de Souza cerrito, dated between approximately 2150 and 2050 years BP, the recorded fauna spectrum is more reduced, with catfish, croakers, black drums and pejerrey (Chim, 2016).

The cerritos of Pontal da Barra and Lagoa Pequena (PSG-02, PSG-07 and PT-02), in the municipality of Pelotas, dated between approximately 2400 and 800 BP, offered up a similar array of species such as the the mammals cavy (*Cavia aperea*), raccoon (*Myocastor coypus*), armadillo (*Dasypus* sp.), armadillo (*Novemcinctus novemcinctus*), domestic dog (*Canis lupus familiaris*), opossum (*Didelphis* sp.), deer (*Mazama* sp.) and water rat (*Holochilus* sp.). Reptiles are represented with specimens of the order *Testudines* (turtles and tortoises). Bony fish included *M. furnieri*, *P. courbina*, *Cynoscion* sp., *Mugil* sp., *Hoplias malabaricus* and *G. barbus; G. genidens; G. planifrons* (Ulguim, 2010, 2018; Sens, 2020; Milheira et al., 2019a). In the biomass calculation performed for the PT-02-Cerrito da Sotéia site, the Osteichthyes class presented the highest values, at 93% of the total, followed by the Mammalia class with 4%, Reptilia with 2% and Malacostraca with 1% (Ulguim, 2010).

The sandbank sites are located between 100 and 9000 m from the Laguna and 600 to 4000 m from the ocean. They are located on depositional barriers from 1 to 10 m in height, formed in the Holocene-Pleistocene ages where currently predominant vegetation of grasses and shrubs is typical of dune formation (Ribeiro et al., 2004). The structures of the cerrito and the three sambaquis are composed of shells of *Erodona mactroides* and *Mesodesma mactroides* (Ribeiro et al., 2004). In two sambaquis, Bager (2013) identified the euryhaline fish *P. courbina, M. furnieri and Genidens* sp. *Mugil* sp., freshwater *Hoplias* sp., *Rhamdia* sp. and marine *Carcharias taurus* and *Chondrichthyes*; the marine molluscs *Mesodesma mactroides, Amiantis purpurata, Phacoides pectinatus, Tivela zonaria, Adelomelon* sp., *Olivancillaria* sp., and the terrestrian *Megalobulimus* sp. In addition to fish and molluscs, he also identified specimens of the deer (*Dasypus* sp.), some unidentified birds, animals of the order *Squamata*, marine mammals, small rodents and scarce fragments of crustacean claws.

Table 8.1 Number of specimens (NISP) identified in the sambaquis and cerritos of the Patos Lagoon, with values adjusted for the volume of 1 m³ of excavation

Archeological site	RS-RG-21	RS-RG-49	RS-RG-20	C-1	RG-8	C-5 II	PSG-02	PSG-07	PT-02	RS-LS-11	RG-48	RS-LC-16	RS-LC-21	RS-LC-42	RS-LC-59	Total (inferred to one cubi meter)	%
Age of the site (years BP)	845 ± 85	2020 ± 50	–	1080 ± 90	–	200 ± 80	Between 1280 ± 20 and 1859 ± 29	Between 1214 ± 22 and 2340 ± 150	Between 1360 ± 40 and 1400 ± 40	Between 2056 ± 84 and 2147 ± 65	1355 ± 45	Between 1350 ± 50 and 2090 ± 60	–	3210 ± 60	Between 1070 ± 60 and 1090 ± 40		
Excavated volume (m3)	0.90	2.70	4.80	1.80	5.40	1.50	0.90	1.00	0.80	5.40	10.80	0.42	2.02	1.20	1.35		
Taxon																	
Genidens sp.	44	1023	64	641	5	359	1309	480	6084	58	578	629	99	620	426	12,417	36.3%
Micropogonias furnieri	40	284	5	9406	51	2269	830	228	3040	672	253	860	72	393	346	18,749	54.9%
Pogonias cromis	3	99	3	36	3	22	249	81	1921	4	5	24	57	134	24	2666	7.8%
Mugil sp.								4	4		1					9	0.0%
Cynoscion leiarchus								1								1	0.0%
Hoplias malabaricus									28		49					76	0.2%
Menticirrhus sp.																0	0.0%
Pimelodella sp.											1					1	0.0%
Pimelodus maculatus																0	0.0%
Rhamdia sp.											250					250	0.7%
Loricariichthys anus											1					1	0.0%

(continued)

Table 8.1 (continued)

Archeological site	RS-RG-21	RS-RG-49	RS-RG-20	C-1	RG-8	C-5 II	PSG-02	PSG-07	PT-02	RS-LS-11	RG-48	RS-LC-16	RS-LC-21	RS-LC-42	RS-LC-59	Total (inferred to one cubi meter)	%
Age of the site (years BP)	845 ± 85	2020 ± 50	–	1080 ± 90	–	200 ± 80	Between 1280 ± 20 and 1859 ± 29	Between 1214 ± 22 and 2340 ± 150	Between 1360 ± 40 and 1400 ± 40	Between 2056 ± 84 and 2147 ± 65	1355 ± 45	Between 1350 ± 50 and 2090 ± 60	–	3210 ± 60	Between 1070 ± 60 and 1090 ± 40		
Excavated volume (m3)	0.90	2.70	4.80	1.80	5.40	1.50	0.90	1.00	0.80	5.40	10.80	0.42	2.02	1.20	1.35		
Taxon																	
Geophagus brasiliensis											8					8	0.0%
Sinbranchus marmoratus											4					4	0.0%
Total	88	1406	71	10,083	59	2651	2388	794	11,076	733	1149	1512	229	1148	796	34,182	

Site ages (BP): Radiocarbon dating intervals

8.6 Faunal Composition of Fishing Resources

The numbers of identified fish (NISP) were collected at 15 archaeological sites located around Patos Lagoon (Schorr, 1975; Schmitz, 1976; Calippo, 2000; Oliveira, 2006; Ulguim, 2010, 2018; Chim, 2016; Milheira et al., 2019a; Sens, 2020). From these sites, 33 radiocarbon datings are available from human bone samples, fish otoliths, charcoals, shells and a domestic dog tooth (Schmitz, 1976; Loureiro, 2008; Chim, 2016; Milheira et al., 2017). Of these, five are still unpublished, and include the base of the Nilton Dutra-RS-LC:42 cerrito (3210 ± 60 years BP/Cal BP 3700 to 3420 – Beta 258667), the sambaquis Farol do Capão da Marca-RS-LC:16 (top: 1350 ± 50 years BP/Cal BP 1390 to 1240 – Beta 25865; base: 2090 ± 60 years BP/Cal BP 2300 to 1970 – Beta 258664) and Capão D'Areia-RS-LC:59 (top: 1070 ± 60 years BP/Cal BP 1180 to 920 – Beta 25866; base: 1090 ± 40 years BP/Cal BP 1220 to 970 – Beta 258668).

Taxonomic identifications of the species compiled in the specialised literature corresponded with the originals, except in two cases. Due to their similarity, it was not possible to discriminate among the otoliths of catfish of the genus *Genidens* present in the region (*G barbus; G genidens, G. planifrons and G. machadoi*) (Marceunik, 2005). In the set of sites, five euryhaline genera and seven genera of freshwater fish were recorded (Table 8.1). Secondly, *Umbrina* sp. (cited in Schorr, 1975) was not included as Oliveira (2006) recently reanalyzed the same collection of cerrito RG-48 and showed it did not include *Umbrina* sp., thus correcting the previous research bias.

The numbers of identified specimens (NISP) adjusted for a standard 1 m^3 excavation volume were calculated for each of the 15 sites (Table 8.1). Among all standardised pooled records, the most frequent eurialine fishes were *M. furnieri* 54.9%, *Genidens* sp. 36.3%, and *P. courbina* 7.8%, and together they were dominant in 14 of the 15 sites. *Rhamdia* sp. 0.7% and *H. malabaricus* 0.2% were the most frequent freshwater fish and were dominant in site RG-48 (Table 8.1).

The historical sequence of NISP in the 15 sites analysed around Patos Lagoon shows that catfish, croaker and, to a lesser extent, black drum were the most conspicuous fish caught throughout the time period under purview (Fig. 8.3). The lowest relative abundances occurred between 3200 and 1800 years BP. The maximum was reached between 1100 and 400 years BP, decreasing again until 200 years BP. Catfish were the most abundant in 1400 years BP and croakers between 1000 and 200 years BP.

8.7 Size Composition of Caught Fish

In the sambaquis (RS-LC-16 and RS-LC-59) and cerritos RS-LC-42 and RS-LC-21 (Calippo, 2000), located on the sandbank, between Laguna dos Patos and the ocean, it was possible to calculate the total lengths (TL) and total weights (TW) of croaker,

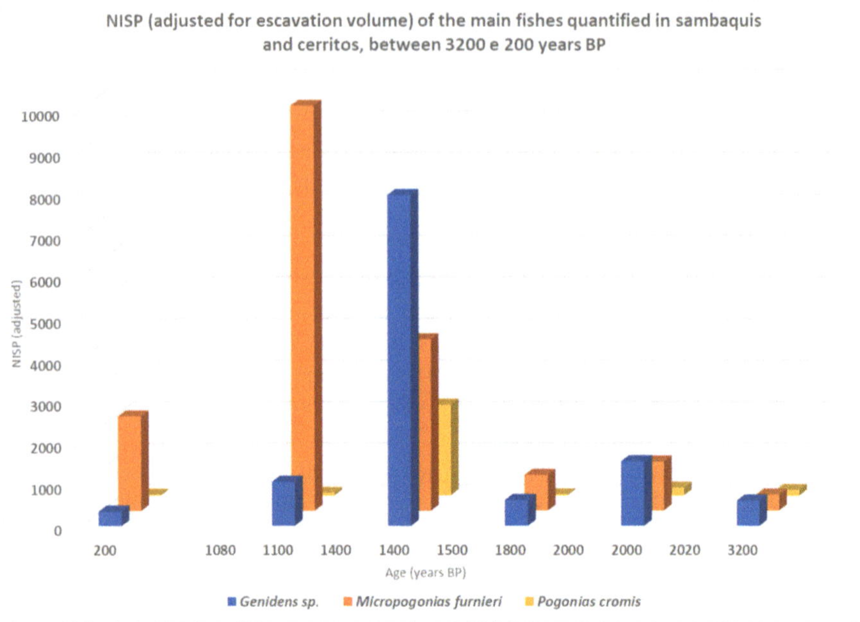

Fig. 8.3 Total number of NISP specimens from each analyzed site, adjusted for a 1 m^3 of excavation

catfish and black drum based on the otolith sizes (OS) of 1446 *M. furnieri* (13.6–77.0 cm), 1783 *Genidens* sp. (11.3–115.8 mm) and 320 *P. courbina* (13.6–104.9 cm) (Fig. 8.4), according to Bastos (1990), Calippo (2000), and Haimovici and Velasco (2000).

Figure 8.4 shows the estimated length compositions for all three species at the four sites. Croaker from the RS-LC-16 and RS-LC-21 sites showed shorter average lengths in the range of 20–25 cm, while in the RS-LC-42 and RS-LC-59 sites the average occured in the range of 30–35 cm (sites where the largest sizes for croakers were also identified). At cerritos RS-LC-21 and RS-LC-42, the most frequent catfish lengths are between 25 and 30 cm. In sambaquis, the most frequent catfish occur between 30 and 45 cm (lacustrine sambaqui RS-LC-16) and 30–35 cm in the marine site RS-LC-59. Black drum of the entire length spectrum occurred in the RS-LC-21 and RS-LC-42 and RS-LC-59 sites, while in the RS-LC-16 site only smaller black drums (from 20 to 45 cm) were present. Considering these fish as a whole, it is observed that the catches predominantly include small fish (which are included in the size spectrum of catfish and croaker caught in contemporary fisheries). In contrast, large black drums, measuring more than 120 cm and weighing over 20 kg, quite frequent in fisheries until the 1970s, were completely absent from the sites.

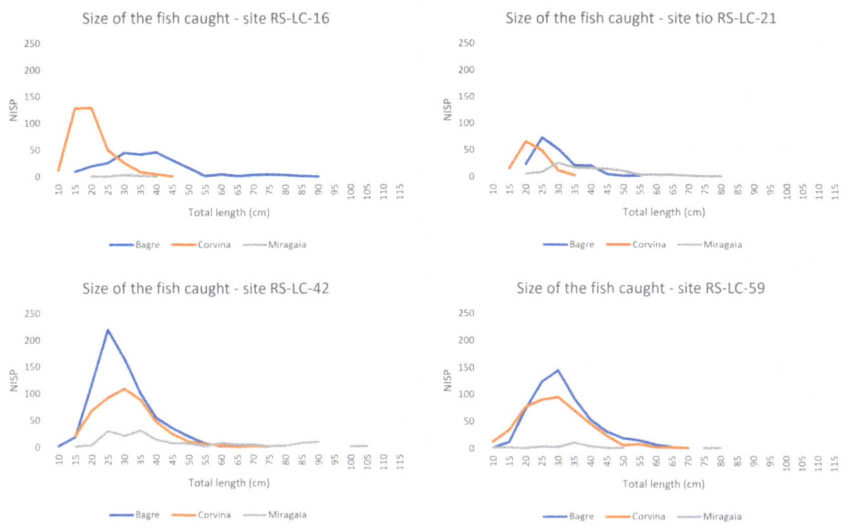

Fig. 8.4 Number and size estimate of catfish, croaker and black drum identified in archaeological sites RS-LC-16, 21, 42 and 59

8.8 Discussion

The spectrum of fauna identified in the cerritos and sambaquis of the estuary and sandbank of the Patos Lagoon indicates that the animal resources captured and recorded in the archaeological sites represent the fauna in local wetland environments such as swamps, lakes and rivers. The data indicates that fish are the most significant fauna in terms of abundance, representing up to 93% of the biomass consumed among the moundbuilders, as indicated in the case of the PT-02-Cerrito da Sotéia cerrito (Ulguim, 2010). The remaining 7% would be represented by the other species of mammals, turtles and birds. The relative calculation of abundance shows that there are two most representative taxa at the sites: the croaker represents around 49.9% of the total weighted NISP in all the sites and the catfish (*Genidens* sp.) with approximately 36.3%. Black drum occurs in only 7.8% of the total weighted NISP. The other taxa do not reach even 0.5%.

The interpretation that croakers, catfish and black drums are the most important estuarine-dependent species in the diet of the populations that inhabited the Patos Lagoon is also supported by isotopic analyses. Chanca et al. (2021) demonstrated that there is a predominance of marine and estuarine fauna captured and consumed by cerriteiros in the region, whose values of collagen samples from human individuals of $\delta^{13}C$ and $\delta^{15}N$ range from −21.3‰ to −10.7‰ and from +5.8‰ to +18.3‰ respectively, which coincides with the zooarchaeological data compiled in the present work.

As a hypothesis, it is suggested that the other species with low frequency may have been consumed occasionally or used as bait. It is common among the fishermen of Patos Lagoon, for example, to use shrimp meat and small fish species as bait. As an alternative hypothesis, the low frequency of some fish species, shrimp and cartilaginous species could have happened due to taphonomic, preservation and/or sampling. An interesting topic that should be considered in future research is the role of the large sized black drum in the diet which, despite its lower frequency (NISP below 10%) may have been important in the diet, due to its large amount of meat-to-size ratio.

From the point of view of the almost 3000 years of fishing history of catfish, croaker and black drum in the area, it is clear these have been the most exploited key-species in the estuarine and sandbank environment, from the pre-colonial period until the present day. The predominance of these species can be inferred considering four non-exclusive questions:

1. They were the most captured species due to food choices inherent to the socio-cultural systems of both indigenous groups. Catfish and croaker are the most abundant taxa in several archaeological sites of fishermen-hunters-gatherers on the south Atlantic coast, such as in the sambaquis located in Babitonga Bay, on the northern coast of Santa Catarina (Fossile et al., 2019), in the sambaqui Jabuticabeiras II (Klokler, 2010), at the Lagoa dos Freitas site (Pavei et al., 2019), on the south coast of Santa Catarina, and the sambaqui Sereia do Mar (Wagner et al., 2020), on the north coast of Rio Grande do Sul. The frequency of catfish and croaker on a large coastal scale suggests standardised consumption behaviour and interaction between different coastal fishing groups, which lasted over millennia of occupation of the lagoon and sandbank environments. Note that local frequency variations may be linked to opportunistic behaviours driven by the ecological availability of species, typical of mixed-economy societies, whose food choices oscillate between animal, and plant species complementary (Iriarte, 2006; Scheel-Ybert & Boyadjian, 2020).
2. Fishing tackle were preferably used to capture these species, masking the frequency of smaller species. It is interesting to note that the most captured species occurred in shallow waters close to the coast, which may indicate that they could be captured with fishing gear that does not necessarily require the use of boats. The use of boats would facilitate fishing of species along their migratory circuits, such as mullet, which occur less frequently in the archaeological record. However, although there are no direct records of vessels in the archaeological contexts of the Patos and Mirim Lagoons, the occurrence of traditional indigenous canoes made of a tree trunk (monoxyls) in other locations along the Pampa, such as the Uruguay River, is well-documented. In addition, there are a considerable number of chroniclers' who mention the use of boats by indigenous populations, which strongly suggests that vessels were a common-place technology (D'Olwer, 1963; Saccone, 2022). Carvalho (2004), when analysing the modification of the insertion structures of the muscle bundles and the irregular bone wear present in individuals buried in sambaquis in Rio de Janeiro, identified anomalies associated with activities related to swimming, handling harpoons or

long spears (the latter also involved in propulsion and canoes and rafts), the use of oars (standing and sitting) and axes (which, indirectly, could also be linked to the manufacture of canoes and rafts) (Calippo, 2010, 2011). The fishing and canoe-using behaviour was also evidenced by physical pathologies identified in human burials found in earthen mounds located in the *cerrito* RS-LS-85: Oscar Erocildo Abreu, located in Rio Grande. There, three individuals' osteoarchaeological analysis showed skeletal pathologies associated with repetitive and intensive activities such as trauma to the head of the ulna (affecting the elbow), flattening of the tibia (indicative of squatting for long periods), and clavicle bones with enthesopathy and arthrosis (Ferreira, 2012). Exploitation of aquatic resources is also evidenced by human remains from the *cerritos* of Pontal da Barra, which show dental wear potentially related to daily activities such as the handling of nets and fibres for the production and use of fishing implements (Ulguim & Milheira, 2017). Such a diagnosis led these researchers to interpret these as chronic symptomatic consequences of the intensive canoe-rowing related to the fishing and transportation activities of the aquatic lifestyle.

3. The choice of these species could have been determined by their ready availability in the estuarine and sandbank environment. It is worth remembering, in this sense, that the sea level would be around 0.5–1 m above current levels, offering an environment with a greater marine influence (Angulo et al., 2006). Bearing in mind that the coastal dynamics and the estuary have changed over the last few thousand years, if the location of the sites were the result of environmental availability alone, there would be a more homogeneous distribution of sites around the lagoon. If the location of the sites followed the environmental conditions (of greater availability or biomass of fish) that would form as a result of the lowering of the sea level, the result would be the occurrence of sites of shorter duration, in which the composition of sizes would be very similar. On the other hand, the data show that fishing over these periods was also not just a direct response to ecological availability. The interpretation of the quantities and sizes of black drums showed that, even though they were fished in smaller numbers than catfishes and croakers, the capture of the largest specimens (verified in three of the four sites in the sandbank) is also associated with specific points within the Patos Lagoon and consumption requires specific/differentiated fishing gear (such as the slingshot, harpoon, spear or bow and arrow), it occurs in all sites throughout the analysed period. The capture of black drums may not have been predominant for the selection of settlement sites, but their presence in the archaeological record certainly indicates a cultural bias, indicated by the need to perceive and identify their capture sites; by the energy supply provided by a single specimen; and by the symbolic aspects and status involved in fishing a larger animal.
4. Analysis of the otoliths sampled in the sambaquis of the sandbank of the Patos Lagoon indicates that fishing occurred throughout the year. The otoliths of many fish, including the croaker, show alternating, more opaque or more translucent growth bands that form at different times (Schwingel & Castello, 1990; Cotrina, 1998). Both types of bands were observed in the sections of croaker otoliths (90 opaque edges and 30 translucent) of the two sites. At the site RS-LC:42 41 (46%) opaque edges and 49 (54%) translucent edges were found. In the RS-LC:59 23

(77%) were opaque and 7 (23%) translucent. Analysing the actual specimens the numbers were 39 (59%) opaque and 26 (41%) translucent specimens in all. This data seems to indicate that fishing, at least for *M. furnieri*, was not seasonal, happening throughout the year, including spring and summer, when these species were more abundant in the estuarine environment (Calippo, 2000). It suggests that the territorial practices of the groups that build cerritos and sambaquis were more stable than the high mobility system suggested by Schorr (1975) and Schmitz (1976). Annual fishing would probably include mullet in the spectrum of captured and consumed fauna. However, little direct archaeological evidence of this species has been identified, which raises the question whether there was a social phenomena regarding food taboo or whether it is simply a question of conservation of bone remains or *in situ* sampling due to the reduced size of the otoliths.

Although the choice of fishing sites and the techniques and technologies were responsible for the selection of species captured and the size of the fish, the importance of the availability of croaker, catfish and black drum must be emphasised. The exploitation of these resources must have been one of the attractions for the establishment and continuity of the occupation of this territory. When comparing the fishing carried out by cerriteiros and sambaquieiros with the data from the artisanal fisheries carried out between 1945 and 1975 that focused on the most abundant resources (Thykjaer et al., 2020), in both the past and present these three species represent the bulk of captured fishing resources. This suggests that over the last few millennia fishing has been directed towards the most abundant species (Fig. 8.5).

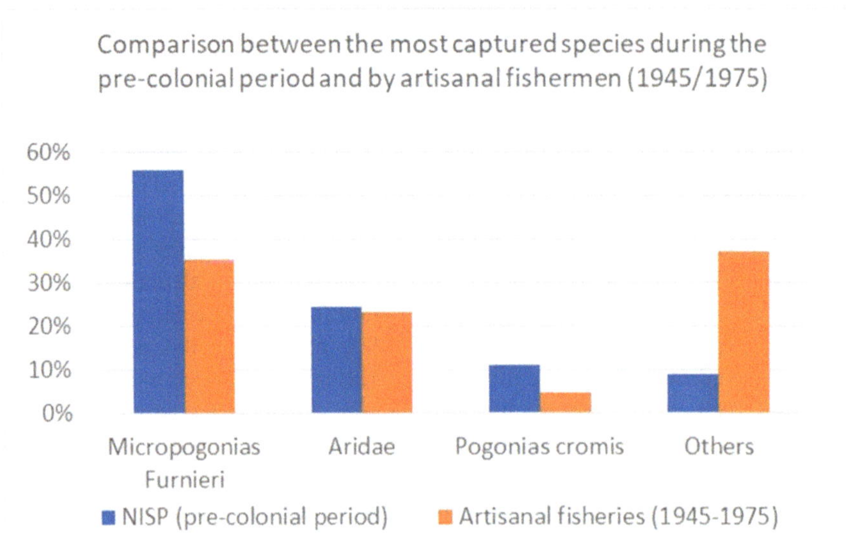

Fig. 8.5 Percentage of species caught by pre-colonial fisheries and artisanal fisheries in the Patos lagoon between 1945 and 1975. (Thykjaer et al., 2020)

In the case of artisanal fishing, the larger presence of other species is justified by the diversification of vessels, the fishing gear used, and contemporary commercial demands, offering access to resources that would not have been used by indigenous peoples for this purpose. As most of the fish recorded in the sambaquis were smaller than the fish landed in the recent artisanal fisheries in the Patos Lagoon, we estimate that the equipment available made it difficult to fish for larger specimens.

To discuss the predominance of catfish, croaker and black drum, it has to be taken into account that the Patos Lagoon has relatively little planktonic productivity and its trophic webs are mostly based on detritus and benthic algae, the food source for all three species. The life strategy of these species under moderate exploitation ensure the maintenance of their. Croakers and black drums that spawn high numbers of eggs in the marine environment or in thermohaline fronts (assuring planktonic food for the pelagic larval stages), which later penetrate estuaries where they find abundant food at the bottom. This explains why these species are abundant in estuarine environments and why, in turn, the location may have been attractive to human occupation. Catfish of the Aridae family, despite their low fecundity, are mouth breeders with large eggs and developed parental care, which reduces juvenile mortality.

Data from nearly 3000 years of exploitation of lagoon resources by indigenous populations show a pattern of sustainable fishing for croaker, black drum and catfish in the Patos Lagoon which has continued to the present day. It is noteworthy that the fishing carried out by the indigenous populations was not based on the commercialization of the surplus; they maintained, for more than three thousand years, an exploitation of fishing resources that did not impact diversity and abundance. Many of the species caught in the past are still the same ones consumed and traded by contemporary populations. However, the decrease in production in recent decades, as pointed out by Thykjaer et al. (2020), demonstrates that commercial fishing, within only a few decades, has dramatically altered the ecological balance and that the resilience of the estuarine and lagoon environments of Patos Lagoon may be at its limit.

8.9 Conclusion

The compilation of ichthyoarchaeological faunal data from 15 archaeological sites of the sambaquis and cerritos, located around the Patos Lagoon and dated between 3200 and 200 years BP, indicates that the most frequently captured and consumed species by pre-colonial indigenous populations are *M. furnieri* (49.9%), *Genidens* sp. (36.3%) and *P. Courbina* (7.8%). Although the predominance of temporal variations, these estuarine species are the basis of the diet, as was also indicated through isotopic studies in human individuals from the sites studied by Chanca et al. (2021), and are also the most consumed species in other wetlands environments as the coastal and estuarine lagoons of the warm temperate region of the south-western Atlantic (Fossile et al., 2019; Klokler, 2010; Pavei et al., 2019; Wagner et al., 2020).

Despite the lack of quantitative zooarchaeological data of the Guarani groups in the Patos Lagoon, the same three species of fish were captured and registered in the sites PS-02-Camping and PS-03-Totó, representing the main aquatic resources in the daily diet, which seems to be the tendency of the coastal indigenous economy on the estuary (Milheira & Ulguim, 2008).

The Total Lengths of the species, measured by otolithometrics, indicate that relatively small specimens of these species predominated in fishing hauls (young and young adults, probably), suggesting limitations in the efficiency of fishing technology to capture larger specimens. In addition, the average size also indicates that fishing could have been carried out from the beach, without the use of fishing gear or capture techniques in the deeper waters of the Patos Lagoon.

The species composition and sizes of the captured fish were influenced both by the greater ecological availability of these species and by specific aspects of the sociocultural systems of the sambaquis and cerritos peoples (environmental knowledge about the species, mastery of preparation techniques and artefacts used in the fishing), which denotes a dialectical relationship between environment and society. The lagoon environment has been economically exploited in a sustainable way for more than three thousand years by indigenous populations, and studies on recent fisheries show changes in species and sizes caught, which may be demonstrating that, according to Thykjaer et al. (2020), the resilience of the Patos Lagoon may be at a tipping point.

Acknowledgments We are grateful to Carla Bica, Eliane Nunes Chim, Leonardo Silva Sens, and Kelly de Oliveira for clarifying information on their published papers and monographs. The research was supported by the Brazilian Council for Scientific and Technological Development (CNPq) through grants 3007994/2020-1, 303266/2018-0, 408208/2021-0 and 305353/2021-7 as well as by the Fundação de Amparo à Pesquisa do Estado do Rio Grande do Sul. The final edition in English version of this text was made by John O'Donnell.

References

Angulo, R. J., Lessa, G. C., & de Souza, M. C. (2006). A critical review of mid- to late-Holocene sea-level fluctuations on the eastern Brazilian coastline. *Quaternary Science Reviews, 25*, 486–506.

Avigliano, E., de Carvalho, B. M., Leisen, M., Romero, R., Velasco, G., Vianna, M., et al. (2017a). Otolith edge fingerprints as approach for stock identification of Genidens barbus. *Estuarine, Coastal and Shelf Science, 194*, 92–96.

Avigliano, E., Leisen, M., Romero, R., Carvalho, B., Velasco, G., Vianna, M., Barra, F., & Volpedo, A. V. (2017b). Fluvio-marine travellers from South America: Cyclic amphidromy and freshwater residency, typical behaviors in *Genidens barbus* inferred by otolith chemistry. *Fisheries Research*. https://doi.org/10.1016/j.fishres.2017.04.011

Bager, F. J. (2013). *Análise do sítio arqueológico Nílton Dutra RS-LC: 42*. Trabalho de Conclusão de Curso, Universidade Federal do Rio Grande – FURG.

Balée, W. (2006). The research program of historical ecology. *Annual Review of Anthropology*. https://doi.org/10.1146/annurev.anthro.35.081705.123231

Bastos, G. C. C. (1990). *Morfologia de otólito de algumas espécies de Perciformes (Teleostei) da costa Sudeste-sul do Brasil.* Dissertação (Mestrado), Instituto Oceanográfico – Universidade de São Paulo.

Bracco, R., del Puerto, L., & Inda, H. (2008). Prehistoria y Arqueología de la Cuenca de Laguna Merín. In D. Loponte, & A. Acosta (Comp.), *Entre la Tierra y el Agua* (Arqueología de Humedales de Sudamérica, pp. 1–60). AINA.

Burns, M. D. de M, Velasco, G., & Cheffe, M. M. (2019). São Gonçalo channel as an ecological corridor for the movement of migratory fishes: Environmental history and perspectives for fishery management in the Mirim lagoon, South Brazil. *Revista Costas, 1*(1), 147–164. https://doi.org/10.26359/costas.0108

Calippo, F. R. (2000). *Análise da ocorrência de otólitos de bagres da Família Ariidae (Teleostei), Micropogonias furnieri e Pogonias cromis (Teleostei: Scianidae) em sítios arqueológicos da restinga da Laguna dos Patos, Rio Grande do Sul, Brasil. Rio Grande.* Trabalho de Conclusão de Curso, Universidade Federal do Rio Grande – FURG.

Calippo, F. R. (2010). *Sociedade Sambaquieira, Comunidades Marítimas.* Tese de Doutorado em Arqueologia, Museu de Etnologia e Arqueologia, Universidade de São Paulo.

Calippo, F. R. (2011). O surgimento da navegação entre os povos dos sambaquis: argumentos, hipóteses e evidências. *Revista do Museu de Arqueologia e Etnologia, 21*, 31–49. Universidade de São Paulo, São Paulo.

Carvalho, C. R. (2004). *Marcadores de estresse ocupacional em populações sambaquieiras no litoral fluminense.* Tese de Doutorado em Saúde Pública, Escola Nacional de Saúde Pública.

Castello, H. P., & Pinedo, M. C. (1977). Botos na Lagoa dos Patos. *Natureza em Revista, 12*, 46–49.

Chanca, I., Caroline, B., Colonese, A. C., Macario, K., Toso, A., Fontanals-Coll, M., dos Anjos, R., Muniz, M., Pereira, R., Talamo, S., & Milheira, R. G. (2021). Food and diet of the pre-Columbian mound builders of the Patos Lagoon region in Southern Brazil with stable isotope analysis. *Journal of Archaeological Science, 133*(2), 105439.

Chao, L., Le Pereira, H., & Vieira, J. P. (1985). Chapter 20: Estuarine fish community of the dos Patos Lagoon, Brazil. A baseline study. In A. Yanez-Arancibia (Ed.), *Fish community ecology in estuaries and coastal lagoons: Towards an ecosystem integration* (pp. 429–450). Univ Nac Aut Mexico Press.

Chim, E. N. (2016). Análise de otólitos do cerrito RS-LS-11 e reconstrução do tamanho de croaker, *Micropogonias furnieri* demarest 1823 (osteichthyes, sciaenidae). *Tessituras: Revista de Antropologia e Arqueologia, 4*, 188–207.

Colonese, A., DeBlasis, P., & Craig, O. (2014). Dietary variability among South American mound-builders: Isotopic and molecular perspectives from Brazilian Sambaquis. *PLoS One, 9*, 1–13.

Costa, M. D. D. P., Muelbert, J. H., Moraes, L. E., Vieira, J. P., & Castello, J. P. (2014). Estuarine early life stage habitat occupancy patterns of whitemouth croaker Micropogonias furnieri (Desmarest, 1830) from the Patos Lagoon, Brazil. *Fisheries Research, 160*, 77–84.

Cotrina, C. P. (1998). Estudio sobre el borde de los otolitos sagita de croaker (Micropogonias furnieri) Frente Marítimo. *Montevideo, 17A*, 39–42.

D'Incao, F., Valentini, H., & Rodrigues, L. F. (2002). Avaliação da pesca de camarões nas regiões sudeste e sul do Brasil, 1965–1999. *Atlantica, 24*, 103–116.

D'Olwer, L. N. (1963). *Cronistas de las culturas precolombinas*. Fondo de Cultura Económica.

DeBlasis, P., Kneip, A., Scheel-Ybert, R., Giannini, P. C., & Gaspar, M. (2007). Sambaquis e paisagem: dinâmica natural e arqueologia regional no litoral do sul do Brasil. *Arqueología Suramericana, 3*, 28–61.

Del Puerto, L. (2015). *Interrelaciones humano-ambientales durante el Holoceno tardío en el este del Uruguay: cambio climático y dinámica cultural*. PhD Thesis, PEDECIBA-Biología, Facultad de Ciencias, Universidad de la República.

Ferreira, M. (2012). Uma análise osteobiográfica dos remanescentes ósseos do sítio arqueológico RS-LS-85: cerrito Oscar Erocildo Abreu. Monografia (Bacharel em Arqueologia) - Instituto de Ciências Humanas e da Informação, Universidade Federal do Rio Grande. Rio Grande-RS, p. 226.

Figuti, L. (1992). *Les sambaquis COSIPA (4200 à 1200 ans BP): Étude de la subsistance chez les peuples de pecheurs-ramasseurs de bivalves de la côte centrale de l'état de São Paulo, Brésil*. Tese de Doutoramento, Institut de Paléontologie Humaine – Museum National d'Histoire Naturelle.

Fossile, T., Ferreira, J., Bandeira, D. R., Figuti, L., Dias-da-Silva, S., Hausmann, N., Robson, H. K., Orton, D., & Colonese, A. C. (2019). Pre-Columbian fisheries catch reconstruction for a subtropical estuary in South America. *Fish and Fisheries, 20*(6), 1124–1137.

Gaspar, M. (2000). *Sambaqui: arqueologia do litoral Brasileiro*. Zahar.

Haimovici, M. (1997). *Recursos Pesqueiros Demersais da Região Sul*. Avaliação do Potencial Sustentável de Recursos Vivos da Zona Econômica Exclusiva (Revizee), editado pela Fundação de Estudos do Mar (FEMAR), Rio de Janeiro.

Haimovici, M., & Cardoso, L. G. C. (2017). Long-term changes in the fisheries in the Patos Lagoon estuary and adjacent coastal waters in Southern Brazil. *Marine Biology Research, 13*, 135–150. https://doi.org/10.1080/17451000.2016.1228978

Haimovici, M., & Velasco, G. R. (2000). Length-weight relationship of marine fishes from Southern Brazil. *Naga, 23*(1), 19–23.

Haimovici, M., Cavole, L. M., Cope, J. M., & Cardoso, L. G. (2021). Long-term changes in population dynamics and life history contribute to explain the resilience of a stock of Micropogonias furnieri (Sciaenidae, Teleostei) in the SW Atlantic. *Fisheries Research, 237*, 105878. https://doi.org/10.1016/j.fishres.2021.105878

Heckenberger, M. (2001). Estrutura, história e transformação: a cultura xinguana na *longue durée*, 1000–2000 d.C. In M. Heckenberger, & B. Franchetto (Orgs.), *Os Povos do Alto Xingu. História e Cultura* (pp. 21–62). UFRJ.

Iriarte, J. (2006). Transformation, mounded villages and adopted cultigens: The rise of early formative communities. In: South-Eastern Uruguay. *World Archaeology, 38*(4), 644–663.

Klokler, D. M. (2010). Comida Para O Corpo E Alma. *Revista de Arqueologia*. https://doi.org/10.24885/sab.v23i1.293

López Mazz, J. M. (2001). Las Estructuras Tumulares (cerritos) Del Litoral Atlantico Uruguayo. *Latin American Antiquity*. https://doi.org/10.2307/971631

Loureiro, A. G. (2008). *Sítio PT-02-Sotéia: Análise Dos Processos Formativos de Um Cerrito Na Região Sudoeste Da Laguna Dos Patos/RS*. https://doi.org/10.11606/d.71.2008.tde-29042008-151446

Marceunik, A. P. (2005). Redescrição de *Genidens barbus* (Lacépède, 1803) e *Genidens machadoi* (Miranda-Ribeiro, 1918), bagres marinhos (Siluriformes, Ariidae) do Atlântico Sul Ocidental. *Papéis Avulsos de Zoologia, 45*(11), 111–125.

Mentz Ribeiro, P. A., Borges Pestana, M., Pereira Penha, M. A., & Rizzi Calippo, F. R. (2004). Levantamentos arqueológicos na porção central da planície costeira do Rio Grande do Sul, Brasil. *Revista de Arqueologia, 17*(1), 85–99. https://doi.org/10.24885/sab.v17i1.194

Milheira, R. G., & Gianotti, C. (2018). The earthen mounds (Cerritos) of Southern Brazil and Uruguay. *Encyclopedia of Global Archaeology*. https://doi.org/10.1007/978-3-319-51726-1_3025-1

Milheira, R. G., & Ulguim, P. F. (2008). Uma contribuição para a Zooarqueologia em sítios Guarani do litoral sul do Brasil, Laguna dos Patos, Pelotas-RS: estratégias de assentamento, aspectos alimentares e função de sítio. *Clío, 23*(1), 84–107.

Milheira, R. G., & Wagner, G. P. (2014). *Arqueologia Guarani no litoral sul do Brasil*. Editora Appris.

Milheira, R. G., Macario, K. D., Chanca, I. S., & Alves, E. Q. (2017). Archaeological earthen mound complex in Patos Lagoon, Southern Brazil: Chronological model and freshwater influence. *Radiocarbon*. https://doi.org/10.1017/rdc.2017.5

Milheira, R. G., Atorre, T., & Borges, C. (2019a). Construtores de cerritos na Laguna dos Patos, Pontal da Barra, sul do Brasil: lugar persistente, território e ambiente construído no holoceno recente. *Latin American Antiquity, 30*(1), 35–54.

Milheira, R. G., Souza, J. G., & Iriarte, J. (2019b). Water, movement and landscape ordering: A GIS-based analysis for understanding the mobility system of late Holocene mound-builders in southern Brazil. *Journal of Archaeological Science, 111*, 105014.

Moreno, F. (2016). La gestión animal en la Prehistoria del Este de Uruguay: de la economía de amplio espectro al control de animales salvajes. *Tessituras, Pelotas, 4*(1), 161–187.

Oliveira, K. (2006). Um sítio de pesca na margem ocidental da lagoa dos Patos: RS-RG-48. *Pesquisas, 63*, 307–336.

Oliveira, A. F., & Bemvenuti, M. A. (2006). O ciclo de vida de alguns peixes do estuário da Lagoa dos Patos, RS. Informações para o ensino fundamental e médio. *Cadernos de Ecologia Aquática, 1*(2), 16–29.

Pavei, D., Borges, C., César, M., Campo, J. B., & Rosa, A. M. (2019). Análise zooarqueológica do Sambaqui Lagoa dos Freitas, extremo sul de Santa Catarina, Brasil: primeiros resultados e tendências de exploração de ecossistemas litorâneos durante o Holoceno final. *Archaeofauna, 28*, 10.15366/archaeofauna2019.28.010.

Pereyra, P. E. R., Mont'Alverne, R., & Garcia, A. M. (2016). Carbon primary sources and estuarine habitat use by two congeneric ariid catfishes in a subtropical coastal lagoon. *Zoologia 33*, 1–7. https://doi.org/10.1590/S1984-4689zool-20150075

Pinedo, M. C. (1990). Ocorrência de pinípedes na costa brasileira. *Gracia de Orla Série Zoologia, 15*(2), 37–48.

Reis, E. G. (1986a). *A Pesca Artesanal de Bagres Marinhos (Siluriformes, Ariidae) no Estuário da Lagoa dos Patos, RS, Brasil* (Documentos Técnicos 05, p. 21). FURG.

Reis, E. G. (1986b). Age and growth of the marine catfish, Netuma barba (Siluriformes, Ariidae), in the estuary of the Patos Lagoon (Brazil). *Fishery Bulletin, 84*, 679–686.

Reis, E. G. (1986c). Reproduction and feeding habits of the marine catfish, Netuma barba (Siluriformes, Ariidae), in the estuary of the Patos Lagoon (Brazil). *Atlantica, 8*, 35–55.

Ribeiro, P. A. M., & Calippo, F. R. (2000). Arqueologia e História Pré-colonial. In P. R. A. Tagliani, P. A. M. Ribeiro, L. H. Torres, & F. das Neves Alves (Eds.), *Arqueologia, História E Socioeconomia Da Restinga Da Lagoa Dos Patos: Uma Contribuição Para O Conhecimento E Manejo Da Reserva Da Biosfera* (pp. 13–40). Fundação Universidade Federal do Rio Grande.

Saccone, E. (2022). Four dugout canoes in Uruguayan collections: Our early maritime heritage. *International Journal of Nautical Archaeology*. https://doi.org/10.1080/10572414.2021.2015914

Santos, P. R. S., Paiva, B., & Velasco, G. (2019). The biggest or the most abundant? Predation of the miragaia Pogonias cromis (Perciformes, Sciaenidae) on benthic organisms in southern Brazil. *Neotropical Biology and Conservation, 14*(4), 431–438. https://doi.org/10.3897/neotropical.14.e48493

Scheel-Ybert, R., & Boyadjian, C. (2020). Gardens on the coast: Considerations on food production by Brazilian shellmound builders. *Journal of Anthropological Archaeology*, 60, 101211.

Schmitz, P. I. (1976). *Sítios de Pesca lacustre em Rio Grande, RS, Brasil*. UNISINOS, (Tese de Livre Docência).

Schorr, M. (1975). *Abastecimento indígena na área alagadiça lacustre de Rio Grande, Rio Grande do Sul, Brasil* (Cadernos, n. 1). ASPE, Departamento de Pesquisas e Projetos.

Schwingel, P. R., & Castello, J. P. (1990). Validación de la edad y el crecimiento de la corvina (Micropogonias furnieri) en el sur de Brasil. *Frente Marítimo, Montevideo, 7*, 19–24.

Sens, L. S. (2020). *Pesca, Comida e Identidade: Um Estudo Zooarqueológico em Cerritos de Índio no Sul do Brasil*. Trabalho de conclusão de curso, UFPEL, Pelotas.

Simon, A. L. H., & Silva, P. F. (2015). Análise geomorfológica da planície lagunar sob influência do canal São Gonçalo – Rio Grande do Sul – Brasil. *São Paulo, UNESP, Geociências, 34*(4), 749–767.

Thykjaer, V., Rodrigues, L., Haimovici, M., & Cardoso, L. G. (2020). Long-term changes in fishery resources of an estuary in Southwestern Atlantic according to local ecological knowledge. *Fisheries Management and Ecology, 27*(2), 185–199.

Tomazelli, L. J., Dillenburg, S. R., & Villwock, J. A. (2017). Late Quaternary geological history of Rio Grande do Sul coastal plain, southern Brazil. *Revista Brasileira de Geociencias, 30*(3), 474–476.
Ulguim, P. F. (2010). *Zooarqueologia e o estudo dos grupos construtores e cerritos: um estudo de caso no litoral da Laguna dos Patos – RS, sítio PT-02 cerrito da Sotéia*. Trabalho de conclusão de curso, UFPEL, Pelotas.
Ulguim, V. F. (2018). *Espinhos, Esporões e Especulações: Análise Zooarqueológica do Cerrito PSG-02 Valverde, Pelotas-RS*. Trabalho de conclusão de curso, UFPEL, Pelotas.
Ulguim, P., & Milheira, R. (2017). Remanescentes humanos em sítios cerritos no sul do Brasil: uma análise osteoarqueológica e bioarqueológica. Cadernos do LEPAARQ, Vol. XIV, n°27.
Wagner, G., Silva, L. A., & Hilbert, L. (2020). O Sambaqui do Recreio: geoarqueologia, ictioarqueologia e etnoarqueologia. *Boletim do Museu Paraense Emílio Goeldi Ciências Humanas, 15*(2). https://doi.org/10.1590/2178-2547-bgoeldi-2019-0084
Wesolowski, V., Mendonça de Souza, S. F., Reinhard, K., & Cecantinni, G. (2007). Grânulos de amido e fitólitos em cálculos dentários humanos: contribuição ao estudo do modo de vida e subsistência de grupos sambaquianos do litoral sul do Brasil. *Revista do Museu de Arqueologia e Etnologia, 17*, 191–210.

Chapter 9
The Onset of Deep-Water Fishing in Southern Brazil

Thiago Fossile, Júlio César de Sá, Jessica Ferreira, and André Carlo Colonese

Abstract Coastal fisheries contributed substantially to population growth and complex forms of social organisation in eastern South America in pre-European times. In the Atlantic Forest coast of Brazil, these societies produced shell mounds (locally known as *sambaquis*) and shallow archaeological sites with abundant fish and shellfish remains that offer snapshots of past species diversity and distribution. Babitonga Bay hosts the largest concentration of coastal archaeological sites in Brazil, playing a prominent role in our understanding of long-term human-ocean interaction in subtropical and tropical South America. Here we reviewed pre-European fish catch composition and associated fishing artefacts from archaeological sites in this region spanning the last 6000 years. Our results reveal changes in fishing practices and an increased capture of pelagic and high trophic level species from 2000 to 1500 years ago. This "fishing up the marine food web" coincided with the appearance of single-piece baited fish hooks and evidence for the colonisation of oceanic islands. The results indicate an expansion of fishing to deep waters and invite us to reconsider the antiquity of the human footprint on ocean ecosystems in the region.

Keywords Coastal adaptation · Late Holocene · Southern Brazil · Fishing technology · Offshore fishing

T. Fossile (✉) · A. C. Colonese
Department of Prehistory & Institute of Environmental Science and Technology (ICTA-UAB), Universitat Autònoma de Barcelona, Barcelona, Spain
e-mail: thiago.fossile@uab.cat; andrecarlo.colonese@uab.cat

J. C. de Sá · J. Ferreira
National Museum, Universidade Federal do Rio de Janeiro, São Cristóvão, Rio de Janeiro, Brazil
e-mail: jcsarqueo@gmail.com; jessferreira.f@gmail.com

A. C. Colonese, R. G. Milheira (eds.), *Historical Ecology and Landscape Archaeology in Lowland South America*, Interdisciplinary Contributions to Archaeology, https://doi.org/10.1007/978-3-031-32284-6_9

9.1 Introduction

Humans have taken advantage of the biological diversity and other ecosystem services of the Brazilian coast for at least 6000 years (Toso et al., 2021). Early evidence of marine fishing along the Brazilian coast for subsistence and ritual practices is documented mainly in the form of shell mounds, locally known as *sambaquis*, and shallow archaeological sites with abundant faunal remains (Prous, 1992; Gaspar et al., 2008). Sambaquis are the most visible and well-known coastal sites, constructed through various cultural activities resulting in the vertical accumulation of faunal remains together with complex sedimentary matrices, numerous human burials, and stone, bone and plant artefacts (Lima, 2000; Villagran et al., 2011; Wagner et al., 2011; Colonese et al., 2014; Fossile et al., 2019). These sites constitute a cultural heritage of national and international interest and expand our understanding of pre-European Indigenous interaction with, and transformation of, coastal environments.

Faunal remains (Klokler, 2014, 2016; Fossile et al., 2019) and stable isotope (Toso et al., 2021; Bastos et al., 2022) analysis indicate that aquatic organisms, and notably fish, were the primary source of dietary proteins to pre-European Indigenous populations. Beyond their nutritional value, fish and shellfish also played a role in the symbolic and ritual spheres (Prous, 2006; Gaspar et al., 2008; Klokler, 2014), as alluded to by the production of a very particular class of stone (zooliths) and bone (zoosteos) artefacts representing a wide variety of terrestrial and aquatic animals (Milheira, 2014; Prous, 2018), and sometimes associated with human burials (Garcia, 2018). Fish were also a common item in funerary rituals, again reinforcing their relevance in the social fabric and long-term identity of these communities (Klokler, 2014, 2016; Fossile et al., 2019). Some regional variability can be observed in the funerary record. For example, the zooarchaeological data from Babitonga Bay (Santa Catarina state) indicate that most of the catches used in funerary rituals were medium-high trophic level species (Fossile et al., 2019, 2020), which contrasts with the view that lower trophic level species were preferentially used in these ceremonial activities (Klokler, 2014, 2016).

Recently, Brazilian archaeologists have started to explore the untapped potential of sambaquis and other coastal sites as archives of biological diversity, species abundance and distribution prior to European contact (Mendes et al., 2018; Fossile et al., 2019, 2020). This is of particular significance in biodiversity hotspots, which are regions undergoing exceptional loss of habitats and biodiversity (Myers et al., 2000), such as the Atlantic Forest of Brazil. In fact, it is becoming increasingly clear that environmental issues can only be effectively addressed if anthropogenic and natural drivers of change are understood from a historical perspective (Jackson, 2001, 2010; Roberts, 2003). The lack of such long-term view can cause misconceptions about the extent and rate of ecosystem alterations (Soga & Gaston, 2018). Within this broad conceptual framework, faunal remains become valuable sources of ecological baselines (Erlandson et al., 2009; Erlandson & Rick, 2010; Braje et al., 2017), and when associated with archaeological artefacts such as fishing and hunting gear, they enable the exploration of the origin and changing nature of historical anthropogenic footprints on past ecosystems (Erlandson et al., 2009; Bettinger, 2015).

Babitonga Bay, located on the northeast coast of the state of Santa Catarina, has over 200 archaeological sites, hosting the largest concentration of sambaquis on the Brazilian coast (DeBlasis et al., 1998; Fish et al., 2000; Lopes et al., 2016; Bandeira et al., 2018). The region holds a prominent position in studies regarding early coastal adaptations in eastern South America (Fig. 9.1). Independent lines of evidence point to fundamental changes in fishing strategies from ca. 2000 cal BP in Babitonga Bay and adjacent coastal areas. Firstly, studies involving bulk collagen stable isotope analysis of human remains show an increase in consumption of high trophic level marine organisms from ca. 2000 cal BP, notably in human individuals associated with Taquara-Itararé ceramics (Toso et al., 2021). Secondly, single-piece baited fishing hooks appear in numerous sites containing ceramic artefacts of the Taquara-Itararé tradition (Tiburtius & Bigarella, 1953; Bryan, 1993; Schmitz et al., 1993). Interestingly, the extent to which the emergence and/or spread of new fishing technology allowed past coastal societies to pursue new marine resources remains a matter of debate. For example, the relation between baited fishing hooks and species of high trophic level in the region has never been explored.

Here, we address some of these enquiries by reviewing the evidence of portable fishing gear and faunal remains in Babitonga Bay from the Middle to the Late Holocene. Our review of the literature shows an apparent association between

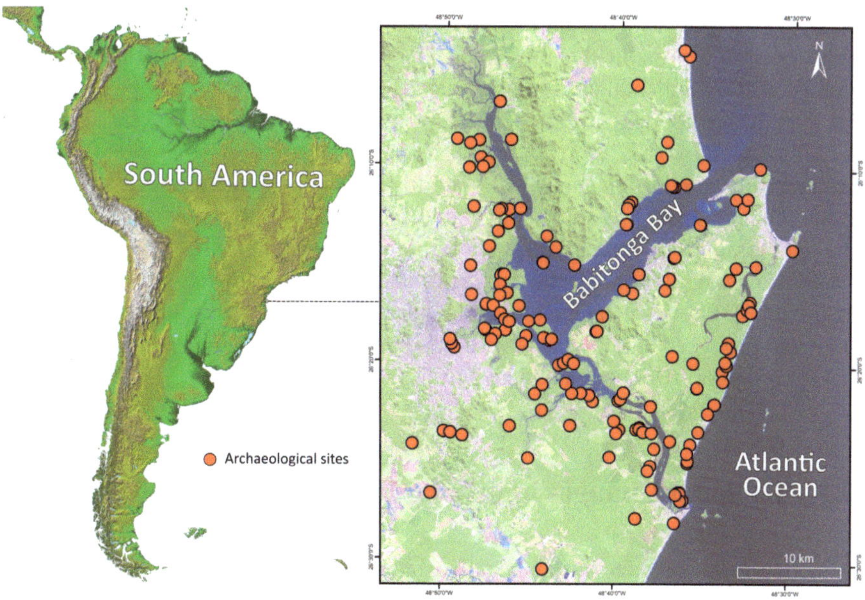

Fig. 9.1 Distribution of pre-Columbian archaeological sites in Babitonga Bay, Santa Catarina, Brazil

fishing technology and marine fish catch composition in Babitonga Bay. The results expand our knowledge about marine resource acquisition by Indigenous populations that inhabited the southern coast of Brazil before European contact, while reinforcing the hypothesis that from ca. 2000 years ago Indigenous groups were "fishing up the food web", and exploiting deeper marine habitats of the southwestern Atlantic Ocean.

9.2 The Study Area: Babitonga Bay

Pre-Columbian occupations in Babitonga Bay are dated between ca. 6600 and 400 cal BP, with most of the sites formed between ca. 5500 and 2200 cal BP (Toso et al., 2021). These groups depended on coastal systems affected by variations in sea level and coastal morphology (Martin et al., 1988; Vieira, 2015; Sá, 2017) occurring since the Middle Holocene, which are thought to have affected resource procurement strategies and settlement patterns (Toso et al., 2021). During the Holocene Climate Optimum (ca. 5500 BP) the sea level of the southern Brazilian coast was ca. 3 m above the current level (Angulo et al., 2006; Toniolo et al., 2020). During its regression towards modern values, ecosystems such as mangroves and coastal lagoons were formed and/or expanded, likely affecting the distribution of coastal and estuarine fauna (Angulo & Müller, 1990; Angulo et al., 2009). This time interval coincides with an increase in the number of archaeological sites, suggesting a period of greater human population density that lasted until 2200 years ago (Toso et al., 2021), after which shell mound formation was minimal and many sites were abandoned. It is possible that pre-European population growth at this time suddenly exceeded environmental carrying capacity, forcing major socio-ecological reorganisations (Lima, 2000; Toso et al., 2021). From ca. 1200 years ago several sambaquis were occupied by groups producing, or using through exchange/adoption, ceramic artefacts attributed to the Taquara-Itararé tradition (also defined as Proto-Jê) of the Brazilian highlands. The spread of ceramics to the coast is possibly associated with the demographic increase of Proto-Jê groups in the southern Brazilian highlands (Robinson et al., 2018) and the decline of shell mound cultures on the coast (Toso et al., 2021). Groups using ceramic artefacts attributed to the Taquara-Itararé tradition also produced shallow sites with abundant faunal remains (Toso et al., 2021; Gilson & Lessa, 2021b) and human burials (Lessa, 2006).

9.3 Regional Changes in Fishing Technology

Although few sites were systematically excavated in Babitonga Bay, a total of 781 artefacts potentially associated with fishing have been recovered from Middle and Late Holocene archaeological contexts. These included projectile points (n = 589, 75.4%), single-piece baited fish hooks (n = 104, 13.3%), fishing net weights (n = 88,

11.3%) and cordage made of plant fibres (Tiburtius et al., 1951, 1954; Tiburtius & Bigarella, 1953; Beck, 1972, 2007; Goulart, 1980; Fossari, 1984; Bryan, 1993; Afonso & DeBlasis, 1994; Tiburtius, 1996; Bandeira, 2004; Peixe et al., 2007; Bandeira et al., 2009; Sá, 2015) (Fig. 9.2a–d). Arguably, some of these artefacts may not have been exclusively used for fishing. Projectile points could have been used for hunting and/or in warfare (Tiburtius & Bigarella, 1953; Fossari, 1984), while cordage could have been employed in a range of activities (Ribeiro, 1986; Gaspar, 1998; Peixe et al., 2007; Bandeira et al., 2009). However, their close association with abundant fish remains suggests a strong link between these portable technologies and fishing practices.

Fishing artefacts were documented in eight sites. Of these, projectile points were recovered in all sites with the exception of Espinheiros II; fish hooks were found mainly in sites near the coast (Forte Marechal Luz, Enseada I, Bupeva II, and Rio Pinheiros). An exception is the recovery of dozens of single-piece fish hooks from Itacoara, located ca. 30 km from the current coastline, but still at close distance to the estuary and small rivers. Net weights were recovered from sites in the inner part of the bay (Morro do Ouro and Itacoara) (Fig. 9.3a).

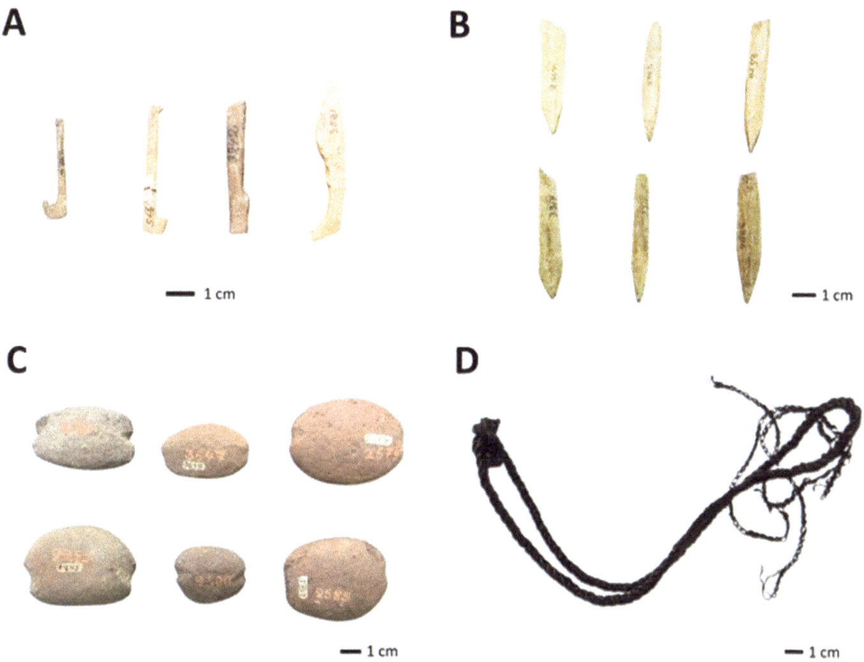

Fig. 9.2 Artefacts recovered in sambaquis from Babitonga Bay, – *Museu Arqueológico de Sambaqui de Joinville* (MASJ) collection. (**a**) Single-piece baited fish hooks from Itacoara; (**b**) Projectile points from various sites; (**c**) Net weights from various sites; (**d**) Cordage (Araceae – *Philodendron* sp.) from Sambaqui Cubatão I. Cordages are exceptionally well preserved in Cubatão I and Espinheiros II due to anaerobic conditions. (**a**, **b** and **c** were adapted from Ferreira et al. (2019), and **d** by the authors)

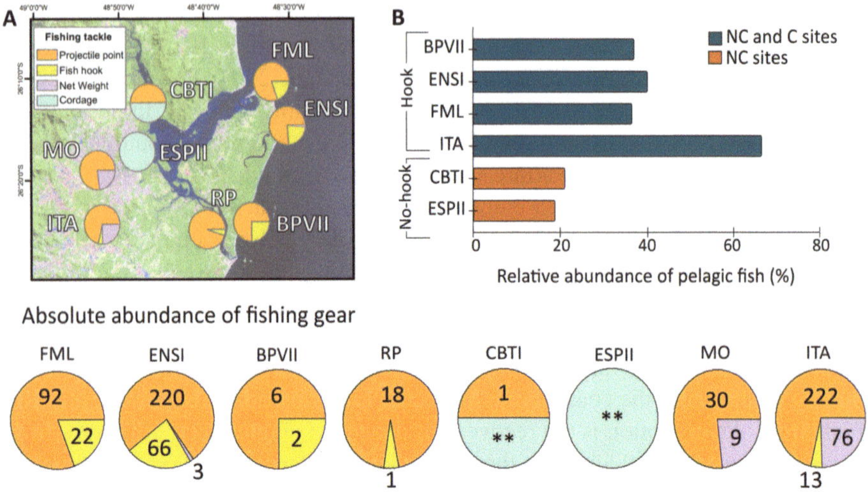

Fig. 9.3 (**a**) Distribution and relative abundance of fishing technology; (**b**) Relative abundance of pelagic fish according to the presence and absence of single-piece baited fish hooks in non-ceramic (NC) and ceramic (C) archaeological sites. MO and RP are not included as they lack systematic faunal analysis. Quantitative data is not available for cordage (**). Abbreviation of sites: *RP* Rio Pinheiros, *MO* Morro do Ouro, *FML* Forte Marechal Luz, *ENSI* Enseada I, *CBTI* Cubatão I, *ESPII* Espinheiros II, *BPVII* Bupeva II, *ITA* Itacoara

Significantly, 97% of single-piece baited fish hooks (n = 93) and 86% of projectile points (n = 295) were recovered from sites with Taquara-Itararé ceramics (Enseada I, Itacoara, and Forte Marechal Luz), in deposits dated between 1390 and 620 BP (Table 9.1). Using faunal data from Fossile et al. (2019, 2020) from 110 archaeological sites in Babitonga Bay, we derived that high trophic level species (Trophic Level value ≥4.0), including *Trichiurus lepturus* (largehead hairtail), *Lagocephalus laevigatus* (smooth puffer) and *Carcharias taurus* (sand tiger shark), were relatively more abundant in ceramic sites with fishing hooks (91% of NISP), compared to non-ceramic sites with no fishing hooks (44% of NISP). Similarly, pelagic species (e.g. *Galeocerdo cuvier* (tiger shark), *Sphyrna* sp. (hammerhead sharks), *Trichiurus lepturus* (largehead hairtail), *Pomatomus saltatrix* (bluefish) and *Lagocephalus laevigatus* (smooth puffer)) were relatively more abundant in sites containing ceramic artefacts and fishing hooks than in non-ceramic sites (Fig. 9.3b). At Bupeva II, elasmobranchs (Rajiformes (rays), *Carcharhinus isodon* (finetooth shark), *Galeocerdo cuvier* (tiger shark), *Carcharodon carcharias* (great white shark), *Isurus oxyrinchus* (shortfin mako), *Carcharias taurus* (sand tiger shark), and *Sphyrna* sp. (hammerhead sharks)) represented 27% of the fish remains in non-ceramic contexts, while in ceramic deposits it rose to 42% (Bandeira, 2004). At Enseada I, elasmobranchs (Rajiformes (rays), *Galeocerdo cuvier* (tiger shark), *Prionace glauca* (blue shark), *Carcharodon carcharias* (great white shark), *Carcharias taurus* (sand tiger shark), and *Sphyrna* sp. (hammerhead sharks)) contributed to 1.6% and 3% in non-ceramic and ceramic contexts, respectively

Table 9.1 Conventional radiocarbon dating of pre-Columbian archaeological sites with fishing artefacts in Babitonga Bay

Archaeological site	Key cultural feature	Lab. code	Material	Conventional date (BP)
Rio Pinheiros II	Ceramic	Beta_443872	Bone (unidentified species)	860 ± 30
Rio Pinheiros II	Non-ceramic (sambaqui)		Marine Shell	3850 ± 140
Rio Pinheiros II	Non-ceramic (sambaqui)		Marine Shell	4580 ± 120
Morro do Ouro	Non-ceramic (sambaqui)		Human bone collagen	3870 ± 40
Morro do Ouro	Non-ceramic (sambaqui)	AA_104770	Human bone collagen	3938 ± 55
Morro do Ouro	Non-ceramic (sambaqui)	Beta_93152	Human bone collagen	4030 ± 40
Morro do Ouro	Non-ceramic (sambaqui)	AA_104768	Human bone collagen	4086 ± 42
Morro do Ouro	Non-ceramic (sambaqui)	Beta_444034	Human bone collagen	4200 ± 30
Morro do Ouro	Non-ceramic (sambaqui)		Human bone collagen	4300 ± 50
Morro do Ouro	Non-ceramic (sambaqui)	AA_104767	Human tooth	4425 ± 39
Forte Marechal Luz	Ceramic	M_1203	Charcoal	620 ± 100
Forte Marechal Luz	Ceramic	M_1200	Charcoal	640 ± 100
Forte Marechal Luz	Ceramic	M_1202	Charcoal	880 ± 100
Forte Marechal Luz	Non-ceramic (sambaqui)	M_1205	Charcoal	850 ± 100
Forte Marechal Luz	Non-ceramic (sambaqui)	M_1204	Seed	1100 ± 100
Forte Marechal Luz	Non-ceramic (sambaqui)	M_1206	Charcoal	1440 ± 110
Forte Marechal Luz	Non-ceramic (sambaqui)		Human bone collagen	1550 ± 40
Forte Marechal Luz	Non-ceramic (sambaqui)	M_1207	Sea mammal bone collagen	2060 ± 120
Forte Marechal Luz	Non-ceramic (sambaqui)	M_1208	Charcoal	3660 ± 130
Forte Marechal Luz	Non-ceramic (sambaqui)		Sea mammal bone collagen	4290 ± 130
Enseada I	Ceramic		Human bone collagen	1390 ± 40
Enseada I	Non-ceramic (sambaqui)		Human bone collagen	3920 ± 40
Cubatão I	Non-ceramic (sambaqui)	Beta_268526	Charcoal	2250 ± 40
Cubatão I	Non-ceramic (sambaqui)	Beta_268518	Human bone collagen	2430 ± 40

(continued)

Table 9.1 (continued)

Archaeological site	Key cultural feature	Lab. code	Material	Conventional date (BP)
Cubatão I	Non-ceramic (sambaqui)	Beta_268523	Human bone collagen	2460 ± 40
Cubatão I	Non-ceramic (sambaqui)	Beta_268525	Human bone collagen	2460 ± 40
Cubatão I	Non-ceramic (sambaqui)	Ly_ 4524	Human bone collagen	2460 ± 30
Cubatão I	Non-ceramic (sambaqui)	Ly_4527	Human bone collagen	2495 ± 30
Cubatão I	Non-ceramic (sambaqui)	Beta_268524	Human bone collagen	2510 ± 40
Cubatão I	Non-ceramic (sambaqui)	Beta_259519	Human bone collagen	2520 ± 40
Cubatão I	Non-ceramic (sambaqui)	Ly_4528	Human bone collagen	2520 ± 40
Cubatão I	Non-ceramic (sambaqui)	Beta_259823	Marine shell	2560 ± 40
Cubatão I	Non-ceramic (sambaqui)	Ly_4526	Human bone collagen	2620 ± 30
Cubatão I	Non-ceramic (sambaqui)	Beta_259821	Human bone collagen	2630 ± 40
Cubatão I	Non-ceramic (sambaqui)	Beta_259824	Marine Shell	2660 ± 40
Cubatão I	Non-ceramic (sambaqui)	Beta_259820	Human bone collagen	2670 ± 40
Cubatão I	Non-ceramic (sambaqui)	Beta_259827	Charcoal	2890 ± 70
Cubatão I	Non-ceramic (sambaqui)	Beta_259825	Charcoal	2970 ± 60
Cubatão I	Non-ceramic (sambaqui)	Ly_4525	Charcoal	2975 ± 30
Cubatão I	Non-ceramic (sambaqui)	Beta_259829	Charcoal	3040 ± 60
Cubatão I	Non-ceramic (sambaqui)	Beta_259826	Charcoal	3110 ± 70
Cubatão I	Non-ceramic (sambaqui)	Beta_259828	Marine shell	3480 ± 60
Espinheiros II	Non-ceramic (sambaqui)	Gif_9415	Charcoal	1160 ± 45
Espinheiros II	Non-ceramic (sambaqui)		Charcoal	1270 ± 60
Espinheiros II	Non-ceramic (sambaqui)	Gif_9416	Charcoal	2970 ± 60
Bupeva II	Non-ceramic (sambaqui)	Kia_22262	Sea mammal bone collagen	2325 ± 25
Itacoara	Ceramic		Human bone collagen	1250 ± 30
Itacoara	Non-ceramic (sambaqui)	Kia_21796	Wood and burned seed	1570 ± 20

(Bandeira, 2004). At Itacoara, elasmobranchs are found only in the ceramic context, representing 14% of the remains (Bandeira, 2004). At Forte Marechal Luz, the remains of elasmobranchs accounted for 38% and 35% of fish remains in non-ceramic and ceramic contexts, respectively (Bryan & Gruhn, 1993). It is worth noting though, that the later non-ceramic phases of Forte Marechal Luz have radiocarbon dates comparable with the earliest sites with Taquara-Itararé ceramics in this region (e.g. Itacoara, Enseada I; Table 9.1).

Results from Babitonga Bay are broadly supported by zooarchaeological studies from adjacent coastal areas. Remains of top predators such as fur seals (*Arctocephalus australis* (South American fur seal)), dolphins (*Delphinus delphis* (short-beaked saddleback dolphin), *Tursiops truncatus* (bottlenose dolphin)) and sharks (*Carcharias taurus* (sand tiger shark), *Carcharodon carcharias* (great white shark), *Galeocerdo cuvier* (tiger shark)), among others, were found in archaeological deposits containing Taquara-Itararé ceramics at Praia das Laranjeiras II (Camboriú), which also preserved remains of single-piece baited fish hooks (Schmitz et al., 1993). At the Taquara-Itararé site of Rio do Meio (Florianópolis), a large variety of fish, including a large diversity of sharks (*Carcharhinus plumbeus* (sandbar shark), *C. obscurus* (dusky shark), *C. leucas* (bull shark), *C. brachyurus* (copper shark), *Rhizoprionodon* sp. (sharpnose sharks), *Carcharodon carcharias* (great white shark), etc.) were exploited by groups using single-piece baited fish hooks (Gilson & Lessa, 2020, 2021a, b).

The observed correlations between the capture of pelagic and high trophic level species and the presence of fishing hooks and spears has been documented in other archaeological records in the Americas. Erlandson et al. (2009) have shown a substantial increase of fish bones in California's northern Channel Islands after the appearance of curved, single-piece fish hooks from about 2500 years ago. Remains of pelagic fishing including large species such as tunas, swordfish, and sharks increased notably after 1500 years ago. This trend resembles the one documented in Babitonga Bay after the introduction of single-piece baited fish hooks.

Single-piece baited fish hooks and spears are particularly effective for catching larger predators and pelagic species (Misund et al., 2002; He et al., 2021). Moreover, fish hooks can be effectively employed in deep water and areas with low visibility (Toso et al., 2021). In support of the arguments presented above is the presence of Taquara-Itararé ceramic artefacts in an archaeological site on Arvoredo Island, an oceanic island located more than 10 km from the southern Brazilian coast (Rohr, 1984; Segal et al., 2017). Moreover, stable carbon and particularly the nitrogen isotopic values of bone collagen of humans found at several coastal sites containing Taquara-Itararé ceramic artefacts broadly support the idea of "fishing up the marine food web" (Toso et al., 2021). The results presented here thus support the growing evidence that offshore fishing was practised by Late Holocene Indigenous groups who used Taquara-Itararé ceramics on the southern coast of Brazil.

9.4 Implications for Biological Conservation

Tracking the origin and changing nature of offshore fishing has implications beyond the boundaries of archaeology (Erlandson et al., 2009; Erlandson & Rick, 2010; Braje et al., 2017). Data from Fossile et al. (2019) suggest that top marine predators must have been abundant in the region in the past to permit their recurrent exploitation (high encounter rate) through relatively simple fishing technology. For example, species with trophic levels ranging between 3.5 and 4.5, including bony (e.g. *Chaetodipterus faber* (Atlantic spadefish), *Trichiurus lepturus* (largehead hairtail)) and cartilaginous fish (e.g. *Carcharias taurus* (sand tiger shark)) were relatively abundant and widespread in the archaeological record. Cartilaginous fish (n = 14 taxa), such as *Carcharias taurus* (sand tiger shark), rays (Batoidea), and *Carcharodon carcharias* (great white shark) occurred in ca. 60% of the sites with fish remains in Babitonga Bay (n = 7). Given the ecological and social importance of these species – some as key functional players in the ecosystem and others as sources of food and livelihood to local coastal communities – an understanding of species and population responses to long-term fishing pressure is critical for their modern-day sustainable management.

Several species documented by Fossile et al. (2019), and revised in this study, are currently threatened, falling into the Critically Endangered (CR), Endangered (EN), or Vulnerable (VU) categories of the International Union for Conservation of Nature – IUCN (IUCN, 2022) and the Instituto Chico Mendes de Conservação da Biodiversidade – ICMBio (Brasil, 2018a). These include *Pomatomus saltatrix* (bluefish), *Cynoscion acoupa* (acoupa weakfish), *Alopias vulpinus* (common thresher shark), *Isurus oxyrinchus* (shortfin mako), *Hyporthodus niveatus* (snowy grouper), *Pogonias cromis* (black drum), *Carcharias taurus* (sand tiger shark), *Carcharodon carcharias* (great white shark) and *Genidens barbus* (white sea catfish). Moreover, *Carcharhinus isodon* (finetooth shark), found at the sites of Itacoara and Bupeva II, is considered regionally extinct (RE) on the Brazilian coast (Brasil, 2018b). Other species lacking sufficient data (DD) for extinction risk assessment are also documented in the archaeological record, including for example *Pagrus pagrus* (red porgy), *Menticirrhus americanus* (southern kingfish), *Gymnothorax ocellatus* (Caribbean ocellated moray), *Pomacanthus paru* (French angelfish), *Mugil liza* (blueback mullet) and *Anisotremus surinamensis* (black margate).

9.5 Final Considerations

The archaeological sites of Babitonga Bay provide unique insights into pre-European Indigenous knowledge, and trends in marine species composition and relative abundance in the Atlantic Forest coast of Brazil. The archaeological faunal record, although still deficient in chronological resolution and often affected by discrepancies in recovery techniques, show that pre-European indigenous groups intensified the capture of pelagic and high trophic level species from *ca.* 2000 years ago. Key to this process was the adoption of efficient and sophisticated fishing (baited hooks) and maritime technologies that allowed them to pursue resources offshore, marking the onset of deep-water fishing in the region. The archaeological faunal record also holds significant information on past species distribution and relative abundance, which can be used to track changes in ecosystem structure, composition and function through time. The information gained can potentially contribute to establishing more accurate reference baselines and aid in ambitious and more informed conservation initiatives.

Acknowledgements The authors thank Alice di Muro and Krista McGrath for their helpful suggestions and comments on the manuscript. TF and ACC thank the TRADITION project that has received funding from the European Research Council (ERC) under the European Union's Horizon 2020 research and innovation programme under grant agreement N° 817911. This work contributes to the ICTA-UAB "María de Maeztu" Programme for Units of Excellence of the Spanish Ministry of Science and Innovation (CEX2019-000940-M). This work was also funded by EarlyFoods (Evolution and impact of early food production systems), 2021 SGR 00527.

References

Afonso, M. C., & DeBlasis, P. A. D. (1994). Aspectos da formação de um grande sambaqui: alguns indicadores em Espinheiros II, Joinville. *Revista do Museu de Arqueologia e Etnologia (Universidade de Sao Paulo Sistema Integrado de Bibliotecas – SIBiUSP), 4*, 21.

Angulo, R. J., & de Müller, A. C. P. (1990). Preliminary characterization of some tidal flat ecosystems on the State of Paraná Coast, Brazil. *Simpósio Sobre Ecossistemas Da Costa Sul e Sudeste Brasileira, 2*, 158–168.

Angulo, R. J., Lessa, G. C., & de Souza, M. C. (2006). A critical review of mid- to late-Holocene sea-level fluctuations on the eastern Brazilian coastline. *Quaternary Science Reviews, 25*, 486–506.

Angulo, R. J., de Souza, M. C., & Lessa, G. C. (2009). The Holocene barrier systems of paranaguá and northern Santa Catarina coasts, southern Brazil. In S. R. Dillenburg & P. A. Hesp (Eds.), *Geology and geomorphology of Holocene coastal barriers of Brazil* (Lecture notes in earth sciences) (pp. 135–176). Springer Berlin Heidelberg.

Bandeira, D. R. (2004). *Ceramistas Pré-coloniais da Baía da Babitonga – Arqueologia e Etnicidade*. Doctor in History, Universidade Estadual de Campinas.

Bandeira, D. R., Oliveira, E. L., & Santos, A. M. P. (2009). Estudo estratigráfico do perfil nordeste do Sambaqui Cubatão I. Joinville/SC. *Revista do Museu de Arqueologia e Etnologia, 19*, 119–142.

Bandeira, D. R., Alves, M. C., de Almeida, G. T., de Sá, J. C., Ferreira, J., Vieira, C. V., da Costa do Amaral, V. M. C., Bartz, M. C., & de Melo, J. C. F., Jr. (2018). Resultados preliminares da pesquisa no sambaqui sob rocha Casa de Pedra, São Francisco do Sul, Santa Catarina, Brasil. *Boletim do Museu Paraense Emílio Goeldi. Ciências Humanas (MCT/Museu Paraense Emílio Goeldi), 13*, 207–225.

Bastos, M. Q., Guida, V., Rodrigues-Carvalho, C., Toso, A., Santos, R. V., & Colonese, A. C. (2022). Elucidating pre-columbian tropical coastal adaptation through bone collagen stable isotope analysis and bayesian mixing models: Insights from Sambaqui do Moa (Brazil). *Revista de Antropología del Museo de Entre Ríos, 7*, 1–10.

Beck, A. (1972). *A variação do conteúdo cultural dos Sambaquis do Litoral de Santa Catarina.* Thesis, Universidade de São Paulo (USP).

Beck, A. (2007). *A variação do conteúdo cultural dos sambaquis do litoral de Santa Catarina* (Sociedade de Arqueologia Brasileira, Ed.). Habilis.

Bettinger, R. L. (2015). *Orderly anarchy: Sociopolitical evolution in aboriginal California.* University of California Press.

Braje, T. J., Rick, T. C., Szpak, P., Newsome, S. D., McCain, J. M., Elliott, E. A., Smith, M. G., & Hamilton, S. L. (2017). Historical ecology and the conservation of large, hermaphroditic fishes in Pacific Coast kelp forest ecosystems. *Science Advances, 3*, e1601759.

Brasil, Instituto Chico Mendes de Conservação da Biodiversidade. (2018a). *Livro Vermelho da Fauna Brasileira Ameaçada de Extinção: Volume VI – Peixes*. Instituto Chico Mendes de Conservação da Biodiversidade (ICMBio).

Brasil, Instituto Chico Mendes de Conservação da Biodiversidade. (2018b). *Livro Vermelho da Fauna Brasileira Ameaçada de Extinção: Volume I (v.2. Mamíferos – v.3. Aves – v.4. Répteis – v.5. Anfíbios – v.6. Peixes – v.7 Invertebrados)*. Instituto Chico Mendes de Conservação da Biodiversidade (ICMBio).

Bryan, A. L. (1993). *The Sambaqui at Forte Marechal Luz, State of Santa Catarina, Brazil* (A. L. Bryan, & R. Gruhn, Eds.). Oregon State University.

Bryan, A. L., & Gruhn, R. (1993). *The Sambaqui at Forte Marechal Luz, State of Santa Catarina, Brazil: Archaeological research at Six Cave or Rockshelter Sites in Interior Bahia, Brazil.* Center for the Study of the First Americans, Oregon State University.

Colonese, A. C., Collins, M., Lucquin, A., Eustace, M., Hancock, Y., de Almeida Rocha Ponzoni, R., Mora, A., et al. (2014). Long-term resilience of late holocene coastal subsistence system in Southeastern South america. *PLoS One, 9*, e93854.

Sá, J. C. de (2015). *Etnoarqueologia e Arqueologia Experimental: desatando informações sobre nós e amarrações no Sambaqui Cubatão I, em Joinville – SC* (A. M. P. Santos, & D. R. Bandeira, Eds.). Undergraduate in Archaeology, Universidade da Região de Joinville (Univille).

Sá, J. C. de (2017). *Sambaquis, patrimônio arqueológico na costa leste de São Francisco do Sul/ SC: reflexões sobre o território, variações do nível relativo do mar (NRM) no quaternário e tensões atuais*. Dissertação (Mestrado em Patrimônio Cultural) – Universidade da Região de Joinville, Mestrado em Patrimônio Cultural e Sociedade, Joinville.

DeBlasis, P., Fish, S. K., Gaspar, M. D., & Fish, P. R. (1998). Some references for discussion of complexity among the Sambaqui moundbuilders from the southern shores of Brazil. *Revista de Arqueología Americana (Pan American Institute of Geography and History), 15*, 75–105.

Erlandson, J. M., & Rick, T. C. (2010). Archaeology meets marine ecology: The antiquity of maritime cultures and human impacts on marine fisheries and ecosystems. *Annual Review of Marine Science, 2*, 231–251.

Erlandson, J. M., Rick, T. C., & Braje, T. J. (2009). Fishing up the food web?: 12,000 years of maritime subsistence and adaptive adjustments on California's Channel Islands. *Pacific Science (University of Hawai'i Press), 63*, 711–724.

Ferreira, J., da Rocha Bandeira, D., Bartz, M. C., Fossile, T., & Mayorka, F. (2019). Reflexões sobre a pesca pré-colonial na Baía da Babitonga, litoral norte de Santa Catarina, Brasil. *Cadernos do LEPAARQ (UFPEL), 16*, 138–155.

Fish, S. K., De Blasis, P., Gaspar, M. D., & Fish, P. R. (2000). Incremental events in the construction of sambaquis, southeastern Santa Catarina. *Revista do Museu de Arqueologia e Etnologia*, 69–87. https://doi.org/10.11606/issn.2448-1750.revmae.2000.109378

Fossari, T. D. (1984). A *Indústria Óssea na Arqueologia Brasileira: Estudo-Piloto do Material de Enseada-SC e Tenório-SP* (U. T. B. de Meneses, Ed.). Universidade de São Paulo (USP).

Fossile, T., Ferreira, J., Bandeira, D. R., Figuti, L., Dias-da-Silva, S., Hausmann, N., Robson, H. K., Orton, D., & Colonese, A. C. (2019). Pre-Columbian fisheries catch reconstruction for a subtropical estuary in South America. *Fish and Fisheries, 20*, 1124–1137.

Fossile, T., Ferreira, J., Bandeira, D. R., Dias-da-Silva, S., & Colonese, A. C. (2020). Integrating zooarchaeology in the conservation of coastal-marine ecosystems in Brazil. *Quaternary International*. https://doi.org/10.1016/j.quaint.2019.04.022

Garcia, J. B. (2018). *Patrimônio Arqueológico: Os Artefatos Zoomorfos e Antropomorfos Sambaquieiros do Estado de Santa Catarina* (D. da Rocha Bandeira, Ed.). Master in Cultural Heritage and Society, Universidade da Região de Joinville (Univille).

Gaspar, M. D. (1998). Considerations of the sambaquis of the Brazilian coast. *Antiquity (Cambridge University Press), 72*, 592–615.

Gaspar, M. D., DeBlasis, P., Fish, S. K., & Fish, P. R. (2008). Sambaqui (shell mound) societies of coastal Brazil. In H. Silverman & W. H. Isbell (Eds.), *The handbook of South American archaeology* (pp. 319–335). Springer New York.

Gilson, S.-P., & Lessa, A. (2020). Ocupação tardia do litoral norte e central catarinense por grupos pescadores-caçadores-coletores. *Revista de Arqueologia, 33*, 55–77.

Gilson, S.-P., & Lessa, A. (2021a). Arqueozoologia do sítio Rio do Meio (SC). *Revista de Arqueologia, 34*, 217–248.

Gilson, S.-P., & Lessa, A. (2021b). Capture, processing and utilization of sharks in archaeological context: Its importance among fisher-hunter-gatherers from southern Brazil. *Journal of Archaeological Science: Reports, 35*, 102693.

Goulart, M. (1980). *Tecnologia e padrões de subsistência de grupos pescadores-coletores pré-históricos, habitantes do sambaqui do Morro do Ouro – Joinville, Santa Catarina*. Museu Arqueológico de Sambaqui de Joinville (MASJ).

He, P., Chopin, F., Suuronen, P., Ferro, R. S. T., & Lansley, J. (2021). *Classification and illustrated definition of fishing gears* (Paper nº 672, Food and Agriculture Organization-FAO, Ed.). Food and Agriculture Organization – FAO.

IUCN. (2022). *The IUCN Red List of Threatened Species*. Version 2022-1.

Jackson, J. B. C. (2001). What was natural in the coastal oceans? *Proceedings of the National Academy of Sciences of the United States of America, 98*, 5411–5418.

Jackson, J. B. C. (2010). The future of the oceans past. *Philosophical Transactions of the Royal Society of London. Series B, Biological Sciences, 365*, 3765–3778.

Klokler, D. (2014). A ritually constructed shell mound: Feasting at the Jabuticabeira II Site. In M. Roksandic, S. M. Souza, S. Eggers, M. Burchell, & D. Klokler (Eds.), *The cultural dynamics of shell-matrix sites* (pp. 151–162). University of New Mexico Press.

Klokler, D. (2016). Animal para toda Obra: fauna ritual em sambaquis. *Revista Habitus – Revista do Instituto Goiano de Pré-História e Antropologia, 14*, 21–34.

Lessa, A. (2006). Reflexões preliminares sobre paleoepidemiologia da violência em grupos ceramistas litorâneos: (I) Sítio Praia da Tapera – SC. *Revista do Museu de Arqueologia e Etnologia, 15–16*, 199–207.

Lima, T. A. (2000). Em busca dos frutos do mar os pescadores-coletores do litoral centro-sul do Brasil. *Revista Usp*. periodicos.usp.br.

Lopes, M. S., Bertucci, T. C. P., Rapagnã, L., de Almeida Tubino, R., Monteiro-Neto, C., Tomas, A. R. G., Tenório, M. C., et al. (2016). The path towards endangered species: Prehistoric fisheries in Southeastern Brazil. *PLoS One, 11*, e0154476.

Martin, L., Suguio, K., Flexor, J., & Azevedo, A. E. G. (1988). *Mapa geológico do Quaternário costeiro dos Estados do Paraná e Santa Catarina*. Departamento Nacional da Produção Mineral do Brasil (DNPM).

Mendes, A. B., Duarte, M. R., & Silva, E. P. (2018). Biodiversity of Holocene marine fish of the southeast coast of Brazil. *Biota Neotropica (Programa BIOTA/FAPESP), 18*. https://doi.org/10.1590/1676-0611-bn-2017-0394

Milheira, R. G. (2014). Zoólitos: algumas reflexões sobre as esculturas sambaquieiras. In J. J. Zocche, J. B. Campos, N. J. O. Almeida, & C. Ricken (Eds.), *Arqueofauna e Paisagem* (pp. 198–207). Habilis Press Editora.

Misund, O. A., Kolding, J., & Fréon, P. (2002). Fish capture devices in industrial and artisanal fisheries and their influence on management. In *Handbook of fish biology and fisheries* (Vol. 2, pp. 13–36). Blackwell Science Ltd.

Myers, N., Mittermeier, R. A., Mittermeier, C. G., da Fonseca, G. A., & Kent, J. (2000). Biodiversity hotspots for conservation priorities. *Nature, 403*, 853–858.

Peixe, S. P., Melo, J. C. F., Jr., & da Rocha Bandeira, D. (2007). Paleoetnobotânica dos macrorestos vegetais do tipo trançados de fibras encontrados no sambaqui Cubatão I, Joinville – SC. *Revista do Museu de Arqueologia e Etnologia, 17*, 211–222.

Prous, A. (1992). *Arqueologia brasileira* (UnB, Ed.). UnB.

Prous, A. (2006). *O Brasil antes dos brasileiros: a pré-história do nosso país*. Zahar.

Prous, A. (2018). As esculturas de pedra (zoólitos) e de osso dos sambaquis do Brasil meridional e do Uruguay. *Revista Memorare, 5*, 197–217.

Ribeiro, B. G. (1986). *Suma etnológica brasileira* (D. Ribeiro, Ed., Vol. 1). Vozes.

Roberts, C. M. (2003). Our shifting perspectives on the oceans. *Oryx: The Journal of the Fauna Preservation Society (Cambridge University Press), 37*, 166–177.

Robinson, M., De Souza, J. G., Yoshi Maezumi, S., Cárdenas, M., Pessenda, L., Prufer, K., Corteletti, R., et al. (2018). Uncoupling human and climate drivers of late Holocene vegetation change in southern Brazil. *Scientific Reports, 8*, 7800.

Rohr, J. A. (1984). Sítios Arqueológicos de Santa Catarina. *Anais do Museu de Antropologia da UFSC, XVI*, 77–167.

Schmitz, P. E., Verardi, I., de Masi, M. A., Rogge, J. H., & Jacobus, A. L. (1993). Escavações Arqueológicas do Pe. João Alfredo Rohr. O Sítio da Praia das Laranjeiras II. Uma Aldeia de Tradição Ceramista Itararé. *Pesquisas, Antropologia, 49*, 181.

Segal, B., Freire, A. S., Lindner, A., Krajewski, J. P., & Soldateli, M. (2017). *MAArE: Monitoramento Ambiental da Reserva Biológica Marinha do Arvoredo e Entorno* (B. Segal, A. S. Freire, A. Lindner, J. P. Krajewski, & M. Soldateli, Eds.). Universidade Federal de Santa Catarina (UFSC).

Soga, M., & Gaston, K. J. (2018). Shifting baseline syndrome: Causes, consequences, and implications. *Frontiers in Ecology and the Environment (Wiley), 16*, 222–230.

Tiburtius, G. A. E. (1996). *Arquivos de Guilherme Tiburtius I* (Fundação Cultural de Joinville (FCJ) e Museu Arqueológico de Sambaqui de Joinville (MASJ), Ed.). Editora Soluções.

Tiburtius, G., & Bigarella, I. K. (1953). Nota sobre os anzóis de osso da jazida páleo-etnográfica de Itacoara, Santa Catarina. *Revista do Museu Paulista, Nova Série, 7*, 381–387.

Tiburtius, G., Bigarella, I. K., & Bigarella, J. J. (1951). Nota prévia sobre a jazida paleoetnografica de Itacoara (Joinville estado de Santa Catarina). *Separata dos arquivos de Biologia, 5*, 315–346.

Tiburtius, G., Bigarella, I. K., & Bigarella, J. J. (1954). Contribuição ao estudo dos Sambaquis do litoral norte de Santa Catarina II: Sambaqui do Rio Pinheiros (n. 8). *Arquivos de Biologia e Tecnologia, 9*, 141–197.

Toniolo, T. de F, Giannini, P. C. F., Angulo, R. J., de Souza, M. C., Pessenda, L. C. R., & Spotorno-Oliveira, P. (2020). Sea-level fall and coastal water cooling during the Late Holocene in Southeastern Brazil based on vermetid bioconstructions. *Marine Geology, 428*, 106281.

Toso, A., Hallingstad, E., McGrath, K., Fossile, T., Conlan, C., Ferreira, J., da Rocha Bandeira, D., et al. (2021). Fishing intensification as response to Late Holocene socio-ecological instability in southeastern South America. *Scientific Reports, 11*, 23506. https://doi.org/10.1038/s41598-021-02888-7

Vieira, C. V. (2015). *Evolução paleogeográfica da planície costeira do extremo norte da ilha de São Francisco do Sul, Santa Catarina, Brasil* (N. O. Horn-Filho, & E. G. Barboza, Eds.). PhD in Geography, Universidade Federal de Santa Catarina.

Villagran, X. S., Klokler, D., Peixoto, S., DeBlasis, P., & Giannini, P. C. F. (2011). Building coastal landscapes: Zooarchaeology and geoarchaeology of Brazilian shell mounds. *The Journal of Island and Coastal Archaeology (Routledge), 6*, 211–234.

Wagner, G., Hilbert, K., Bandeira, D., Tenório, M. C., & Okumura, M. M. (2011). Sambaquis (shell mounds) of the Brazilian coast. *Quaternary International: The Journal of the International Union for Quaternary Research, 239*, 51–60.

Chapter 10
Exploitation, Management and Cultivation of Plants by Holocene Populations of the Cerrado of Brazil (South America)

Daniela Dias Ortega, Marco Madella, Jonas Gregorio de Souza, and Ximena Suarez Villagran

Abstract In the Cerrado, a Savannah-like environment of Brazil, ancient indigenous peoples have been exploiting, managing, and cultivating plants for thousands of years. This chapter presents an archaeological perspective with a summary of what is known of the archaeobotanical evidence in the region, both in terms of macro remains – for example, charcoal and seed fragments – and micro remains – such as starch, phytoliths, and pollen. We review the evidence from the first hunter-gatherer societies of the Early Holocene (around 12 kya) to the transition and development of ceramist groups that implemented diverse strategies for plant cultivation. The archaeological evidence suggests that a marked change in the occupation of the Cerrado took place around 4 kya, when settlements shifted from rock shelters to open-air sites or villages, with the first appearance of ceramics. In the eastern Cerrado, abundant desiccated macro-botanical remains have been documented after ca. 4 kya, including domesticated plants, such as maize. In recently excavated rock shelters, the evidence for ceramic production appears later, after ca. 2 kya, with pottery associated with storage pits containing plant remains of several domesticated and wild plants, including maize, manioc, and beans. The current summary of archaeobotanical evidence shows that, since the Early Holocene, the

D. D. Ortega · J. G. de Souza
CASEs – Department of Humanities, Universitat Pompeu Fabra (UPF), Barcelona, Spain
e-mail: daniela.d.ortega@upf.edu; jonas.gregorio@upf.edu

M. Madella
CASEs – Department of Humanities, Universitat Pompeu Fabra (UPF), Barcelona, Spain

ICREA, Barcelona, Spain

School of Geography, Archaeology, and Environmental Studies, University of the Witwatersrand, Johannesburg, South Africa
e-mail: marco.madella@upf.edu

X. S. Villagran (✉)
Museu de Arqueologia e Etnologia, Universidade de São Paulo, São Paulo, Brazil
e-mail: villagran@usp.br

A. C. Colonese, R. G. Milheira (eds.), *Historical Ecology and Landscape Archaeology in Lowland South America*, Interdisciplinary Contributions to Archaeology, https://doi.org/10.1007/978-3-031-32284-6_10

indigenous peoples of the Cerrado used wild plants typical of this biome with the subsequent addition of a diversity of cultivated and domesticated species by the Late Holocene. The diversity of species that have been identified in many archaeological sites highlights, not only the diversity of plant use by early human populations in the Cerrado, but also the enormous potential of future archaeobotanical studies in Central Brazil.

Keywords Archaeobotany · Plant management · Indigenous peoples · Holocene · South America

10.1 Introduction

The tropics of Central and South America have been the cradle of domestication of plants that play a major role in global economies (Piperno & Pearsall, 1998; Cassino et al., 2021). Pre-Columbian cultivation in the Neotropics was a heterogeneous phenomenon, with variable importance of cultivated plants depending on the region and period (Iriarte et al., 2020). Large-scale landscape transformations have been documented in forests and savannas, some of which show profound modifications such as ponds, canals, and raised fields (Erickson, 2008; McKey et al., 2010; de Souza et al., 2018; Colonese et al., 2020). Furthermore, growing archaeobotanical evidence attests to the antiquity of such a diversity of domesticated plants in South America, with maize, for example, present since ca. 6000–4000 BP in different parts of the continent (Hilbert et al., 2017; Iriarte et al., 2004; Maezumi et al., 2018; Lombardo et al., 2020; Colonese et al., 2020).

The Cerrado (Fig. 10.1), a biome covering 2 million km^2 in central and eastern Brazil, is a mosaic of woodland savannas and riverine forest corridors which connect the Amazon with other tropical rainforests on the Atlantic coast and La Plata Basin (Marques et al., 2020). The Cerrado climatic, geological, and sedimentary settings show variable conditions for the preservation of botanical macro-remains, with the numerous rock shelters of its easternmost regions having excellent preservation (Freitas & Rodet, 2010; Strauss et al., 2016; Villagran et al., 2017; Cassino et al., 2021) while the open-air sites having less than ideal conditions due to the chemical characteristics of the sediments.

Most of the debate about plant cultivation, domestication, and dispersal in lowland South America has focused on the Amazon rainforest and its margins. However, the Cerrado, a major vegetation system of the continent, still must be better understood both in terms of what plants were utilized in economic and dietary terms by the different societies since the Early Holocene, and the nature of ancient agriculture by the ceramist villagers of the Late Holocene.

To synthesize our understanding of the past plant use and management by the Holocene populations of the Cerrado, we present here a review of the archaeobotanical and paleoenvironmental data arising from the analysis of botanical macro and micro-remains from the biome's archaeological sites (Fig. 10.2). To

Fig. 10.1 Visual aspect of the Cerrado biome showing the open, savanna-like vegetation with interspersed trees and shrubs, palms, and herbaceous species (bottom right, *Ananas* cf. *ananassoides*). Serra do Espinhaço, Minas Gerais (Brazil). (Images by M Moore and M Madella)

complement such information, we also summarize the results of subsistence studies based on fauna and material culture.

10.2 Early Human Cultures in the Cerrado: Hunter-Gatherer Groups

The earliest archaeological sites of Central Brazil date from the Pleistocene/Holocene transition and from the Early Holocene (Miller, 1983; Schmitz & Barbosa, 1985; Vilhena-Vialou & Vialou, 1994; Neves et al., 2013; Viana et al., 2016; Strauss et al., 2020). The human populations of these sites had a foraging subsistence strategy, and the diet was based on the consumption of plants and small-to-middle-size animals (Bueno et al., 2013; Scheel-Ybert, 2001; Kipnis, 2002). Rock shelters are the most common type of settlement, and they were used for both domestic and funerary activities (Vilhena-Vialou & Vialou, 1994; Schmitz et al., 2004; Neves et al., 2013; Viana et al., 2016; Strauss et al., 2020). Open-air sites from this period

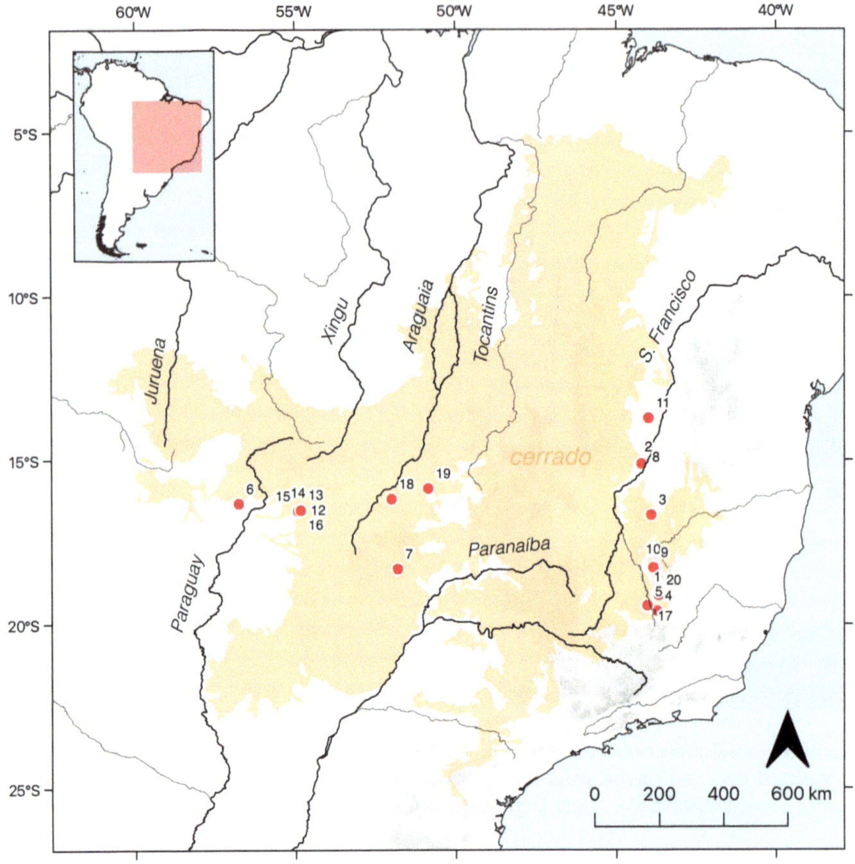

Fig. 10.2 Archaeological sites in the Cerrado of Central Brazil, where botanical remains have been identified. (1) BA-RC-28; (2) Lapa do Boquete; (3) Lapa dos Bichos; (4) Lapa Pintada; (5) Lapa do Caboclo; (6) Lapa da Hora; (7) Santana do Riacho; (8) Vereda III; (9) Lapa das Boleiras; (10) Cerca Grande; (11) Lapa do Santo; (12) Lapa Mortuária; (13) Harold Walter; (14) Lapa da Amoreira; (15) Lapa das Carrancas; (16) Lapa Grande de Taquaraçu; (17) GO-JA-01; (18) Ferraz Egreja; (19) Morro Solteiro; (20) Antiqueira; (21) Pacífico; (22) Santa Elina. (Map prepared by the authors in the open software QGIS 3.16 (https://qgis.org/) using free data from Natural Earth (https://www.naturalearthdata.com/) and WWF (https://www.worldwildlife.org/publications/terrestrial-ecoregions-ofthe-world))

have been rarely discovered in Central Brazil, except for the Sumidouro Site in Lagoa Santa (Araujo et al., 2013).

The rock shelter site **BA-RC-28**, located in the southwest area of the state of Bahia, is dated to the terminal Pleistocene, with a chronology starting at least at 15000 BP and an anthropic deposit containing limestone flakes, bone fragments, and mollusks (Schmitz et al., 1996; Bitencourt & Schmitz, 1994). In the later deposits dating from 6805 BP to the present, plant remains were recovered from several pits (Schmitz et al., 1996; Bitencourt & Schmitz, 1994) and included *Zea mays* (maize) cobs and straw from unidentified species; fruits and seeds of *Spondias tuberosa* (umbu tree); seeds from *Caryocar brasiliense* (pequi), *Talisia esculenta* (pitomba), fragmented seeds of *Syagrus oleracea* (gueroba palm) – probably intentionally broken to extract the edible parts; rinds from the fruits of *Hymenaea stigonacarpa* (jatobá); gourd fragments and *Gossypium* sp. (cotton), among others (Schmitz et al., 1996; Bitencourt & Schmitz, 1994). Excavators also found textiles (made from cotton) and a bag made of plant fibers (Schmitz et al., 1996; Bitencourt & Schmitz, 1994).

In the region of Minas Gerais, the study of rock shelter sites such as **Santana do Riacho**, **Lapa do Boquete**, **Lapa dos Bichos**, **Lapa do Santo**, and **Lapa Grande de Taquaraçu** (Bissaro, 2008; Perez, 2009; Mingatos & Okumura, 2016; Strauss et al., 2016, 2020; Chim, 2018), which were occupied from the Early to the Late Holocene, suggests that the hunter-gatherer populations exploited a diversity of roots and fruits complemented by small and middle size animals (Kipnis, 2002; Neves & Kipnis, 2004). There is no evidence for the exploitation of megafauna species, such as tiger (*Smilodon* sp.) and giant sloth (*Catonyx cuvieri*), although they were available in the ecosystem (Neves & Piló, 2003; Hubbe et al., 2013). Pleistocene megafauna was not the dietary staple of these populations, as opposed to other contemporaneous hunter-gatherers from the southern lowlands of South America, it is possible that they were sporadically hunted (Kipnis, 2002; Neves & Piló, 2003; Neves & Kipnis, 2004).

The **Lapa do Santo** rock shelter site, in Minas Gerais, contains more than 30 human burials exhibiting complex funerary rituals, and it was occupied from the Early to the Late Holocene (Strauss et al., 2016, 2020). A preliminary study of phytoliths and starch grains in sediments and lithic artefacts from the Early Holocene levels of the site indicates the good preservation of botanical micro-remains (Ortega, 2019). The phytoliths identified the presence of several plant groups such as Arecaceae (palms), Bambusoideae (bamboos), Poaceae (grasses), Aristidoideae, Chloridoideae, Cyperaceae, Zingiberales, as well as dicotyledonous plants (mostly trees and shrubs) (Ortega, 2019). A pollen study from the same site, with samples from a column dating between 2100 cal BP to the present day, identified a mix of plants typical of the flora of the Cerrado (Table 10.1) (Morrow, 2014 apud Ortega, 2019).

Starch analysis of the lithic instruments from the Early Holocene contexts of the **Lapa Grande de Taquaraçu** rock shelter revealed the exploitation of *Dioscorea* sp. (yams) and Arecaceae (palms) and it was able to demonstrate that the instruments were used to process both plant and animal resources (Angeles Flores, 2015; Angeles Flores et al., 2016; 2023).

In the rock shelter **Santana do Riacho**, botanical macro-remains highlighted the use of fruits, nuts, and fibers from a diversity of plants such as *Astrocaryum* sp. (palm), *Caryocar brasiliense*, *Hymenaea courbaril* (jatobá), *Symplocos* sp. (shrubs and trees). These remains were all associated with the human burials and often accompanied by artefacts made from plant materials, like beads from the fruits of *Scleria* sp. (pente de macaco), or vegetable braids, ropes, and tree bark parts that were used to swaddle newborns (Resende & Prous, 1991).

Table 10.1 Botanical taxa identified in the discussed archaeological sites

BA-RC-28	**6805 to present** *Caryocar brasiliense* (pequi), *Hymenaea stigonocarpa* (jatobá), *Talisia esculenta* (pitomba), *Spondias tuberosa* (umbu tree), Syagrus oleracea (gueroba palm), Fabaceae (legumes), Malvaceae (shrubs and trees), Poaceae (Schmitz et al., 1996; Bitencourt & Schmitz, 1994)
Santana do Riacho	**10000–8000 BP** *Caryocar brasiliense* (pequi), *Hymenaea courbaril* (jatobá), *Astrocaryum* sp. (palms), *Syagrus coronata* (licuri), *Symplocos* sp. (shrubs and trees) (Resende & Prous, 1991)
Lapa do Boquete	**1500–570 BP** *Zea mays* (maize), *Manihot esculenta* (manioc) (Freitas, 2004) *Bixa Orellana*, *Caryocar brasiliense*, *Eugenia dysenterica* (cagaita), *Fevillea trilobata* (Fava Santo Inácio), *Guazuma ulmifolia*, *Lagenaria* vulgaris (bottle gourd), *Spondias tuberosa*, *Spondias lutea* (cajá), *Syagrus coronata*, *Syagrus oleracea*, *Talisia esculenta*, *Zea mays*, *Annona* sp., *Byrsonima* sp., *Cnidoscolus* sp. (cansanção), *Enterolobium* sp. (tamboril), *Gossypium* sp. (cotton), *Hymenaea* sp., *Manihot* sp., *Passiflora* sp., *Phaseolus* sp. (beans), *Piptadenia* sp. (angico), *Scleria* sp. (pente de macaco), *Solanum* sp. (jurubeba), *Struthanthus* sp. (herb), *Sterculia* sp. (tree) (Resende & Cardoso, 1996; Freitas, 2003, 2004; Freitas & Rodet, 2010)
Lapa dos Bichos	**750–150 BP** *Arachis hypogaea* (peanut), *Manihot* cf. *esculenta*, *Phaseolus vulgaris* (beans), *Zea mays*., *Bixa orellana*, *Caryocar brasiliense*, *Guazuma ulmifolia* (mutamba), *Mucuna sloanei* (tree), *Spondias mombin* (hog plum), *Spondias tuberosa*, *Chromolaena* sp. (herb), *Cucurbita* sp. (squash), *Lagenaria* sp. (bottle gourd), *Passiflora* sp. (passion flower), *Lasiacis* sp. (herb), *Trichogonia* sp. (herb), Solanaceae (herb), Eupatorieae (herb), Gnaphalieae (herb), Vernonieae (herb), Asteraceae (herb), Bromeliaceae (herbs), Poaceae (grasses), *Alibertia edulis* (puruí), *Syagrus* cf. *oleracea* (gueroba palm), *Alibertia* sp. (tree), *Annona* sp. (fruit trees), *Byrsonima* sp. (fruit trees), *Ficus* sp. (trees), *Hymenaea* sp. (jatobá) **2000–750 BP** *Manihot* cf. *esculenta*, *Zea mays*, *Spondias mombin*, *Spondias tuberosa*, *Lagenaria* sp., *Passiflora* sp., *Syagrus* cf. *oleracea*, *Hymenaea* sp., Solanaceae **4250–2000 BP** *Caryocar brasiliense*, *Spondias tuberosa*, *Manihot* cf. *esculenta*, cf. *Phaseolus lunatus*, *Lagenaria* sp., *Syagrus* cf. *oleracea*, *Hymenaea* sp., Solanaceae **6500–4250 BP** *Hymenaea* sp., *Syagrus* cf. *oleracea*, Poaceae (Shock, 2010; Shock et al., 2013)
Lapa do Santo	**12700–8000 BP** *Ipomoea batatas* (sweet potato), *Dioscorea* sp., Bambusoideae (bamboos), Zingiberales (herbs), Arecaceae (Ortega, 2019) **2100 cal. BP to present** 6–12 cm – Arecaceae, Asteraceae, Chenopodiaceae/Amaranthaceae, Malpighiaceae, Melastomataceae, Myrtaceae, Poaceae 12–30 cm – Euphorbiaceae, Malpighiaceae, Malvaceae, Melastomataceae, Myrtaceae, Nyctagenaceae, Poaceae (Morrow, 2014 apud Ortega, 2019)
Lapa Grande de Taquaraçu	**Early Holocene** *Dioscorea* sp. (yams), Arecaceae (palms) (Angeles Flores, 2015; Angeles Flores et al., 2016; 2023)

(continued)

Table 10.1 (continued)

Lapa Pintada	**1200–800 BP** *Zea mays*, *Caryocar brasiliense*, *Guazuma ulmifolia*, *Manihot* cf. *esculenta*, *Lagenaria* sp., *Lasiacis* sp., *Passiflora* sp., *Syagrus* cf. *oleracea*, *Annona* sp., *Byrsonima* sp., *Hymenaea* sp., *Myrciaria* sp. (fruit shrubs and trees), Vernonieae, Asteraceae, Solanaceae **4400–4300 BP** *Caryocar brasiliense*, *Manihot* cf. *esculenta*, *Syagrus* cf. *oleracea*, *Lagenaria* sp., *Passiflora* sp., *Hymenaea* sp. **7000 BP** *Hymenaea* sp., *Syagrus* cf. *oleracea* (Shock, 2010)
GO-JA-01 (Gruta do Diogo)	**Early Holocene** (10.600–9000 BP) *Byrsonima crassifolia* (murici), *Cordia trichotoma* (louro-pardo), *Syagrus oleracea*, *Attalea speciosa* (babaçu palm), *Anacardium* sp. (fruit trees), *Hymenaea* sp., *Syagrus* sp. (palms), Fabaceae **Middle Holocene** (around 8500) *Arachis hypogaea*, *Attalea speciosa*, *Butia capitata* (palm), *Syagrus oleracea*, *Guazuma ulmifolia*, *Hymenaea stigonocarpa*, *Syagrus flexuosa* (acumã palm), *Anacardium* sp., *Syagrus* sp. **Late Holocene** (1500–200 BP) *Arachis hypogaea*, *Annona montana* (araticum), *Byrsonima crassifolia*, *Cordia trichotoma*, *Hymenaea stigonocarpa*, *Luehea divaricate* (açoita-cavalo), *Licania tomentosa* (oiti tree), *Prunus myrtifolia* (peach), *Syagrus flexuosa*, *Syagrus oleracea*, *Syagrus romanzoffiana* (jerivá palm), *Zea mays*, *Anacardium* sp., *Gossypium* sp. (cotton), Lecythidaceae (chestnut), *Orbignya* sp. (palm), *Oryza* sp. (rice), *Lagenaria* sp., *Passiflora* spp. (passion fruit), Fabaceae (Schmitz et al., 1989, 2004)
Santa Elina	**400 cal BP** *Licania* sp. (trees), *Sebastiania* sp. (trees), Apocynaceae (shrubs and trees), Asteraceae, Combretaceae (trees), Fabaceae (legumes), Mimosoideae (legumes), Melastomataceae (shrubs and trees), Myrtaceae (shrubs and trees), Rubiaceae (shrubs and trees), Rutaceae (shrubs and trees), Poaceae **4000–2000 cal BP** *Acrocomia aculeata* (macauba palm), *Anacardium humile* (cashew), *Attalea phalerata* (uricuri), *Attalea vitrivir* Zona (palm), *Caryocar brasiliense*, *Dipteryx alata* Vogel (baru), *Guazuma ulmifolia*, *Guettarda viburnoides* (tree), *Hymenaea stigonocarpa* (jatobá), *Hymenaea courbaril*, *Protium heptaphyllum* (fruit tree), *Manihot* sp., *Ficus* sp., *Inga* spp. (ice-cream-bean), *Lecythis* sp. (shrubs and trees), *Matayba* sp. (shrubs and trees), *Neea* sp. (shrubs and trees), *Spondias* sp. (shrubs and trees), Anacardiaceae (fruit trees), Annonaceae (fruit trees), Apocynaceae (shrubs and trees), Arecaceae, Bromeliaceae, Fabaceae, Myrtaceae, Rubiaceae, Sapotaceae (shrubs and trees), Poaceae (Chaves, 2005; Scheel-Ybert & Solari, 2005; Bachelet, 2014; Scheel-Ybert & Bachelet, 2020)

From the sites of **Lapa dos Bichos** and **Lapa Pintada**, also in Minas Gerais, many plant taxa were identified from the abundant macro-remains recovered (Shock, 2010; Shock et al., 2013). The later period of occupation of Lapa dos Bichos, with a chronology of 750–150 BP, had a diverse assemblage of plants, presented in Table 10.1, among them domesticated species such as *Arachis hypogaea* (peanut), *Manihot* cf. *esculenta* (manioc), *Phaseolus vulgaris* (beans), *Zea mays*, *Cucurbita* sp. (squash), *Lagenaria* sp. (bottle gourd), and *Bixa orellana* (achiote).

Typical Cerrado plants were also identified, such as *Caryocar brasiliense* and *Hymenaea* sp. (jatobá), also palms, fruit trees and other arboreal and herbaceous species (Table 10.1) (Shock, 2010; Shock et al., 2013).

From deposits of Lapa dos Bichos dated at 2000–750 BP, the excavations recovered *Zea mays*, *Manihot* cf. *esculenta*, *Syagrus* cf. *oleracea*, *Spondias mombin*, *Spondias tuberosa*, *Hymenaea* sp., *Lagenaria* sp., *Passiflora* sp. and Solanaceae (Shock, 2010; Shock et al., 2013). The layers with a chronology of 4250–2000 BP were characterised by the presence of cf. *Manihot esculenta* and *Caryocar brasiliensis*, among others (Table 10.1) (Shock, 2010; Shock et al., 2013). The most ancient deposits, dated at 6500–4250 BP, had a less rich assemblage, with *Syagrus* cf. *oleracea*, *Hymenaea* sp., and wild grasses (Poaceae) (Shock, 2010; Shock et al., 2013).

At Lapa Pintada, the later period, dated at 1200–800 BP, had the presence of *Zea mays*, *Manihot* cf. *esculenta*, among others (Table 10.1). Also, there were typical Cerrado plants such as *Caryocar brasiliense*, *Guazuma ulmifolia* and *Hymenaea* sp. as well as fruit trees and herbs (Table 10.1) (Shock, 2010). In the deposits with a chronology of 4400–4300 BP the excavations recovered *Manihot* cf. *esculenta*, *Caryocar brasiliense*, *Syagrus* cf. *oleracea*, *Lagenaria* sp., *Passiflora* sp. and *Hymenaea* sp. (Shock, 2010) while the oldest sediments, dated at around 7000 BP, were much less diverse with macro-remains only from *Hymenaea* sp. and *Syagrus* cf. *oleracea* (Shock, 2010).

In the southwest of Goiás, three episodes of occupation with macrobotanical remains were identified in the rock shelter **GO-JA-01** (Schmitz et al., 1989, 2004). The oldest phase of occupation, the Paranaíba phase, covers part of the Early Holocene, with dates from 10600 to 9000 BP (non-calibrated). These deposits are characterized by lithic technology with plan-convex unifacial tools (Schmitz et al., 1989, 2004; Ramos & Viana, 2019). The human groups frequenting the rock shelter during the Early Holocene used native Cerrado plants, including the fruits from different palm species (see Table 10.1), *Hymenaea* sp. and Fabaceae (Schmitz et al., 2004). During the Middle Holocene, in GO-JA-01 a new lithic technology appears (phase Serranópolis) that is developed on flakes recycled from the debitage of the previous occupation phase (Ramos & Viana, 2019). In this period, the botanical macro-remains show the presence of *Arachis hypogaea* (peanut), *Anacardium* sp., (cashew), and *Hymenaea stigonocarpa*, palms and other native Cerrado plants (Table 10.1) (Schmitz et al., 2004).

In the Late Holocene, in GO-JA-01, the so-called Jataí phase, dated at approximately 900 BP and characterized by lithic instruments and ceramics, is then followed by the most recent occupation in connection to Tupiguarani ceramics (Schmitz et al., 1989, 2004; Ramos & Viana, 2019). These occupations are characterized by groups cultivating several crops like *Arachis hypogaea*, *Oryza* sp. (rice), *Zea mays*, *Lagenaria* sp., and *Gossypium* sp. (cotton) (Schmitz et al., 2004). Although cultivated plants are common, native wild plants are still very important in the resource exploitation strategies of such groups as suggested by the recovery of macro remains from several other species as well as a variety of palms and other arboreal species (Table 10.1) (Schmitz et al., 2004).

In Mato Grosso, the rock shelter **Santa Elina** was occupied from the Pleistocene to the Late Holocene and the anthropic deposits yielded lithic industry, body adornments, braided fibers and well-preserved botanical macro-remains (Vilhena Vialou, 2011; Scheel-Ybert & Bachelet, 2020; Ceccantini & Gusella, 2001). No ceramic, even in the later occupation layers, was found at the site (Vialou et al., 2017). Anthracological and carpological analysis in Santa Elina identified in the oldest layers dated at 4000–2000 cal BP many plant taxa, such as *Manihot* sp. and Cerrado species (Table 10.1) but also Fabaceae (legumes), Bromeliaceae and Poaceae (Scheel-Ybert, 2001; Bachelet & Scheel-Ybert, 2017; Scheel-Ybert & Bachelet, 2020; Chaves, 2005; Scheel-Ybert & Solari, 2005). The more recent contexts of Santa Elina (dated from 400 cal BP to the present) preserved macro-remains from shrubs and trees (Table 10.1) as well as Mimosoideae (legumes) and grasses (Poaceae) (Scheel-Ybert, 2001; Bachelet & Scheel-Ybert, 2017; Scheel-Ybert & Bachelet, 2020; Chaves, 2005; Scheel-Ybert & Solari, 2005).

Isotopic analysis of human and faunal remains from many of the above sites also supports the hypothesis that these human populations had an omnivore, broad spectrum diet, but with an important intake of plant food (Hermenegildo, 2009). A study of health and lifestyle conducted on human remains from the sites of **Lapa do Santo**, **Lapa das Boleiras**, **Cerca Grande**, **Harold Walter**, **Lapa da Amoreira**, **Lapa de Carrancas**, **Lapa Mortuária**, and **Santana do Riacho** suggests low mobility, the occurrence of infections and important levels of stress, all of which are characteristic normally associated with agricultural societies even though the evidence from all the other archaeological data does not support an economy based on agriculture for such an early period (Da-Gloria & Larsen, 2014). It has been therefore suggested that these populations would have had a high carbohydrate consumption – higher than other forager societies – and that such carbohydrates were sourced from wild tubers, seeds, and fruits (Da-Gloria & Larsen, 2014), as clearly highlighted by the archaeobotanical record.

10.3 The Villagers and Cultivation/Agriculture in the Cerrado

In Central Brazil, the first evidence of pottery appears with the Una Tradition, (Schmitz et al., 1989) and it is hypothesized that Una ceramics were adopted or developed by hunter-gatherer groups that already practised some form of cultivation (Schmitz et al., 1986; Wüst, 1983, 1998; Moreira, 1981). This hypothesis is supported by the location of settlements in the landscape, predominantly in areas of forest and Cerrado with access to a diversified set of resources and associated with clayey soils, which are more suited for cultivation (Schmitz & Barbosa, 1985; Wüst, 1983). The cultivation of plants in Central Brazil is seen as the agent that would have eventually pushed for the technological development of pottery production (Schmitz et al., 1986; Wüst, 1983; Moreira, 1981). In the same way, the cultivation of plants could be seen as the cause behind the changes observed in the lithic assemblage, with the assemblage shifting from abundant knapped and *façonnated* instruments related to hunting, cutting and scraping (both animal and vegetal tissues), to polished stone tools and other instruments related to grinding (Schmitz & Barbosa, 1985).

The **Una Tradition** appears in Central Brazil at around 2500 BP, in the regions of the Upper Araguaia and Middle Tocantins river basins (Schmitz et al., 1986; Vilhena-Vialou & Vialou, 1994; Robrahn-González, 1996). Una sites are found also in the low Paranaíba and Rio Vermelho river basins, in Goiás, Mato Grosso, Mato Grosso do Sul, Bahia, Tocantins and Minas Gerais states (Schmitz & Barbosa, 1985; Robrahn-González, 1996). Una sites tend to be in areas with rugged relief and Cerrado biome, as rock shelters for the more ancient settlements and open-air sites for the more recent ones (Schmitz & Barbosa, 1985; Schmitz et al., 1989). The ceramic contains *caraipé* (the bark of Bignoniaceae trees) as temper and the containers are smaller (with capacities of less than a liter) and thinner than those of later ceramist traditions; they have simple or inflected contours, and the forms are shallow plates, bowls, and small vessels – which rarely have plastic decorations (Schmitz & Barbosa, 1985; Schmitz et al., 1989; Robrahn-González, 1996). The pottery is not found in high density in the archaeological sites suggesting, together with its typological characteristics, that the vessels had a domestic use. This is also supported when comparing Una pottery to the abundance and dimensions of the containers of later ceramic traditions, which were also used for more diverse social activities (Robrahn-González, 1996; Schmitz et al., 1989).

Rock shelters such as **Lapa do Boquete**, in the eastern Cerrado, were initially investigated in the 1970s when excavations unveiled deep stratigraphic sequences with dates as early as ca. 10.000 BP (Bird et al., 1991). A marked change in the occupation of the sites happens at around 4000 BP when ceramics of the Una tradition make their first appearance together with abundant desiccated remains of plants and plant parts (Bird et al., 1991).

In Lapa do Boquete, botanical macro-remains were found in storage pits lined with basketry, wood, or palm leaves (Resende & Cardoso, 2009). The plant remains included *Zea mays* cobs, *Syagrus oleracea* nuts, *Manihot esculenta*, *Arachis hypogaea*, *Phaseolus* sp. (beans), *Gossypium* sp., tobacco, and chilies (Resende & Cardoso, 2009; Freitas & Rodet, 2010). In the pits there were also plants that are known to have insecticide properties, such as *Annona* and *Styrax*, and which could have been used for plague control (Resende & Cardoso, 2009). The materials have been dated from around 1500 to 500 BP, with the dates obtained directly on maize cobs being later than ca. 1000 BP (Freitas & Rodet, 2010). Other plants identified by macro remains are *Caryocar brasiliense*, *Lagenaria vulgaris* (bottle gourd), *Bixa Orellana*, *Syagrus coronata*, *Fevillea trilobata* (fava Santo Inácio), *Spondias lutea* (cajá), *Guazuma ulmifolia*, *Talisia esculenta*, *Spondias tuberosa*, *Hymenaea* sp., *Enterolobium* sp. (tamboril), *Solanum* sp. (jurubeba), *Scleria* sp. (pente de macaco), *Piptadenia* sp. (angico), *Eugenia dysenterica* (cagaita), *Byrsonima* sp., *Passiflora* sp., *Cnidoscolus* sp. (cansanção), *Struthanthus* sp. and *Sterculia* sp. (Resende & Cardoso, 1996; Freitas, 2003, 2004; Freitas & Rodet, 2010).

Table 10.2 Botanical taxa identified in each site

Lapa da Hora	*Zea mays*, *Manihot esculenta* (Freitas, 2003, 2004)
Lapa do Caboclo	*Zea mays* (Freitas, 2003)
Ferraz Egreja	5460 ± 40 BP to 200 ± 70 BP *Curatella americana*, *Humiria balsamifera*, *Anacardium* sp., *Anadenanthera* sp., *Aspidosperma* sp., *Albizia* sp., *Allophylus* sp., *Astronium* sp., *Bauhinia* sp., *Brosimum* sp., *Byrsonima* sp., *Cespedesia* sp., *Copaifera* sp., *Cochlospermum* sp., *Cupania* sp., *Chrysophyllum* sp., *Dasyphyllum* sp., *Eschweilera* sp., *Eugenia* sp., *Ficus* sp., *Himatanthus* sp. (trees), *Hymenaea* sp., *Hirtella* sp., *Mabea* sp., *Tachigali* sp., *Myrsine* sp., *Magonia* sp., *Nectandra/Ocotea*, *Pterodon* sp., *Pouteria* sp., *Qualea* sp., *Sterculia* sp., *Spondias* sp., *Sweetia* sp., *Terminalia* sp., *Vochysia* sp., *Virola* sp., Arecaceae, Fabaceae, Sapindaceae, cf. *Xylopia*, cf. *Tococa*, cf. *Siparuna* (Bachelet, 2014, 2016)
Antiqueira	1300–500 BP *Curatella americana*, *Goupia glabra*, *Aspidosperma* sp., *Allophylus* sp., *Anadenanthera* sp., *Bagassa* sp., *Copaifera* sp., *Chrysophyllum* sp., *Hirtella* sp., *Nectandra* sp., *Pterodon* sp., *Peschiera* sp., *Pouteria* sp., *Sweetia* sp., *Sterculia* sp., *Luehea* sp., *Virola* sp., Bambusoideae, Fabaceae (Bachelet, 2014, 2016)
Morro Solteiro	1000–320 BP *Curatella americana*, *Astronium* sp., *Aspidosperma* sp., *Anadenanthera* sp., *Andira/Ormosia*, *Byrsonima* sp., cf. *Cecropia* (embaúba), *Hymenaea* sp., *Sweetia* sp. (Bachelet, 2014, 2016)
Pacífico	1620 ± 30 BP to 1000 BP *Curatella americana*, *Dipteryx alata*, *Aspidosperma* sp., *Bagassa* sp., *Bauhinia* sp., *Byrsonima* sp., *Chrysophyllum* sp., *Heteropterys* sp., *Inga* sp., *Machaerium* sp. (trees), *Rollinia* sp., Sapindaceae (Bachelet, 2014, 2016)
Vereda III	*Zea mays*, *Ipomoea batatas* (Gardiman, 2014; Rodrigues & Gardiman et al., 2016)

In **Lapa do Boquete**, **Lapa do Caboclo**, **Lapa da Hora**, **Malhador**, and **Tikão** rock shelters, ceramics appears only after ca. 2000 BP, accompanied by storage pits with botanical macro-remains, which include maize, manioc, beans, and several other domesticated and wild plants (Table 10.2) (Freitas, 2003; Freitas & Rodet, 2010). *Zea mays* and *Manihot esculenta* were also identified by starch analysis at Lapa do Boquete and Lapa da Hora (Freitas, 2004), and by a DNA analysis at Lapa do Caboclo (Freitas, 2003).

The **Aratu Tradition** appears at around 1000 BP and is found from the coast of the states of Pernambuco, Bahia, and Espírito Santo, to the Araguaia, Tocantins, and Paranaíba river basins in Mato Grosso, Mato Grosso do Sul, Goiás, Minas Gerais, and São Paulo states (Calderón, 1971; Dias Júnior, 1971; Wüst, 1983; Perota, 1974; Wüst & Barreto, 1999). The origins of this cultural tradition are probably in the southern region of Amazonia (Brochado, 1981; Robrahn-González, 1996). The sites are open-air settlements, mostly in a forested setting of the Cerrado (Oliveira & Viana, 1999; Soares, 2013) and consist of large ring villages (Wüst & Barreto, 1999).

The pottery of the Aratu Tradition is characterised by mineral non-plastics in the paste, and in some contexts, an increase in the use of *caraipé* through time (Schmitz & Barbosa, 1985). The vessels are of bigger dimensions in comparison to other contemporary ceramist traditions, with conical, spherical, or ellipsoid pear-shaped forms used for the storage of liquids and, in secondary use, as urns for burials (Schmitz & Barbosa, 1985; Wüst & Barreto, 1999). Geminated-double form and artifacts classified as stamps and pipes have been also found (Schmitz & Barbosa, 1985). It is assumed that manioc was not cultivated by the Aratu ceramists, due to the absence of the typical pottery plates used for tuber processing (Schmitz & Barbosa, 1985). Thus, and based on ethnographic similitudes, it has been hypothesized that maize could have had a central role in their subsistence (Schmitz & Barbosa, 1985; Soares, 2013).

The transition from the Una to the Aratu tradition involves a change in the settlement system, from ephemeral rock shelter camps to large and permanent ring villages. This is frequently interpreted as reflecting population growth, a more sedentary lifestyle and greater reliance on agriculture (Oliveira & Viana, 1999; Robrahn-González, 1996; Wüst & Barreto, 1999). The transformation in artefact assemblages (large, conical vessels, suggested to have been used for fermentation; Fernandes, 2003) and settlement patterns (large circular villages with central plazas, reaching 50 hectares; Wüst & Barreto, 1999) of the Aratu tradition is suggestive of large, sedentary populations, who potentially relied more on plant cultivation than their predecessors, the makers of Una ceramics. Ethnographic analogies indicate that Aratu villages would have supported a population of about 150–2000 people (Wüst, 1983; Wüst & Barreto, 1999).

The **Aratu-Sapucaí Tradition** relates to groups in North East and Central Brazil, dated between 1650 and 150 BP, with ceramic assemblages very similar to the ones of the Aratu Tradition (Dias Júnior, 1975; Schmitz & Barbosa, 1985; Prous, 1992; Schmitz & Rogge, 2008; Afonso, 2009; Alves, 2018). Evidence of cultivation in the Aratu-Sapucaí Tradition was recovered from the site of **Vereda III**, in Minas Gerais. A study of plant micro-remains from large-sized ceramics identified starch and evidence of fermentation, suggesting the production of an alcoholic beverage made from maize (Gardiman, 2014; Rodrigues & Gardiman, 2016). Starch of sweet potatoes (*Ipomoea batatas*) was also identified in other vessel types (Gardiman, 2014; Rodrigues & Gardiman, 2016).

The **Uru Tradition** of Mato Grosso and Goiás chronologically covers from 1150 to 250 BP and is hypothesized to be of Amazonian origin (Wüst, 1975, 1998; Oliveira & Viana, 1999; Wüst & Barreto, 1999; Robrahn-Gonzalez, 1996). This tradition survived until current times and is represented in the ceramics produced by the indigenous Iny people – commonly known as "Karajá" (Wüst, 1975; Nimuendajú, 1981). The former Uru settlements are found around the Tocantins and Araguaia river basins, in open and relatively plain areas of Cerrado with soils characterised by low fertility (Schmitz & Barbosa, 1985; Wüst, 1983; Rubin et al., 2019). The ceramics of the Uru Tradition include the widespread use of *caraipé* as temper (Schmitz & Barbosa, 1985) and the most common typologies are plates and roasters, big shallow bowls with reinforced edges and jars for storage, fermentation and conservation of drinks (Schmitz & Barbosa, 1985; Robrahn-González, 1996; Ortega, 2016; Rubin et al., 2019). Uru subsistence was based on the cultivation of manioc and maize and complemented by fishing (Schmitz et al., 1976; Schmitz & Barbosa,

1985). The largest villages are estimated to have sheltered between 1000 and 2000 people (Wüst, 1992, 1998).

The **Tupiguarani Tradition** is the geographically most widespread cultural tradition spanning from the Amazon to southern Brazil and many other areas of South America (Oliveira & Viana, 1999). The origins of this tradition can be traced back to Amazonia, whence they expanded to forested environments throughout the eastern lowlands of South America (Brochado, 1984; Noelli, 1996; Almeida & Neves, 2015; Souza et al., 2021). In Central Brazil, this tradition is not very represented – in comparison with the Uru and Aratu traditions – and it is dated from 650 BP (Fernsteseifer & Schmitz, 1975; Wüst & Barreto, 1999), while in the region of Serra da Mesa it is dated from 1050 BP (Pontim, 2013). The sites are located in the Cerrado or in gallery forests, in the plains or in gentle slope areas. They have been occupied by many groups through time and are sometimes built on top of earlier settlements of other pottery traditions (Wüst & Barreto, 1999; Oliveira & Viana, 1999; Pontim, 2013). The pottery of the Tupiguarani Tradition is known for its decorations, mostly painted in Goiás and Mato Grosso but largely plastic decorations in Mato Grosso do Sul (Martins & Kashimoto, 1998). The pottery contains *caraipé* and grog (crushed ceramic material) in the paste, and the forms have complex contours and shoulders, convex or plain bases, and reinforced edges (Schmitz & Barbosa, 1985; Robrahn-González, 1996). The subsistence of the Tupiguarani was based on plant cultivation, possibly mainly manioc (Schmitz & Barbosa, 1985).

It is usually assumed in the literature that the Una and Aratu traditions of the Cerrado were more heavily dependent on maize cultivation than their Amazonian neighbors (Tupiguarani and Uru traditions), whose diet was supposedly focused on manioc (Oliveira & Viana, 1999; Prous, 1992; Robrahn-González, 1996). For example, it is suggested that the Aratu Tradition did not cultivate manioc, because plates – forms normally associated with the processing of manioc – were not a typology present in the pottery assemblage of this tradition (Schmitz & Barbosa, 1985). However, such assumptions, based on indirect evidence (repertoire of vessel typology and ethnographic analogy) are still lacking direct confirmation from archaeobotanical or isotopic evidence.

In the region of Cidade de Pedra, Mato Grosso, charred wood analysis (anthracology) was carried out in four archaeological rock shelters (Bachelet, 2014, 2016). The site **Ferraz Egreja**, which was continuously occupied from 6000 to 100 BP, with the first ceramics dated at 3000 BP (Bachelet, 2016; Monteiro, 2005; Vilhena Vialou, 2006). The sites of **Antiqueira**, **Morro Solteiro,** and **Pacifico** were occupied from 1000 BP and the excavation uncovered lithic industries, ceramics, hearths, and carbon concentrations (Bachelet, 2016). Sites whose pottery was analysed show a variety of industries through time, with not all of the vessels fitting in a single "tradition" (Monteiro, 2005). Indeed, the site of Ferraz Egreja has a few vessels with characteristics of the Tupiguarani tradition, while the Antiqueira site has vessels with characteristics of the Uru tradition (Monteiro, 2005).

In the sites **Ferraz Egreja**, **Antiqueira**, **Morro Solteiro and Pacífico**, macroremains from many species listed in Table 10.2 were identified (Bachelet, 2014, 2016). In the Ferraz Egreja site, this evidence comes from contexts dated 5460 ± 40

BP to 200 ± 70 BP (Bachelet, 2014, 2016). In the Antiqueira **site**, the chronology is situated between 1300 and 500 BP; in the Morro Solteiro site, between 1000 and 320 BP; and in the Pacífico site, between 1620 ± 30 BP and 1000 BP. The latter has a plant assemblage including a diversity of tree species (Table 10.2) (Bachelet, 2014, 2016).

10.4 A Discourse Between Plant Processing and Lithic Instruments

Archaeobotany, the analysis of plant macro and micro-remains, is an efficient tool for understanding the plant exploitation strategies of past societies. Notwithstanding pre and post-depositional taphonomic processes, the recovery of botanical remains is routinely carried out from archaeological sites around the world albeit some depositional environments can be more challenging for the preservation of such remains, especially when not carbonised. The acid soils of the Cerrado biome are not very favourable for the preservation of organic remains especially when not carbonized (Schmitz & Barbosa, 1985). However, the drier conditions of some sites – specifically the rock shelters – can result in very good preservation by desiccation (Bird et al., 1991).

The study of lithic industry assemblages can also be used to understand how plants were exploited by earlier societies. Lithics are an important constituent of the Cerrado archaeological record, but they are still awaiting in-depth studies, and therefore their function is often still unclear (Moreno de Sousa, 2014). However, there are some exceptions in the archaeological sites of Itaparica and Lagoa Santa, where the function of the lithic instruments is partially understood (Lourdeau, 2010; Rodet et al., 2011; Bueno, 2012; Pugliese, 2007; Moreno de Sousa & Araujo, 2018).

The Itaparica tradition (Schmitz et al., 1989, 2004) or technocomplex (Lourdeau, 2010), dated from the Early Holocene and found in Central and Northeastern Brazil, is characterized by the presence of unifacial plano-convex tools (nicknamed "slugs") (Schmitz et al., 1989, 2004; Moreno de Sousa, 2016a, b) or, in another definition, unifacially shaped artefacts (Lourdeau, 2012). Although this typology has been used to characterize the tradition, there is substantial technological and morphological heterogeneity among the tools (Fogaça & Lourdeau, 2008; Rodet et al., 2011; Moreno de Sousa, 2016a, b). Within this typology, it was suggested that a variety of "*slug*" was used for processing starchy plants, and plant micro-remains analysis from the active portions of the tools revealed the presence of starch, confirming such assumption (Angeles Flores et al., 2016; 2023).

The lithic industry of Lagoa Santa, dating from the Early to the Late Holocene, has different characteristics from other contemporary industries (Moreno de Sousa, 2014; Moreno de Sousa & Araujo, 2018). The industry is mainly defined by small flakes and cores of hyaline quartz, which were also polished for the production of axes (Pugliese, 2007; Bueno, 2012; Moreno de Sousa, 2014; Moreno de Sousa &

Araujo, 2018). Their functionality is considered to have been related to activities such as scraping and cutting soft materials, like non-woody plants and plant parts, and/or medium to small size animals (Pugliese, 2007; Moreno de Sousa, 2014). The use of such implements for plant processing has been confirmed by ongoing studies of botanical micro-remains in the Early Holocene contexts of Lagoa Santa (Angeles Flores, 2015; Ortega, 2019; Angeles Flores et al., 2023).

At the site of Lapa do Santo, a study of botanical micro-remains from the lithic implements was able to identify well-preserved starch and phytoliths (Ortega, 2019), also highlighting that the preservation of such micro-remains was much higher on the lithic implements than in the surrounding sediments. The micro-remains allowed for the identification of a variety of plants such as Araceae (the arum family), Arecaceae (palms), and Poaceae (grasses) as well as *Ipomoea batatas* (sweet potato) and *Dioscorea* sp. (yams) (Ortega, 2019). The significant presence of starch from several plants with underground starchy organs (known as USOs in the archaeobotany literature) suggests that such parts were cleaned and cut using the various lithic tools.

The Culture-Historical theoretical framework that characterised Brazilian archaeology in the past attempted to correlate ancient pre-colonial indigenous pottery "traditions" with the ceramics produced by contemporary indigenous societies, using direct analogy (Schmitz & Barbosa, 1985; Oliveira & Viana, 1999). Moreover, specific functions were suggested on the basis of similarities in technology, morphology or decoration (Schmitz & Barbosa, 1985; Oliveira & Viana, 1999).

For the Una Tradition, the earliest ceramic tradition in the Cerrado, technological studies indicate that the functionality of the vessels is related to serving and ritual activities because of their thin walls and small dimensions (Schmitz & Barbosa, 1985; Schmitz et al., 2004; Robrahn-Gonzalez, 1996). However, there are no studies about the Una pottery function through direct evidence from micro-botanical remains (Schmitz & Barbosa, 1985; Schmitz et al., 2004). For the Uru Tradition, plates (*prato* or *assador*) and bowls (*tijela*) are pottery typologies that are considered diagnostic of the tradition. Both have a flat base, but plates are more open and shallower than bowls (Chmyz, 1976; Schmitz & Barbosa, 1985). These containers are also considered diagnostic of the processing of manioc for the preparation of flour (Chmyz, 1976; Schmitz & Barbosa, 1985; Robrahn-González, 1996). The same ceramic typologies are present in the Orinoco Valley in Venezuela, where their function is also hypothesized to be related to manioc processing. However, a study on the starch grains recovered from pottery and lithics assumed to have been used for the processing of manioc (flakes for graters and rock-plates; Perry, 2001, 2004) did not show the presence of *Manihot esculenta* starch. Instead, the starch assemblage indicated that such artefacts were used for the processing of a variety of tubers as well as corn and palm fruits (Perry, 2001, 2004). Therefore, the combination of a lack of macro-remains and the use of form-function analogies with contemporary traditional societies hampered a correct interpretation of the use of some artefacts as well as a proper understanding of the plant resources exploited and the diet of the early agricultural groups of the Cerrado. An approach that can help ameliorate such a problem is the recovery and analyses of a diverse array of plant evidence originating from the artefacts and the sediments of the numerous Cerrado sites. These would include starch grains, phytoliths, and pollen (Fullagar, 2006, 2009; Sandweiss, 2007; Lombard & Wadley, 2009; Pearsall, 2015) but also lipid and proteins specifically from the pottery vessels (e.g. Scott et al., 2021).

For the Aratu-Sapucaí tradition, the morphology of the large piriform vessels together with the internal use-wear marks indicated a function related to the fermentation of some liquid substance (Gardiman et al., 2016). This has been then confirmed by starch analysis, which found *Zea mays* (maize) starch in such vessels. Therefore, the function of the piriform vessels of the Aratu-Sapucaí tradition was in connection with the production of maize fermented beverages (Gardiman et al., 2016). At the moment, this is the only study that confirmed a ceramic function by using plant micro-remains and it shows the huge potential of specifically starch analysis to understand vessel function, social economy, and resource use in the archaeology of the Cerrado of Central Brazil. More such studies are needed to test the diverse hypotheses of resource exploitation currently based on a form-function approach (both for pottery and lithics) and ethnographic analogy and to better understand subsistence and the use of plants in the pre-colonial societies of the Cerrado.

10.5 Final Considerations

Evidence for plant exploitation by the early settlers of the Cerrado region of Central Brazil has been found starting from the first occupations of this area dated at the Early Holocene. These groups were hunter-gatherers that foraged the savanna-like environment of the Cerrado for acquiring carbohydrates (starch) and vegetable fats from the available fruits, seeds, and underground organs of the many species composing this vegetation (Cassino et al., 2021). Furthermore, plants were also exploited for non-food resources, such as for fibers, for wood and for fuel. The beginning of cultivation in the Cerrado (by adoption or local innovation) is in connection -as in many other parts of the world– with the emergence of pottery, which would have facilitated the processing and storage of such plant resources.

The earliest evidence for the use of pottery in the area is the Una tradition, with most of the discovered sites consisting of sparse occupations in rock shelters (Robrahn-González, 1996). The Una Tradition is followed by the Aratu Tradition, with typical settlements consisting of large ring villages (Wüst & Barreto, 1999). The subsistence of the Aratu tradition is assumed to have relied on maize cultivation (Robrahn-González, 1996; Wüst & Barreto, 1999). In parallel, the Uru tradition, which has an Amazonian origin (Robrahn-González, 1996; Wüst & Barreto, 1999), also established itself in the Cerrado. The Uru tradition ceramists are considered to have been cultivating manioc well before arriving in central Brazil (Robrahn-González, 1996; Wüst & Barreto, 1999). Also associated with manioc is the Tupiguarani tradition, which expanded from the Amazon (for discussion, see Brochado, 1984; Noelli, 1996; Almeida & Neves, 2015). This tradition was also present in the Cerrado area, but its occupation is less intense than the Aratu and Uru traditions (Robrahn-González, 1996).

Although it is not clear whether the emergence of the Una and Aratu traditions are due to local processes or to migrations, the change from ephemeral rock shelter camps to large, permanent ring villages is frequently interpreted as reflecting population growth, a more sedentary lifestyle, and an overall greater reliance on agriculture (Henriques Jr., 2006; Oliveira & Viana, 1999; Robrahn-González, 1996; Wüst & Barreto, 1999). However, the importance of different cultivars (e.g. maize *versus* manioc), the association of such cultivars with the different ceramic traditions of the Cerrado, and the exploitation of the plant resources in general is often based on indirect evidence, such as the repertoire of vessel shapes and ethnographic analogy. Although such approaches can help formulate initial hypotheses, they need to be tested with further data from the archaeological record, such as plant remains in their different forms of preservation (charred, desiccated, or as phytoliths or starch grains) or chemical signatures from isotopes, lipids, and proteins as well as a DNA. Therefore, future studies on plant proxies, residues from pottery, and genetics should be able to test such hypotheses and help to better understand the plant exploitation strategies and cultivation, and possibly if there were instances of domestication, among the human populations of the Cerrado of Central Brazil.

References

Afonso, M. C. (2009). *Um painel da arqueologia pré-histórica no Estado de São Paulo: os sítios cerâmicos*. UESC.

Almeida, F. O., & Neves, E. G. (2015). Evidências arqueológicas para a origem dos tupi-guarani no leste da amazônia. *MANA, 21*(3), 499–525.

Alves, M. A. (2018). Tradições arqueológicas ceramistas de dois povos Jê: Kayapó meridional e Kaingang da região centro norte de São Paulo. *Revista Do Museu de Arqueologia e Etnologia, 31*, 1–21. https://doi.org/10.11606/issn.2448-1750.revmae.2018.146044

Angeles Flores, R. (2015). *Uso de Recursos Vegetais em Lapa Grande de Taquaraçu: Evidencias Macro e Microscópicas* (p. 308 f). Dissertação de Mestrado – Museu de Arqueologia e Etnologia, Universidade de São Paulo, São Paulo.

Angeles Flores, R., Moreno de Sousa, J. C., Araujo, A. G. de M, & Ceccantini, G. (2016). Before Lagoa Santa: Micro-remain and technological analysis in a lithic artifact from the Itaparica industry. *Journal of Lithic Studies, 3*(1), 6–29. https://doi.org/10.2218/jls.v3i1.1423

Angeles Flores, R., Venezia, M. E., Araujo, A. G. D. M., & Ceccantini, G. (2023). Confirmation of starchy plant processing in Lagoa Santa by paleoindian people: Results of microrremain and use-wear analysis at Lapa Grande de Taquaraçu. *PaleoAmerica, 9*(1), 7–24. https://doi.org/10.1080/20555563.2023.2174828.

Araujo, A. G. M., Strauss, A. M., Feathers, J. K., Paisani, J. C., & Schrage, T. J. (2013). Paleoindian open-air sites in tropical settings: A case study in formation processes, dating methods, and paleoenvironmental models in Central Brazil. *Geoarchaeology, 28*(3), 195–220. https://doi.org/10.1002/gea.21442

Bachelet, C. (2014). Pré-História no Cerrado: Análises antracologicas dos abrigos de Santa Elina e da Cidade de Pedra (Mato Grosso). Fronteiras: Journal of Social, Technological and Environmental. *Science, 3*(2), 96. https://doi.org/10.21664/2238-8869.2014v3i2.p96-110

Bachelet, C. (2016). Use of wood resources by Holocene hunter-gatherers of Cidade de Pedra, Mato Grosso, Brazil. *Cadernos do LEPAARQ, XIII*(25), 387–400. https://periodicos.ufpel.edu.br/ojs2/index.php/lepaarq/article/viewFile/7374/5704

Bachelet, C., & Scheel-Ybert, R. (2017). Landscape and firewood selection in the Santa Elina rock shelter (Mato Grosso, Brazil) during the Holocene. *Quaternary International, 431*, 52–60. https://doi.org/10.1016/j.quaint.2015.12.019

Bird, R. M., Dias, O., Jr., & Carvalho, E. T. (1991). Subsídios para a arqueobotânica no Brasil: o milho antigo em cavernas de Minas Gerais, Brasil. *Revista de Arqueologia, 6*, 14–31.

Bissaro Jr, M. C. (2008). *Tafonomia como ferramenta zooarqueológica de interpretação: viés de representatividade óssea em sítios arqueológicos, paleontológico e etnográfico*. MSc Dissertation, Universidade de São Paulo.

Bitencourt, A. L. V., & Schmitz, P. I. (1994). O paleoambiente do sítio arqueológico BA-RC-28: Um estudo através dos minerais dos sedimentos do abrigo. *Revista de Arqueologia, 8*(1), 255–272. https://doi.org/10.24885/sab.v8i1.478

Boëda, E. (1997). *Technogenèse de systèmes de production lithique au Paléolithique inférieur et moyen en Europe occidentale et au Proche-Orient*. Unpublished Thèse d'Habilitation à diriger des recherches, Université Paris X—Nanterre.

Brochado, J. P. (1981). A tradição cerâmica Tupiguarani na América do Sul. *Cliometrica, 3*, 117–164.

Brochado, J. P. (1984). *An ecological model of the spread of pottery and agriculture into Eastern South America*. PhD dissertation, University of Illinois at Urbana-Champaign, Ann Arbor UMI, Inc.

Bueno, L. (2012). Entre abrigos e lagoas: Tecnologia lítica e territorialidade em Lagoa Santa (Minas Gerais, Brasil). *Revista de Arqueologia, 25*(2), 62–83. https://doi.org/10.24885/sab.v25i2.355

Bueno, L., Dias, A. S., & Steele, J. (2013). The Late Pleistocene/Early Holocene archaeological record in Brazil: A geo-referenced database. *Quaternary International, 301*, 74–93. https://doi.org/10.1016/j.quaint.2013.03.042

Calderón, V. (1971). Breve notícia sobre a arqueologia de duas regiões do Estado da Bahia. *Publicações Avulsas do Museu Paraense Emílio Goeldi, 15*, 163–177.

Cassino, M. F., Shock, M. P., Furquim, L. P., Ortega, D. D., Machado, J. S., Madella, M., & Clement, C. R. (2021). Archaeobotany of Brazilian indigenous peoples and their food plants. In *Local food plants of Brazil* (pp. 127–159). Springer. https://doi.org/10.1007/978-3-030-69139-4_8.

Ceccantini, G. C. T., & Gusselia, L. W. (2001). Plant anatomy and palaeoethnobotany at Santa Elina shelter (Jangada, MT, Brazil). *Revista Do Museu de Arqueologia e Etnologia, 11*, 189. https://doi.org/10.11606/issn.2448-1750.revmae.2001.109417

Chaves, S. A. M. (2005). Novas perspectivas paleoetnobotânicas e paleovegetacionais do Planalto Central do Brasil: uma proposta palinológica. In Á. V. Vialou (Ed.), *Pré-história do Mato Grosso* (Santa Elina) (Vol. 1, pp. 149–154). EDUSP.

Chim, E. N. (2018). *Zooarqueologia da Lapa Grande de Taquaraçu* (p. 265 f). Dissertação de Mestrado – Museu de Arqueologia e Etnologia da Universidade de São Paulo, São Paulo.

Chmyz, I. (1976). Terminologia Arqueológica brasileira para a cerâmica. Cadernos De Arqueologia. *Ano, 1*(1). Paranaguá: Museu de Arqueologia e Artes Populares, UFPR.

Colonese, A. C., Winter, R., Brandi, R., Fossile, T., Fernandes, R., Soncin, S., McGrath, K., Von Tersch, M., & Bandeira, A. M. (2020). Stable isotope evidence for dietary diversification in the pre-Columbian Amazon. *Scientific Reports, 10*, 16560. https://doi.org/10.1038/s41598-020-73540-z

Da-Gloria, P., & Larsen, C. S. (2014). Oral health of the Paleoamericans of Lagoa Santa, Central Brazil. *American Journal of Physical Anthropology, 154*(1), 11–26. https://doi.org/10.1002/ajpa.22467

de Oliveira, J. E., & Viana, S. A. (1999). O Centro-Oeste antes de Cabral. *Revista USP, 44*, 142. https://doi.org/10.11606/issn.2316-9036.v0i44p142-189

de Souza, J. G., Schaan, D. P., Robinson, M., Barbosa, A. D., Aragão, L. E. O. C., Marimon, B. H., Marimon, B. S., da Silva, I. B., Khan, S. S., Nakahara, F. R., & Iriarte, J. (2018). Pre-Columbian earth-builders settled along the entire southern rim of the Amazon. *Nature Communications, 9*(1). https://doi.org/10.1038/s41467-018-03510-7

Dias Júnior, O. F. (1971). Breves notas a respeito das pesquisas no sul de Minas Gerais. Resultados preliminares do quarto ano de pesquisas: PRONAPA 4: 1968-1969. *Publicações Avulsas do Museu Emílio Goeldi, 15*, 133–148.

Dias Júnior, O. F. (1975). Pesquisas arqueológicas no sudeste brasileiro. Boletim do Instituto de Arqueologia Brasileira. *Série Especial, 1*, 3–17.

Erickson, C. L. (2008). Amazonia: The historical ecology of a domesticated landscape. In H. Silverman & W. H. Isbell (Eds.), *The handbook of South American archaeology* (pp. 157–183). Springer.

Fensterseifer, E., & Schmitz, P. I. (1975). Fase Iporá. Uma fase Tupiguarani no sudoeste de Goiás. *Anuário de Divulgação Cientifica, 7*(2), 19–27.

Fernandes, H. L. A. (2003). *Os sepultamentos do Sítio Aratu de Piragiba*. Dissertação de mestrado. Universidade Federal da Bahia, Salvador.

Fogaça, E., & Lourdeau, A. (2008). Uma abordagem tecno-funcional e evolutiva dos instrumentos plano-convexos (lesmas) da transição Pleistoceno/Holoceno no Brasil central. *FUMDHAMentos., 7*, 260–347.

Freitas, F. de O. (2003). Áreas de contato de populações pré-históricas, observado através do resgate de DNA de amostras arqueológicas de milho – Zea Mays, L. *Revista de Arqueologia, 16*(1), 47–57. https://doi.org/10.24885/sab.v16i1.178

Freitas, F. de O. (2004). Uso de Material Arqueológico no Estudo de Evolução de Plantas. *Revista de Arqueologia, 17*(1), 33–40. https://doi.org/10.24885/sab.v17i1.190

Freitas, F. de O, & Rodet, M. J. (2010). O que ocorreu nos últimos 2000 anos no vale do Peruaçu? Uma análise multidisciplinar para abordar os padrões culturais e suas mudanças entre as populações humanas daquela região. *Revista do Museu de Arqueologia e Etnologia, 20*, 109–126. https://doi.org/10.11606/issn.2448-1750.revmae.2010.89917

Fullagar, R. (2006). Starch on Artifacts. In H. Barton & R. Torrence (Eds.), *Ancient starch research*. Left Coast Press.

Fullagar, R. (2009). Stones, stories and science. In M. Haslam, G. Robertson, A. Crowther, S. Nugent, & L. Kirkwood (Eds.), *Archaeological science under a microscope: Studies in residue and ancient DNA analysis in honour to Thomas H. Loy* (Vol. 30, pp. 4–7). ANU E Press.

Gardiman, G. G. (2014). *Vereda III: Uma análise paleoetnobotânica*. Monografia. Universidade Federal de Minas Gerais.

Gardiman, G. G., Rodrigues, I. M. M., Cascon, L. M., & Isnards, A. (2016). A morfologia dos vasos Jê na produção de cauim de milho em Vereda III: Uma proposição. *Revista Do Museu de Arqueologia e Etnologia, 27*, 111. https://doi.org/10.11606/issn.2448-1750.revmae.2016.137288

Géminard, L. (1985). Avant-Propos. In Y. Deforge (Ed.), *Technologie et génétique de l'objet industriel*. Ed. Maloine (Collection Université de Compiègne), 196p.

Henriques Jr., G. (2006). *Arqueologia regional da província cárstica do Alto São Francisco: um estudo das tradições ceramistas Una e Sapucaí*. University of São Paulo, São Paulo.

Hermenegildo, T. (2009). *Reconstituição da dieta e dos padrões de subsistência das populações pré-históricas de caçadores-coletores do Brasil Central através da ecologia isotópica* (p. 126 f). Dissertação de Mestrado – Escola Superior de Agricultura "Luiz de Queiroz", Universidade de São Paulo, Piracicaba.

Hilbert, L., Neves, E. G., Pugliese, F., Whitney, B. S., Shock, M., Veasey, E., Zimpel, C. A., & Iriarte, J. (2017). Evidence for mid-Holocene rice domestication in the Americas. *Nature Ecology & Evolution, 1*, 1693–1698. https://doi.org/10.1038/s41559-017-0322-4

Hubbe, A., Hubbe, M., & Neves, W. A. (2013). The Brazilian megamastofauna of the Pleistocene/Holocene transition and its relationship with the early human settlement of the continent. *Earth-Science Reviews, 118*, 1–10. https://doi.org/10.1016/j.earscirev.2013.01.003

Iriarte, J., Holst, I., Marozzi, O., Listopad, C., Alonso, E., Rinderknecht, A., & Montaña, J. (2004). Evidence for cultivar adoption and emerging complexity during the Mid-Holocene in the La Plata Basin, Uruguay. *Nature, 432*, 614–617.

Iriarte, J., Elliott, S., Maezumi, S. Y., Alves, D., Gonda, R., Robinson, M., Gregorio de Souza, J., Watling, J., & Handley, J. (2020). The origins of Amazonian landscapes: Plant cultivation, domestication and the spread of food production in tropical South America. *Quaternary Science Reviews, 248*, 106582. https://doi.org/10.1016/j.quascirev.2020.106582

Kipnis, R. (2002). *Foraging societies of eastern central Brazil: An evolutionary ecology study of subsistence strategies during the terminal Pleistocene and early/middle Holocene*. PhD thesis, University of Michigan.

Leroi-Gourhan, A. (1964). *Le Geste et la parole*. Ed. Albin Michel. 421 p.

Lombard, M., & Wadley, L. (2009). The impact of micro-residue studies on south African middle stone age research. In M. Haslam, G. Robertson, A. Crowther, S. Nugent, & L. Kirkwood (Eds.), *Archaeological science under a microscope: Studies in residue and ancient DNA analysis in honour to Thomas H. Loy* (Vol. 30, pp. 11–28). ANU E Press.

Lombardo, U., Iriarte, J., Hilbert, L., Ruiz-Pérez, J., Capriles, J. M., & Veit, H. (2020). Early Holocene crop cultivation and landscape modification in Amazonia. *Nature, 581*(7807), 190–193. https://doi.org/10.1038/s41586-020-2162-7

Lourdeau, A. (2010). *Le technocomplexe Itaparica: Définition techno-fonctionnelle des industries à pièces façonnées unifacialement à une face plane dans le centre et le nordest du Brésil pendant la transition Pléistocène-Holocène et l'Holocène ancien* (p. 477 f). Tese de Doutorado – Université Paris Ouest Nanterre La Défense, Paris.

Lourdeau, A. (2012). *The Itaparica Technocomplex: The first conspicuous settlement of Central and Northeastern Brazil from a technological perspective* (Current research in the pleistocene special edition, pp. 53–56).

Loy, T. H. (1994). Methods in the analysis of starch residues on prehistoric stone tools. In J. G. Hather (Ed.), *Tropical Archaeobotany: Applications and new developments* (pp. 86–114). Routledge.

Maezumi, S. Y., Alves, D. T., Robinson, M., De Souza, J. G., Levis, C., Barnett, R. L., Oliveira, E. A., Urrego, D., Schaan, D., & Iriarte, J. (2018). The legacy of 4,500 years of polyculture agroforestry in the eastern Amazon. *Nature Plants, 4*, 540–547.

Marques, E. Q., Marimon-Junior, B. H., Marimon, B. S., Matricardi, E. A. T., Mews, H. A., & Colli, G. R. (2020). Redefining the Cerrado-Amazonia transition: Implications for conservation. *Biodiversity and Conservation, 29*, 1501–1517. https://doi.org/10.1007/s10531-019-01720-z

Martins, G. R., & Kashimoto, E. M. (1998). Arqueologia na área impactada pelo Gasoduto Bolívia-Brasil: trecho Terenos Três Lagoas/MS. *Revista do Museu e Arqueologia e Etnologia, 8*, 87–107. São Paulo.

McKey, D., Rostain, S., Iriarte, J., Glaser, B., Birk, J., Holst, I., & Renard, D. (2010). Pre-Columbian agricultural landscapes, ecosystem engineers and self-organized patchiness in Amazonia. *Proceedings of the National Academy of Sciences, 107*, 7823–7828.

Miller, E. T. (1983). *História da cultura indígena do alto-médio Guaporé*. Dissertação (Mestrado em História) – Pontifícia Universidade Católica. Porto Alegre.

Mingatos, G. S., & Okumura, M. (2016). Modelo de Amplitude de Dieta aplicada a restos faunísticos do sítio Lapa do Santo (MG) e suas implicações para o entendimento da dieta em grupos Paleoíndios do Brasil central. *Palaeoindian Archaeology, 1*(1), 13.

Moreira, E. (1981–1984). Análise dos restos de alimentos de origem animal – Programa Arqueológico de Goiás, recolhidos nas escavações do abrigo GO-JA-01. *Anuário de Divulgação Científica, 10*, 98–112. Goiânia.

Moreno de Sousa, J. C. (2014). *Cognição e Cultura no Mundo Material: Os Itaparicas, os Umbus e os "Lagoassantenses"* (pp. 200). Volume 1. MAE-USP. Dissertação de Mestrado.

Moreno de Sousa, J. C. (2016a). Lithic technology of an Itaparica industry archaeological site: The Gruta das Araras rockshelter, Midwest of Brazil. *Journal of Lithic Studies, 3*(1), 87–106. https://doi.org/10.2218/jls.v3i1.1298

Moreno de Sousa, J. C. (2016b). Did Palaeoindian technology persist during the middle or late Holocene in Central Brazil? A review from the Córrego do Ouro 19 site (GO-CP-17), Goiás state. *Palaeoindian Archaeology, 1*(1), 32–49.

Moreno de Sousa, J. C., & Araujo, A. G. de M. (2018). Microliths and polished stone tools during the pleistocene-holocene transition and early holocene in South America: The Lagoa Santa lithic industry. *PaleoAmerica, 4*(3), 219–238. https://doi.org/10.1080/20555563.2018.1531350

Monteiro, L. C. (2005). *Abrigos e aldeias: análise dos contextos tecnológicos das ocupações de ceramistas na Cidade de Pedra, Rondonópolis, Mato Grosso*. Dissertação de Mestrado, Museu de Arqueologia e Etnologia, Universidade de São Paulo. https://doi.org/10.11606/D.71.2006. tde-15092006-151147.

Neves, W. A., & Kipnis, R. (2004). Further evidence of a highly cariogenic diet among late Paleoindians of Central Brazil. *Current Research on Pleistocene, 21*, 81–83.

Neves, W. A., & Piló, L. B. (2003). Solving Lund's dilemma: New AMS dates confirm that humans and megafauna coexisted at Lagoa Santa. *Current Research in the Pleistocene, Orono, 20*, 57–60.

Neves, W. A., Hubbe, M., Bernardo, D., Strauss, A. M., Araujo, A. G. M., & Kipnis, R. (2013). Early human occupation of Lagoa Santa, eastern-Central Brazil: Craniometric variation of the initial settlers of South America. In K. Graf, C. Ketron, & M. Water (Eds.), *Paleoamerican odyssey* (pp. 397–412). Texas A & M University Press.

Nimuendajú, C. (1981). *Mapa Etno Histórico do Brasil e Regiões Adjacentes*. Biblioteca Digital Curt Nimuendajú. Avaliable http://www.etnolinguistica.org/biblio:nimuendaju-1981-mapa. 16 Mar 2016.

Noelli, F. S. (1996). As hipóteses sobre o centro de origem e rotas de expansão dos Tupi. *Revista de Antropologia, 39*(2), 7–53. https://doi.org/10.11606/2179-0892.ra.1996.111642

Ortega, D. D. (2016). *A Cerâmica Arqueológica do Sítio Lago Rico: Questões Sobre Funcionalidade, Funcionamento e Função* (142 pp.). Monografia (Graduação em Arqueologia). Pontifícia Universidade Católica de Goiás, Goiânia.

Ortega, D. D. (2019). *Microvestígios botânicos em artefatos líticos do sítio Lapa do Santo (Lagoa Santa, Minas Gerais)*. Universidade de São Paulo. Retrieved July 30, 2021, from https://doi.org/10.11606/d.71.2019.tde-8092019-104742

Pearsall, D. M. (2015). *Paleoethnobotany: A handbook of procedures* (3rd ed., 513 p.). Academic.

Perez, C. P. (2009). *Paleoecologia de mamíferos viventes como ferramenta na caracterização do ambiente holocênico de Lagoa Santa, MG*. Dissertação de Mestrado, Universidade de São Paulo.

Perry, L. (2001). *Prehispanic subsistence in the Middle Orinoco Basin: Starch analyses yield new evidence*. PhD diss, Department of Anthropology, Southern Illinois university at Carbondale, Carbondale.

Perry, L. (2004). Starch analyses reveal the relationship between tool type and function: An example from the Orinoco valley of Venezuela. *Journal of Archaeological Science, 31*(8), 1069–1081. https://doi.org/10.1016/j.jas.2004.01.002

Perota, C. (1974). Resultados preliminares sobre a arqueologia da região Central do Estado do Espirito Santo. PRONAPA 5. Resultados Preliminares do 5. Ano (1969–1970). *Publicações Avulsas, 26*, 127–140. Museu Paraense Emílio Goeldi.

Piperno, D. R., & Pearsall, D. M. (1998). Background of tropical agricultural origins. In *The origins of agriculture in the Lowland Neotropics* (pp. 1–38). Elsevier. https://doi.org/10.1016/b978-012557180-7/50002-3

Pontim, R. de L. (2013). A Tradição Tupiguarani na Bacia do Alto Tocantins. *Revista Do Museu De Arqueologia E Etnologia. Suplemento*, (supl. 15), 11–113. https://doi.org/10.11606/issn.2594-5939.revmaesupl.2013.113578

Prous, A. (1992). *Arqueologia Brasileira*. EdUnB.

Pugliese, Jr. F. (2007). *Os Líticos de Lagoa Santa: um estudo sobre organização tecnológica de caçadores-coletores do Brasil Central* (p. 154 f). Dissertação de Mestrado – Museu de Arqueologia e Etnologia, Universidade de São Paulo, São Paulo.

Rabardel, P. (1995). *Les Hommes et les technologies: Approche cognitive des instruments contemporains* (193 p.). Armand Colin.
Ramos, M. P., & Viana, S. (2019). Diagnose tecno-funcional de amostragem lítica datada do início do Holoceno Médio no sítio arqueológico GO-JA-01: Características da estrutura de lascamento em presença. *Revista Mosaico – Revista de História, 12*, 135. https://doi.org/10.18224/mos.v12i0.7292.
Resende, E. M. T. P., & Cardoso, J. S. (1996). Estruturas de Armazenamento Vegetal: Os "Silos" do Vale do Rio Peruaçu (MG). *Coleção Arqueologia, Porto Alegre, EDIPUCRS, 1*(2), 249–264.
Resende, E. M. T., & Cardoso, J. S. (2009). Vestígios vegetais: arqueobotânica e técnicas tradicionais de armazenamento. Arquivos do Museu de História Natural e Jardim Botânico – UFMG: Arqueologia do vale do Rio Peruaçu e adjacências. *Belo Horizonte, 19*, 231–260.
Resende, E. T., & Prous, A. (1991). Os vestígios vegetais do Grande Abrigo de Santana do Riacho. *Arq Mus Hist Nat UFMG, 12*, 87–111.
Robrahn-González, E. M. (1996). Os grupos ceramistas pré-coloniais do Centro-Oeste brasileiro. *Rev. do Museu de Arqueologia e Etnologia, São Paulo, 6*, 83–121.
Rodet, M. J., Duarte-Talim, D., & Barri, L. F. (2011). Reflexões sobre as primeiras populações do Brasil Central: "Tradição Itaparica". *Habitus, Goiânia, 9*(1), 81–100.
Rodrigues, I., & Gardiman, G. (2016). Fora das grandes aldeias: o sítio Vereda III e suas informações complementares sobre sistemas de ocupação de grupos portadores de cerâmica associada à tradição Aratu-Sapucaí. *Revista do Museu de Arqueologia e Etnologia, 103*. https://doi.org/10.11606/issn.2448-1750.revmae.2016.137284
Rubin, J. C. R., Da Silva, R. T., Bayer, M., Barberi, M., Barbosa, J. B., Ortega, D. D., et al. (2019). Ocupación precolonial en la cuenca hidrográfica del río Araguaia, estados de Goiás y Mato Grosso, Brasil: síntesis aproximada y dos estudios de casos. *Revista del Museo de La Plata, 4*(2), 401–436.
Sandweiss, D. H. (2007). Small is big: The microfossil perspective on human-plant interaction. *Proceedings of the National Academy of Sciences of the United States of America (PNAS), 104*(9), 3021–3022.
Scheel-Ybert, R. (2001). Man and vegetation in southeastern Brazil during the late holocene. *Journal of Archaeological Science, 28*(5), 471–480. https://doi.org/10.1006/jasc.2000.0577
Scheel-Ybert, R., & Bachelet, C. (2020). A good place to live: Plants and people at the Santa Elina Rock Shelter (Central Brazil) from late pleistocene to the holocene. *Latin American Antiquity, 31*(2), 273–291. https://doi.org/10.1017/laq.2020.3
Scheel-Ybert, R., & Solari, M. E. (2005). Análise dos macrorrestos vegetais do setor oeste: antracologia e carpología. In Á. V. Vialou (Ed.), *Pré-história do Mato Grosso* (Santa Elina) (Vol. 1, pp. 139–147). EDUSP.
Schmitz, P. I. (1977). Arqueologia de Goias em 1976: Projeto Paranaiba. *Estudos Revista da Universidade Católica de Goiás, Goiânia, IV*(5), 19–77.
Schmitz, P. I., & Barbosa, A. S. (1985). *Horticultores pré-históricos do Estado de Goiás*. Instituto Anchietano de Pesquisas.
Schmitz, P. I., & Rogge, J. H. (2008). Um sítio da tradição cerâmica Aratu em Apucarana, PR. *Revista Do Museu de Arqueologia e Etnologia, 18*, 47. https://doi.org/10.11606/issn.2448-1750.revmae.2008.89828
Schmitz, P. I. et al. (1986). Caiapônia. IAPUnisinos, São Leopoldo.
Schmitz, P. I., Barbosa, A. S., Jacobus, A. L., & Ribeiro, M. B. (1989). Arqueologia nos Cerrados do Brasil Central: Serranópolis I. *Pesquisas – Antropologia, 44*, 208.
Schmitz, P. I., Barbosa, A. S., Miranda, A. F., Ribeiro, M. B., & Barbosa, M. O. (1996). Arqueologia nos Cerrados do Brasil Central – Sudoeste da Bahia e Leste de Goiás: O Projeto Serra Geral. *Pesquisas, Antropologia, 52*, 198.
Schmitz, P. I., Rosa, A. O., & Bitencourt, A. L. (2004). Arqueologia nos Cerrados do Brasil Central. SERRANÓPOLIS III. *Pesquisas – Antropologia, 60*, 1–286.
Scott, A., Power, R. C., Altmann-Wendling, V., Artzy, M., Martin, M. A. S., Eisenmann, S., Hagan, R., et al. (2021). Exotic foods reveal contact between South Asia and the Near East

during the second millennium BCE. *Proceedings of the National Academy of Sciences, 118*(2), e2014956117. https://doi.org/10.1073/pnas.2014956117

Shock, M. P. (2010). *Holocene hunter-gatherer plant use and foraging choice: a test from Minas Gerais, Brazil.* PhD thesis, University of California, Santa Barbara.

Shock, M. P., Kipnis, R., Bueno, L., & Silva, F. M. (2013). A chronology of the introduction of domesticated plants in Central Brazil. *Tipiti, 11*(2), 52–59.

Sigaut, F. (2002). Technology. In T. Ingold (Ed.), *Companion encyclopedia of anthropology* (pp. 420–459). Routledge.

Soares, J. (2013). Discutindo a tradição Aratu: proposta de um modelo de dispersão e implantação nas zonas de tensão ecológica. *Revista Do Museu De Arqueologia E Etnologia, 23*, 61–77. https://doi.org/10.11606/issn.2448-1750.revmae.2013.107025

Souza, J. G., Madella, M., & Noelli, F. S. (2021). Reassessing the role of climate change in the Tupi expansion (South America, 5000–500 BP). *Journal of the Royal Society, Interface, 18*, 20210499.

Strauss, A., Oliveira, R. E., Villagran, X. S., Bernardo, D. V., Salazar-García, D. C., Bissaro, M. C., Jr., Pugliese, F., Hermenegildo, T., Santos, R., Barioni, A., de Oliveira, E. C., Moreno de Sousa, J. C., Jaouen, K., Ernani, M., Hubbe, M., Inglez, M., Gratão, M., Rockwell, H., Machado, M., de Souza, G., Chemale, F., Kawashita, K., O'Connell, T. C., Israde, I., Feathers, J., Campi, C., Richards, M., Wahl, J., Kipnis, R., Araujo, A., & Neves, W. (2016). Early Holocene ritual complexity in South America: The archaeological record of Lapa do Santo (East-Central Brazil). *Antiquity, 90*(354), 1454–1473. https://doi.org/10.15184/aqy.2016.220

Strauss, A., Mariano Rodrigues, I. M., Baeta, A., Villagran, X. S., Alves, M., Pugliese, F., Jr., Bissaro, M., Jr., de Oliveira, R. E., de Souza, G. N., Bueno, L., de Sousa, J. C. M., Morrow, J. J., Reinhard, K. J., Hermenegildo, T., Perez, G. C., Chim, E. N., de Oliveira dos Santos, R., de Paiva, M., Kipnis, R., & Neves, W. (2020). The archaeological record of Lagoa Santa (East-Central Brazil): From the late pleistocene to historical times. In *Lagoa Santa Karst: Brazil's Iconic Karst region* (pp. 227–281). Springer. https://doi.org/10.1007/978-3-030-35940-9_12

Vialou, A. Vilhena, org. (2006). *Pré-história do Mato Grosso. Vol. 2 – Cidade de Pedra* (232 p.). Edusp.

Vialou, D., Benabdelhadi, M., Feathers, J., Fontugne, M., & Vialou, A. V. (2017). Peopling South America's centre: The late Pleistocene site of Santa Elina. *Antiquity, 91*(358), 865–884. https://doi.org/10.15184/aqy.2017.101

Viana, S. A. (2005). *Variabilidade tecnológica do sistema de debitagem e de confecção dos instrumentos líticos lascados de sítios lito-cerâmicos da região do Rio Manso/MT* (p. 340 f). Tese de Doutorado – Programa de Pós-Graduação em história, Pontifícia Universidade Católica do Rio Grande do Sul, Porto Alegre.

Viana, S. A., Ribeiro, C. V., & Oliveira, S. D. (2011). Cauixi em Cerâmica Arqueológica: uma Questão de Escolhas Culturais. *Revista de Arqueologia da Sociedade de Arqueologia Brasileira, 24*, 20–39.

Viana, S. A., Ramos, M. P. M., Rubin, J. C. R., Barberi, M., & Boëda, E. (2016). O Complexo Arqueológico de Palestina de Goiás/Brasil – uma avaliação dos conjuntos líticos mais antigos em contextualização macrorregional. *Revista Cadernos Do Ceom, 29*(45), 188. https://doi.org/10.22562/2016.45.07

Vilhena Vialou, A. (2011). *Occupations humaines et faune éteinte du Pléistocène au centre de l'Amérique du Sud: l'abri rupestre Santa Elina, Mato Grosso, Brésil. Peuplements et Préhistoire en Amérique* (pp. 193–208). Comité des travaux historiques et scientifiques.

Vilhena-Vialou, A., & Vialou, D. (1994). Les premiers peuplements préhistoriques du Mato Grosso. *Bulletin de La Société Préhistorique Française, 91*(4), 257–263. https://doi.org/10.3406/bspf.1994.9734

Villagran, X. S., Strauss, A., Miller, C., Ligouis, B., & Oliveira, R. (2017). Buried in ashes: Site formation processes at Lapa do Santo rockshelter, East-Central Brazil. *Journal of Archaeological Science, 77*, 10–34. https://doi.org/10.1016/j.jas.2016.07.008

Wüst, I. (1975). A Cerâmica Karajá de Aruanã. *Anuário de Divulgação Científica, 2*, 95–165.

Wüst, I. (1983). *Aspectos da ocupação pré-colonial em uma área do Mato Grosso de Goiás – tentativa de análise espacial*. Dissertação de Mestrado, USP, São Paulo.

Wüst, I. (1992). Contribuições arqueológicas, etnoarqueológicas e etno-históricas para o estudo dos grupos tribais do Brasil central: o caso Bororo. *Revista do Museu de Arqueologia e Etnologia, 2*, 13–26.

Wüst, I. (1998). Continuities and discontinuities: Archaeology and ethnoarchaeology in the heart of the Eastern Bororo territory, Mato Grosso, Brazil. *Antiquity, 72*(277), 663–675. https://doi.org/10.1017/S0003598X0008707X

Wüst, I., & Barreto, C. (1999). The ring villages of Central Brazil: A challenge for Amazonian archaeology. *Latin American Antiquity, 10*(1), 1–21.

Wüst, I., & Schmitz, P. I. (1975). Fase Jataí: estudo preliminar. *Anuário de Divulgação Científica, II*(2), 71–93. Goiânia.

Chapter 11
Wild Plant Resources and Cerritos de Indio Archaeological Sites at the India Muerta-Paso Barranca Archaeological Locality

Joaquín Mazarino

Abstract This paper presents the results of the analysis of the spatial distribution of earthen mound archaeological sites known as *cerritos de indio* at the *India Muerta-Paso Barranca* archaeological locality in the Merín Lagoon Basin. Starting from a database of 639 archaeological sites in the study area and considering the wild plant resources present in the archaeobotanical record, their location was analysed based on the geomorphological units and plant compositions of the area. The results indicate a strong correlation with the presence of food, medicinal and industrial resources suggesting mound location responds to factors causally related to the economy of the groups that raised them.

Keywords South American wetlands · Merín Lagoon Basin · *Cerritos de indio* · Wild plant resources · Domestic landscape

11.1 Introduction

Wetland environments have traditionally been described as hostile places, the dominant perception for those who do not inhabit the wetland or are not familiar with it has been negative, associating it with diseases, impassable, unproductive places and that ultimately should be avoided at all costs (Brown, 1997). However, different branches of ecology suggest that these highly productive environments can sustain ecosystems rich in diversity of plant and animal species (Finlayson et al., 2015).

The relationship between humans and wetlands dates back to the development of the first hominids (Reed, 1997). These environments have provided human groups

J. Mazarino (✉)
Instituto de Ecología y Ciencias Ambientales (IECA), Facultad de Ciencias, Universidad de la República, Montevideo, Uruguay

A. C. Colonese, R. G. Milheira (eds.), *Historical Ecology and Landscape Archaeology in Lowland South America*, Interdisciplinary Contributions to Archaeology, https://doi.org/10.1007/978-3-031-32284-6_11

with resources such as food, construction materials, tool making, fuel and shelter, as well as different plant species that have been used for their medicinal and psychotropic properties (Krtshnamurthy, 1990; Ramsar, 2008; Wilkinson, 1958). To this day different groups around the world depend on wetlands resources for their food security, access to water and medicines (Finlayson et al., 2015). Wetlands have also been the scenario for the development of different cultivation systems and environmental management practices that have co-evolved with these landscapes and have ultimately contributed to their present-day configuration.

Although wetlands in its broadest definition can be found in any climate they are usually associated with alluvial processes and are a common feature in most floodplains. When it comes to Primary Net Production (PPN) floodplains tend to be more productive under any climatic conditions when compared to geomorphological units where other than alluvial processes are dominant (Brown, 1997).

Despite the central role wetlands have played in human past, "wet" archaeological sites have received considerably less attention than dry land in archaeology, which has skewed the reconstruction of the past (Bernick, 1998). From the mid twentieth century to the late 1990s the prevailing view was that wetlands impose environmental constraints and limitations on socio-cultural development (Meggers, 1954; Tedeschi, 2011). In South America, this bias is best exemplified in the works of the *Programa Nacional de Pesquisa Arqueológica* (PRONAPA) and *Programa de Pesquisa Arqueológica na Bacia Amazônica* (PRONAPABA) from the 1960s (Simões, 1972; Moraes, 2015). Meggers (1971) described the lowlands of the Amazon basin as a green hell and an illusory paradise. From this point of view, all evidence of complexity in the archaeological record was due to the influence of more developed groups typical of the highlands (Miller, 1999). Later, with the development of new theoretical perspectives, these strongly deterministic approaches were called into question.

New theoretical approaches have been gaining more and more traction in the study of the neotropical lowlands of South America, most notably Historical Ecology (see among others: Erickson, 2003, 2006; Fraser et al., 2009; Neves & Petersen, 2006). By integrating perspectives from ecology, anthropology and archaeology, historical ecologists have lowlands as highly productive environments that, far from restricting, can promote socio-cultural development. Historical Ecology emphasises the idea of landscape understood as a multidimensional physical entity with temporal and spatial characteristics that has been modified by human activities in such a way that human actions and intentions can be inferred from its analysis (Balée & Erickson, 2006). Through this new approach, the Amazonian landscape was no longer considered pristine, but its exuberance and biodiversity came to be seen as a contingent product of human-environment interaction during the Holocene (Balée, 2010).

In Uruguay, lowland archaeology has played an instrumental role in the development of professional archaeology. Since the mid-1980s, considerable research has centred on the Merín Lagoon Basin (MLB) wetlands (Fig. 11.1). This region has been the scenario for one of the most notable anthropic modifications of the landscape in pre-Hispanic times (Bracco, 2006; Gianotti & Bonomo, 2013). From

Fig. 11.1 Top: *India Muerta* wetlands; Bottom: earthen mound at *García-Ricci* archaeological site

approximately 5500 to 200 ^{14}C yr. BP, human groups raised thousands of earthen mounds or "*cerritos de indio*", as they are known locally, in the MLB (Bracco et al., 2015). These mounds present a circular or elliptical base with average diameters of 30 to 40 m and heights ranging from a few decimetres to 7 meters and can be found isolated or in groups of up to 50 mounds (Bracco et al., 2011).

The location of these archaeological sites is far from random, they are generally found in floodplains, wetlands or rocky hills that form protruding wedges towards and into wetlands (Bracco et al., 2008). This relationship was noted by researchers early on (Ferrés, 1927), hence culture-environment interactions became part of the archaeology of south-eastern Uruguay since its inception and has served as one of the main conceptual tools in interpreting and explaining mound settlements (Bracco et al., 2011).

For the last 20 years, different archaeological investigations have incorporated theoretical and methodological tools from disciplines such as paleohydrology, palaeontology, paleolimnology, palaeobotany, geomorphology and geology, among others, to the study of these mounds (see among others: del Puerto, 2009; Inda, 2009; del Puerto & Inda, 2005; García-Rodríguez et al., 2010; Mazarino et al., 2015; Bracco et al., 2005, 2011, 2016; Duarte et al., 2017; Campos et al., 2001). The last decade has seen more and more archaeologists approach mound archaeological sites from a Historical Ecology and Landscape Ecology perspectives, shedding new light on the groups that built and occupied them (for example: Bracco et al., 2019; del Puerto, 2015; López et al., 2014).

The aim of the research presented in this paper is to analyse mound location in relation to the different geomorphological units they occupy and their distribution with respect to wetlands wild plant resources from the perspective of Historical Ecology.

11.2 Subsistence and Environmental Change

In 1987, the *Comisión de Rescate Arqueológico de la Laguna Merín* was created with a strong commitment to the positivism of the New Archaeology (Bracco et al., 2015). One of the main goals was the characterization of the adaptative strategies of mound builders, for which environmental characterization and subsistence studies were carried out (del Puerto & Inda, 2005).

Most paleoclimatic reconstructions were based mainly on the paleolimnological record of coastal lagoons from biological (diatoms and phytoliths), geochemical and sedimentological indicators (Bracco et al., 2009; García-Rodríguez et al., 2010; Inda et al., 2006; Inda, 2009; del Puerto et al., 2013; del Puerto, 2011, 2015). Integrating chronologies and core data from the Rocha, Blanca and Negra lagoons, Bracco et al. (2011) propose the following climatic sequence from 7000 ^{14}C yr. BP:

- 7000–5000 ^{14}C yr. BP: more temperate and humid climate, the sea reached its current level and even exceeded it by 5 m around 6000 AP.
- 5000–3000 ^{14}C yr. BP: drier, colder, and highly seasonal conditions begin to settle, sea level decline.
- 3000–1500 ^{14}C yr. BP: more temperate and humid conditions like current ones, the decline in sea level continues to current levels.
- 1500–500 ^{14}C yr. BP: more temperate and humid conditions prevailed.
- 500–200 ^{14}C yr. BP: current conditions.

Starting in the 1990s, bio-siliceous particle analysis began to be applied to the study of subsistence strategy of prehistoric groups in the MLB (del Puerto & Inda, 2005). Subsequently, other indicators such as pollen, starch and charcoal were also considered, which revealed the diversity of plant resources used by prehistoric populations and the incorporation of crops in prehistoric times in southeast Uruguay (Campos et al., 2001; del Puerto & Inda, 2005; del Puerto, 2015; Inda & del Puerto, 2007).

There is evidence of early adaptations to the lowlands in the MLB with an intensive harvesting of the *Butiá capitata* palm fruit around 8000 ^{14}C yr. BP (López-Mazz et al., 2014). However, the greatest evidence of wild plant resource management is found from the middle Holocene onwards (del Puerto, 2015).

Most of the archaeobotanical, archaeofaunistic and isotopic data from southeast Uruguay comes mainly from 8 archaeological sites that can be geographically divided into: Estancia Mal Abrigo, Los Ajos (Los Ajos hill range), Puntas de San Luis (India Muerta wetland), Isla Larga, CH2D01 (San Miguel hill range), Los Indios (Negra lagoon), Cráneo Marcado and Guardia del Monte (Castillos lagoon). It should be noted that the Castillos lagoon is located outside the MLB, on the

Atlantic slope (Fig. 11.2). Based on these records, a model of 3 main periods of consumption and resource management has been proposed (del Puerto et al., 2016):

– 4500–3000 ^{14}C yr. BP: Towards the end of the Middle Holocene there was an intensification of occupations in coastal areas. For this period, the isotopic and archaeofaunal record points to a use of marine resources mainly in relation to the

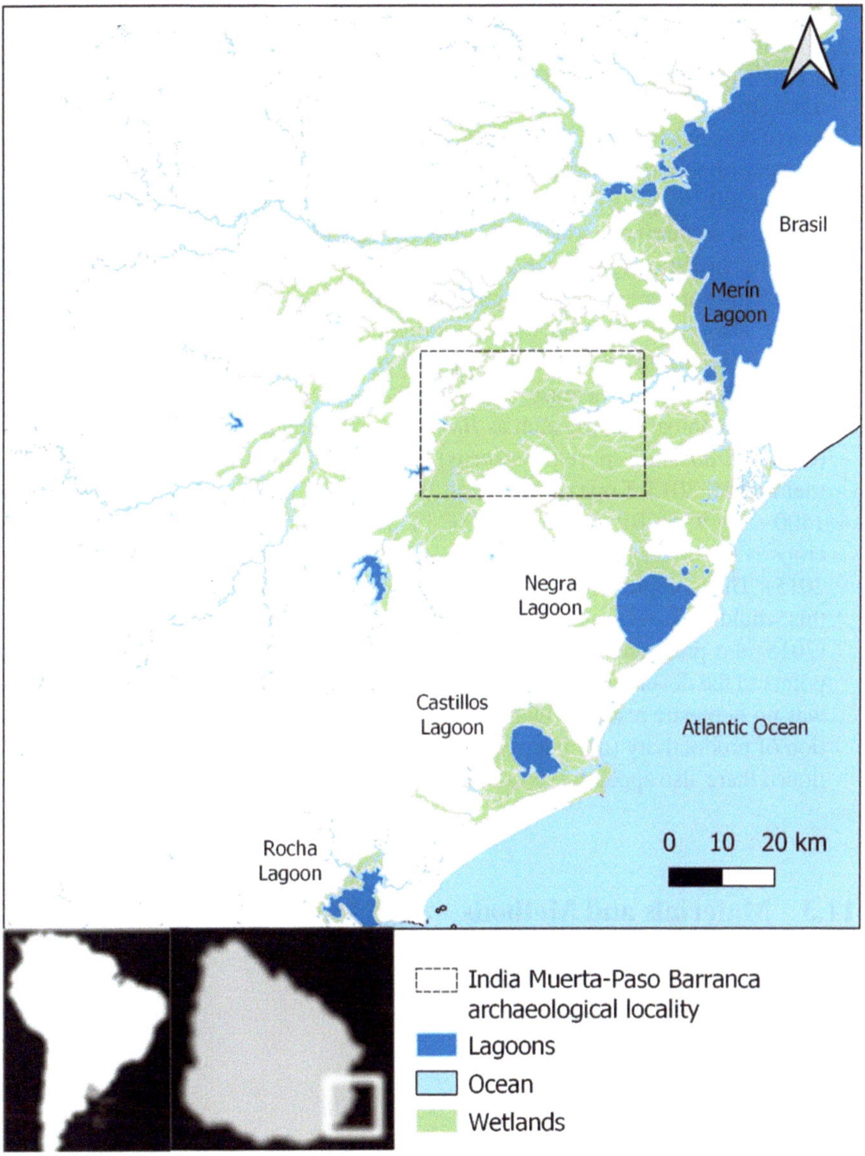

Fig. 11.2 Location of the IMPB locality within the MLB

contribution of proteins to the diet (Inda et al., 2011; Capdepont & Pintos, 2006). Different indicators point to the use of wild plant resources with a strong presence of *Butiá capitata* (butiá palm tree), *Syagrus romanzoffiana* (pindó palm tree), *Scirpus* sp. (junco), *Cyperus* sp. (tiririca), *Canna* sp. (achira) and different types of *cyperaceae* among others (Campos et al., 2001; del Puerto, 2015; Iriarte et al., 2001). The archaeological record for this period highlights a high variability of wild plant resources with evidence of storage and processing (Capdepont & Pintos, 2006). In most archaeological sites there is no significant presence of crops, however in some of them the occurrence of phytoliths and starch grains attributed to *Cucurbita* sp. (squash), *Phaseolus* sp. (beans) and *Zea mays* (corn) have been identified (Iriarte et al., 2001). This could indicate that although there was a consumption of domestic species, their contribution to the diet of mound builder groups was not very important, prioritizing wild resources (del Puerto, 2015).

- 3000–1500 ^{14}C yr. BP: The archaeofaunal record continues to point to a consumption of marine resources (Capdepont & Pintos, 2006; Pintos, 2000), however the isotopic signal of human remains from this period indicate that they would play a secondary role compared to continental resources (del Puerto, 2015). There is evidence of consumption of some potentially managed resources such as *Cucurbita* sp. and *Phaseolus* sp. (del Puerto, 2015; del Puerto & Inda, 2008). There is an increase in the richness of wild plants represented in the archaeobotanical record, such as *Bromelia* sp. (bananita do mato), *Datura ferox* (cardo), *Equisetum* sp. (equisetáceas), *Bambusae* (caña) and *Oryzeae* (wild rice) (del Puerto, 2015; López-Mazz et al., 2014).
- 1500–250 ^{14}C yr. BP: Starting at 1500 ^{14}C yr. BP the first indisputable evidence of crops is observed with *Zea mays*, *Cucurbita* sp. and *Phaseolus* sp. (del Puerto, 2015). The isotopic record of human remains points to consumption of corn, but this would have been a secondary contribution to the diet. According to del Puerto (2015) the presence of corn phytoliths outside the mound archaeological sites points to the development of agricultural practices. This does not imply that there was no domestic resource management in earlier times, but rather an intensification of productivity (del Puerto, 2015). Among the wild species previously mentioned there also appears *Typha domineguensis* (totora) (López-Mazz et al., 2014).

11.3 Materials and Methods

11.4 Study Area

The MLB is located between latitudes 31°–34° S and 52°–54° N, covering an area of 54,000 km^2, 24,000 km^2 of which belong to Brazil and the remaining 30,000 km^2 to Uruguay (Bracco et al., 2011). Its western limit in Uruguay is the Santa Lucía river basin, to the south it is limited by La Angostura hill range, a narrow strip that runs parallel to the Atlantic Ocean and a series of coastal lagoons that accompany the edge of the coast (Panario, 1988). The Negra lagoon is in the interior plains and is the only one that flows into the Merín lagoon. From the structural point of view, the MLB is in a rift valley (Panario, 1988) that has shown signs of recent activity (Panario et al., 2014).

According to Soriano's (1991) phytogeographic classification the area belongs to the "Grasslands of the Río de la Plata". In this region, "the pampas" are differentiated from the "campos", the first one corresponding to the Argentine territory, characterized by sub-humid to semi-arid grasslands with low tree presence, while the latter comprises the entirety of Uruguay's territory and the south of Rio Grande do Sul (Brazil) and is dominated by subtropical or savanna grasslands in the north (northern campos) and humid in the south (southern campos), with an important presence of trees and shrubs (Soriano, 1991). The vegetation develops under a temperate humid climate classified as subtropical, with relatively abundant (1000–1500 mm) and irregular rains corresponding to Cfa in Koeppen's classification (Peel et al., 2007).

Achkar et al. (2012) divide the MLB into three broad landscape units:

Hill ranges: This unit coincides with the edge of the rift valley, with heights ranging from 450-150 meters above sea level with the highest elevations to the southwest of the MLB. The rocky hill ranges with flattened watersheds reflect erosive processes that have acted since the Cretaceous period. Soils are generally shallow, of medium fertility, poorly permeable and well drained. The valleys present fluvial forest, xerophytic communities and open summer meadows, the latter being the predominant ecosystem.

Hills and hillocks: Consist of elevations of 150-50 meters above sea level. Watersheds are flat surfaces developed on crystalline rocks and silty-clay sediments. Soils in this unit are highly fertile, of slow permeability and moderate drainage. The predominant ecosystem is the summer meadow with a dense carpet and minor associated parks. The banks of the river courses are surrounded by dense fluvial forests, and in the valleys between the rocky hills patches of forests may appear.

Plains and Fluvial Plains: These unit extends eastward from the river plains to the Merín lagoon coast with heights between 50-0 meters above sea level, this unit is where wetlands develop due to its flatness and low point in the basin. Soils show low to medium fertility, slow permeability, and poor drainage. The predominant ecosystems are summer meadows with dense carpet, wetlands are made up of hydrophytic communities and palm groves. On the riverbanks fluvial forest accompanies the river courses.

11.5 India Muerta-Paso Barranca Archaeological Locality

The India Muerta-Paso Barranca (IMPB) archaeological locality was delimited based on archaeological and environmental criteria. It includes the plains of the southern sector of the MLB encompassing wetlands and palm grove environments

(Bracco et al., 2015). It extends from the middle course of the India Muerta stream in the west to the headwaters of the San Luis River in the east and from the headwaters of the Pelotas stream to the north to the Los Ajos hill range in the south with an approximate area of 1389 km2 (Fig. 11.2). It includes the Pelotas wetlands, the northern end of the San Miguel and Isla Negra wetlands and the entirety of the India Muerta wetland.

Archaeological research indicates that the India Muerta wetland concentrates the largest number of mound archaeological sites as well as the tallest, radiocarbon and luminescent dating techniques place its elevation between 4500 and 2000 years ^{14}C BP (Bracco et al., 2005, 2008; Duarte et al., 2017).

Most wetlands at IMPB start at the inflection zone of the India Muerta stream (Montaña & Bossi, 1995). The presence of alluvial fans and partially clogged channels points to changes in the configuration of the basin associated with the channels carrying capacity, rainfall, and Holocene climatic changes. Montaña and Bossi (1995) attribute the presence of paleochannels to a fluvial palaeosystem that would have encompassed the current India Muerta basin and the middle basin of the Cebollatí River further north (Montaña & Bossi, 1995).

South of the India Muerta wetlands, at the intersection of Cañada Grande and Channel N°2, logs were found in a living position below a deposit that occupies ¾ of the section of the channel indicating a process of clogging of the river system that was dated to 2580 ± 90 ^{14}C yr. BP (Bracco et al., 2012). To the north of the wetland, an abandoned and clogged palaeochannel was identified on the ravines of the Cebollatí river at the height of Enrique Martínez town. The deposits were identified as indicators of the beginning of the clogging process which dates between 3110 ± 160 OSL BP (Mazarino et al., 2015) and 2730 ± 60 ^{14}C yr. BP (Bracco et al., 2012) (Fig. 11.3). All of this points to a major decrease in the flow received by the wetlands of India Muerta, which could have been offset by the transition to more humid conditions that began in this temporary range (del Puerto, 2009; Inda, 2009).

The dynamics of the river system had great cultural and environmental implications. On the one hand, it would imply that from approximately 5000 to 2500 BP, during a period of greater aridity the India Muerta area would have functioned as a wetland, as it had a basin twice as large, when more humid conditions were installed, its basin was captured by part of the Cebollatí river, however, the increase in rainfall allowed it to sustain a wetland ecosystem. Thus, the India Muerta area would have worked as an eco-refuge since the middle Holocene, gathering a greater supply of resources for human populations (Bracco et al., 2016).

11.6 Wild Plant Resources in the MLB

Resources are dependent on the context and even cultural preferences. For example, the eye of a tuna could be considered disgusting to the average Uruguayan citizen but a delicacy for a Japanese. Therefore, a clarification of what a wild plant resource is in this paper is in order.

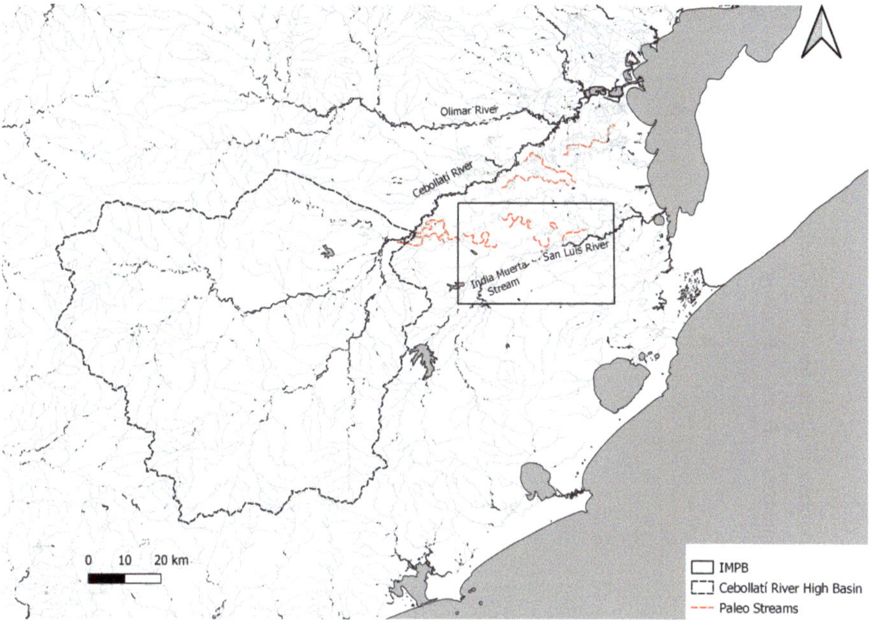

Fig. 11.3 Cebollatí River higher basin and paleo-streams connecting it to IMPB. (Redrawn from Montaña & Bossi, 1995)

Only plant resources identified in the archaeobotanical record of mound builders were considered (del Puerto et al., 2016). Of these resources, only those for which there is ethnohistoric evidence of consumption were taken into consideration mainly: food, fuel, household implements, boats, homes, industries and medicinal (del Puerto, 2011). Finally, only the wild resources present in the study area were included (Table 11.1).

11.7 Spatial Analysis

For the distribution, a database of 639 mounds in IMPB previously published by Bracco et al. (2015) was used. Most of these mounds were identified by aerial photography, satellite images and field observations. Mound distribution was introduced as a shp layer in an ArcGIS work environment.

The geomorphological characterization followed the units proposed by Sombroek (1969) for the Food and Agriculture Organization of the United Nations (Fig. 11.4). This classification was based mainly on the topography of the landscape, edaphic properties and frequency and permanence of floods and was ultimately developed as a vector layer which concentrates specific properties for each spatial object.

Table 11.1 Wild plant resources present at the MLB and their use, "x" marks the presence of the resource in the archaeological record

Family	Species	Common name	Food	Fuel	Domestic supplement	Boats	Housing	Industries	Medicinal
Anacardiaceae	Sp. *Indet.*	Molle						X	X
Arecaceae	*Butia capitata*	Butia	X				X	X	
	Syagrus romanzoffiana	Pindó	X		X		X	X	X
Berberidaceae	Sp. *indet*	Espina Amarilla	X						
Bromeliaceae	*Bromelia* sp.	Bananita do Mato	X						
Cannabaceae	*Celtis* sp.	Tala							
Cannaceae	*Canna* sp.	Achira	X					X	X
Cyperaceae	*Cyperus* sp.	Tiririca					X		X
	Scirpus sp.	Junco					X	X	
Euphorbiaceae	Sp. *indet*	Curupí						X	X
Lauraceae	Sp. *indet*	Laurel		X		X			X
Maranthaceae	*Calathea* sp.	Talia			X				
Myrtaceae	Sp. *indet*	Guayabo	X	X				X	X
Phytolacacae	*Phytolacca dioica L.*	Ombú	X						X
Rhamnaceae	Sp. *indet*	Coronilla						X	
Rosaceae	Sp. *indet*		X						
Salicaceae	Sp. *indet*	Sauce						X	X
Santalaceae	Sp. *indet*								
Sapindaceae	Sp. *indet*	Chirca		X				X	
Sapotaceae	Sp. *indet*	Mataojo	X						X
Solanaceae	Sp. *indet*		X						X
	Datura ferox L.	Cardo	X						X
Thyphaceae	*Thypha* sp.	Totora	X		X	X	X	X	

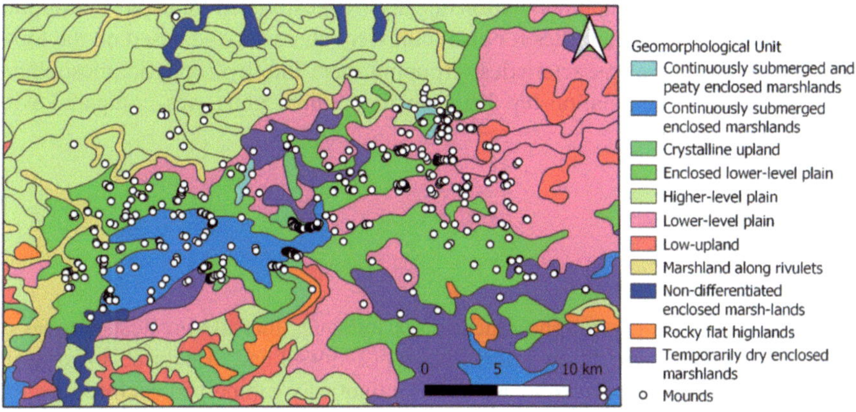

Fig. 11.4 Mound distribution by geomorphological unit

Table 11.2 Plant formation classification

Life mode	Access to water	Plant formation
Forest	Hydrophytic	Hydrophytic Forest
	Mesophytic	Ombu Forest
		Hill Forest
		Riparian Forest
		Palm grove
	Xerophytic	Psammophilic Forest
Shrubland	Hydrophytic	Riparian shrubland
	Mesophytic	Espinillar
	Xerophytic	Psammophilic shrubland
Herbaceous	Hydrophytic	Caraguatal
		Tirirical
		Gramal
		Pajonales
		Totoral
		Juncal
	Mesophytic	Hunquillar
		Uliginous grassland
		Drained grassland
	Xerophytic	Psammophilic grassland

The distribution of plant resources conformed to the mapping and classification of Plant Formations (PF) of south-eastern wetlands proposed by Barreneche and Zarucki (2017). This PF classification combines different attributes on PF including vegetation present and edaphic properties, life mode and access to water (among others) in a shp layer. For the purposes of this research PF were grouped according to their life mode and access to water (Table 11.2).

Surface area and mound population density were estimated for each geomorphological unit. Plant communities for each unit were characterized based on the topographic and drainage characteristics, grouping them into hydrophytic, mesophytic and xerophytic. The relationship between these three units was compared with mound distribution.

11.8 Results

The distribution by geomorphological units indicates that the greatest number of mounds is found in the "Lower-level plain" (27.99%) and "Enclosed lower-level plain" (27.04%). In the lower areas, the "Continuously submerged enclosed marshlands" (16.98%) and "Temporarily dry enclosed marshlands" (7.08%) stand out. The lowest concentration occurs in the high areas of "Higher-level plain" (9.12%) and "Crystalline uplands" (5.97%). However, the highest densities are found in wetlands (1.79 / km^2) and in hills (2.93 /km^2) (Table 11.3).

Mound distribution over geomorphological units were estimated as independent variables. In this way, it would be expected to observe a coincidence between the percentage of the area represented by each unit and the number of mounds present there. Comparing the real distribution against the expected one, it can be observed that the greatest contrasts are found in the "Continuously submerged enclosed marshlands", the "Lower-level plain", the "Enclosed lower-level plain" and the "Crystalline upland" (Fig. 11.5).

Wild plant resources were quantified for each life mode and plant community by access to water. Based on the life mode, forests gather the highest number of species (14) followed by herbaceous (6) and shrubland (4) (Table 11.4). The communities

Table 11.3 Number, percentage and density of mound grouped by geomorphological unit

Geomorphological unit	Code	Area (km^2)	Mounds (n)	Mounds (%)	Density
Marshland along rivulets	Ba	52	15	2.35	0.29
Continuously submerged enclosed marshlands	Bri	85	61	9.55	0.72
Continuously submerged and peaty enclosed marshlands	Brit	5	9	1.41	1.79
Non-differentiated enclosed marshlands	Brnd	27	3	0.47	0.11
Temporarily dry enclosed marshlands	Brs	165	24	3.76	0.15
Crystalline upland	C	40	116	18.15	2.93
Higher-level plain	LA	408	30	4.69	0.07
Lower-level plain	LM	285	172	26.92	0.60
Enclosed lower-level plain	LMb	246	204	31.92	0.83
Low-upland	M	54	2	0.31	0.04
Rocky flat highlands	SR	27	3	0.47	0.11

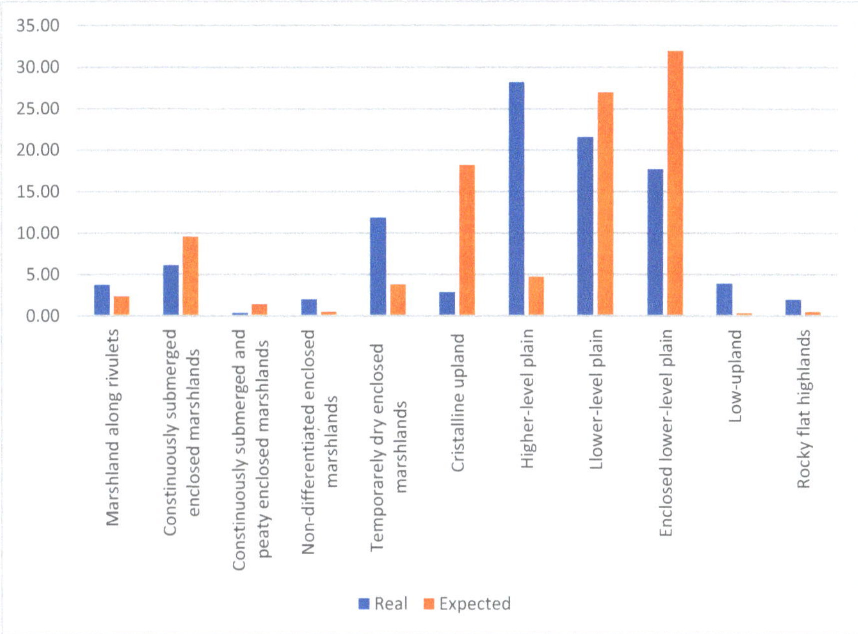

Fig. 11.5 Real mound distribution versus expected grouped by geomorphological unit

Table 11.4 Number of wild plant resources grouped by life mode

Life Mode	Number of species
Forest	14
Shrubland	4
Herbaceous	6

that concentrate the highest number of species are mesophytic (10) and hydrophytic (9) followed by xerophytic (5) (Table 11.5). These numbers reflect the diversity of the offer but the presence of a species in one of these categories does not exclude its occurrence in others.

When considering the documented use in ethnohistoric sources for these resources it can be observed that the mesophytic community has the greatest supply of food, medicinal and industrial resources (Table 11.6). The hydrophytic community follows in terms of supply of resources (Fig. 11.6).

Once the geomorphological units have been grouped (Fig. 11.8) by the type of plant community that they support, it is observed that the highest concentration of mounds occurs in the mesophytic communities (55.03%), followed by the hydrophytic (29.72%) with the lowest concentration in xerophytics (15.25%) (Table 11.7) (Fig. 11.7).

Table 11.5 Number of wild plant resources grouped by access to water

Plant community by access to water	Number of species
Hydrophytic	9
Mesophytic	10
Xerophytic	5

Table 11.6 Use of wild plant resources grouped by access to water

	Food	Fuel	Domestic supplement	Boats	Housing	Industries	Medicinal
Xerophytic	3	0	1	0	0	3	4
Mesophytic	10	3	2	1	2	5	7
Hydrophytic	5	2	1	2	3	7	7

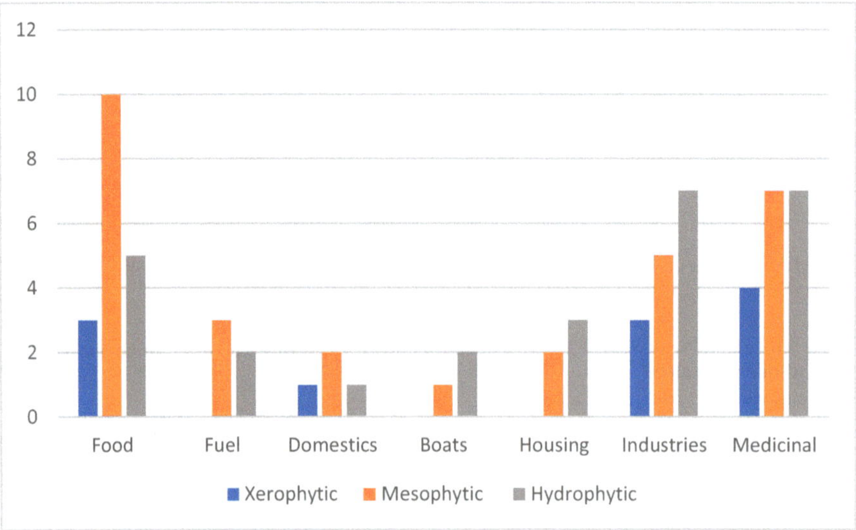

Fig. 11.6 Wild plant resources available by plant community

Table 11.7 Number, percentage and density of mound grouped by plant community unit

	Area (km^2)	Mound (n)	Mound (%)	Density
Hydrophytic	448	189	29.72	0.42
Mesophytic	579	350	55.03	0.60
Xerophytic	362	97	15.25	0.27

11.9 Discussion

Although the results arrived to in this paper should be considered preliminary, the set of spatial data generated is strong enough to support useful interpretations and clarify the direction for future research. The analysis of the data on mound

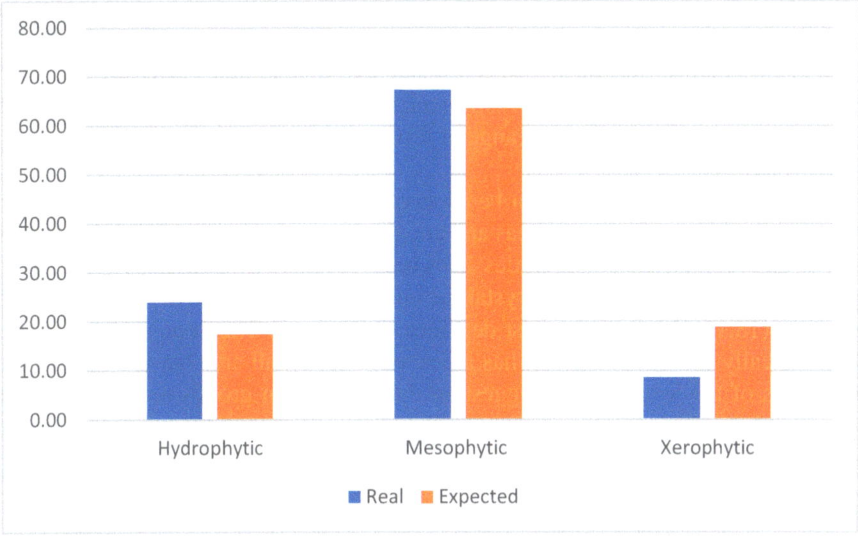

Fig. 11.7 Real mound distribution versus expected grouped by plant community

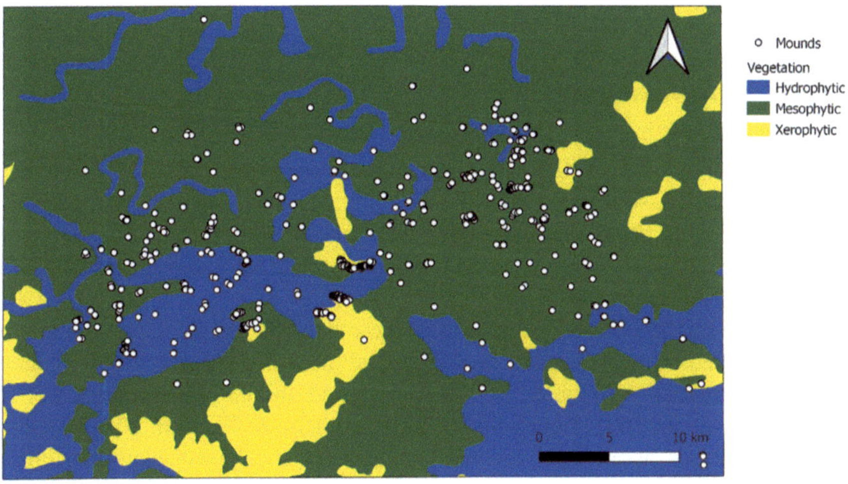

Fig. 11.8 Mound distribution by plant community

distribution and wild plant resources presents in the MLB archaeobotanical record indicate a strong correlation between these archaeological sites and the supply of resources. It should be noted that the evaluation of resources carried out in this work is linked only to the diversity of the offer and not to the relative abundance that each species can offer.

The concentration of mounds in environments that bring together the greatest food, medicinal and industrial offer suggests that their location responds to factors related to the economy of their builders. A higher concentration and density of mounds is occasionally observed in mesophytic and hydrophytic forest and grassland environments. Specifically, species such as *Typha domineguensis*, *Canna* sp., *Scirpus* sp., *Butiá capitata*, *Syagrus romanzoffiana* and *Bromelia* sp. meet in these environments, providing a wide range of resources that would be available for most of the year.

Even for the high areas it can be observed that mound groupings are found in wedges that go into the lower areas and are located on the border between the xerophytic and mesophytic communities. This becomes more evident when considering geomorphological units. The "Crystalline upland" presents relative low numbers of mounds (18.15%) but the highest densities (2.93 /km^2). This topographically and lithologically differentiated unit has an exceedingly small area but is closer to the resources of the lower lands, suggesting that the human groups that settled there sought to exploit the resources available in the vicinity of neighbouring environments.

On the other hand, these results reinforce the idea that the IMPB zone has functioned as a wetland since the middle Holocene. Although the information on plant formations was based on recent observations, the coincidence of mounds with the currently observable wild resources indicates that their distribution has remained relatively stable since the beginning of the elevation of the mounds in the IMPB.

11.10 Conclusions

Earthen mounds in the IMPB span over 5000 years of history and thousands of square kilometres, so it is not possible to define one objective or reason why they were raised, what behaviours originated them and how their function changed of millennia so any generalization could not be more than informed speculations (del Puerto et al., 2021).

Research on the genesis and growth of these mounds has paid special attention to accumulated elements such as sediments and discarded objects, which integrate waste from human activity, as well as human and animal burials (Bracco et al., 2015; Iriarte et al., 2001; López-Mazz, 2001; Milheira & Gianotti, 2018; Schmitz, 1976). Proposals about their function range from occupation platforms during floods, funerary places, spaces of ritual importance, territorial markers, habitation sites, locations for food processing and disposal, places for the preparation, use and disposal of different instruments, controlled burning in the form of earth stoves and ovens, support for crops and multi-functional systems for agricultural purposes (Bracco et al., 2021; Andrade Lima & López-Mazz, 1999; Baeza and Panario, 1999; Bracco, 2006; Iriarte, 2006; Bracco et al., 2016; del Puerto, 2015; Duarte et al., 2017; López-Mazz, 2001).

The goal of this research was not to elucidate either the mode or motive of construction, or the elements that constitute its matrix. Rather than zeroing in on their "purpose" or function this work aimed to assess the relation and distribution of these mounds with respect to environmental variables. After reviewing mound distributions some precisions can be made, mainly that the persistence of these occupations over millennia implies a close relationship between human groups and the environment, especially the wild vegetable resources they provided.

Acknowledgement I am grateful to professors Roberto Bracco and Daniel Panario for facilitating the geomorphological mapping and providing comments on methodological aspects of the work and data analysis. To Professor Ofelia Gutiérrez for her help in setting up the geographic information system. The results presented here are advances of the master's thesis project "Estructuras monticulares en la génesis del paisaje de los humedales del sureste de Uruguay. Un abordaje desde la Ecología del Paisaje" financed by Agencia Nacional de Investigación e Innovación through the Sistema Nacional de Becas.

References

Achkar, M., Domínguez, A., & Pesce, F. (2012). *Cuenca de la Laguna Merín-Uruguay. Aportes para la discusión ciudadana.* IECA-Facultad de Ciencias/Programa Uruguay Sustentable/ REDES-AT.

Andrade Lima, T., Lopez-Mazz J. (1999) La Emergencia de la Complejidad entre los Cazadores Recolectores de la Costa Atlántica Meridional Sudamericana. *Boletín de Antropología Americana*, n°.17, 18 y 19: 128–176.

Baeza, J., Panario D. (1999) La horticultura indígena en las estructuras monticulares. En: Actas de las Primeras Jornadas sobre Cenozoico en Uruguay. Sociedad Uruguaya de Geología and INGEPA, UNCIEP, Facultad de Ciencias, Montevideo.

Balée, W. (2010). *Contingent diversity on anthropic landscapes. Diversity, 2*, 163–181.

Balée, W., & Erickson, C. (Eds.). (2006). *Time and complexity in historical ecology: Studies in theneotropical lowlands*. Columbia University Press.

Barreneche J, Zarucki M (2017) Mapeo y clasificación de formaciones vegetales del sitio Ramsar bañados del este y franja costera. Documentos de trabajo N° 52. Rocha.

Bernick, K. N. (1998). *Hidden Dimensions: The cultural significance of wetland archaeology*. UBC Press.

Bracco Boksar, R. (2006). Montículos de la Cuenca de la Laguna Merín: Tiempo, Espacio y Sociedad. *Latin American Antiquity, 17*, 511–540.

Bracco Boksar, R., Inda, H., Castiñeira, C., Sprechmann, P., & García-Rodríguez, F. (2005). Relationships between Holocene Sea-level variations, trophic development, and climatic change in Negra lagoon, southern Uruguay. *Journal of Paleolimnology, 33*, 252–263.

Bracco Boksar, R., del Puerto, L., & Inda, H. (2008). Prehistoria y Arqueología de la Cuenca de Laguna Merín. In D. Loponte & A. Acosta (Eds.), *Entre la tierra y el agua* (pp. 1–59). Arqueología de humedales del Este de Sudamérica.

Bracco Boksar, R., del Puerto, L., Inda, H., Panario, D., Castiñeira, C., & García-Rodríguez, F. (2011). The relationship between emergence of moundbuilders in SE Uruguay and climate change inferred from opal phytolith records. *Quaternary International, 245*, 62–73.

Bracco Boksar, R., del Puerto, L., Inda, H., Capdepont, I., Panario, D., & García-Rodríguez, F. (2012). *Evolución ambiental y constructores de cerritos en la región de India Muerta: Un replanteo* (pp. 14–15). III Jornadas del Cenozoico. Montevideo.

Bracco Boksar, R., Panario, D., Duarte, C., & Bazzino, A. (2019). Estructuras monticulares y hormigueros en el sur de la cuenca de la Laguna Merín: ¿Ingenieros ambientales y/o la estrategia del bricoleur. *Revista de Antropología del Museo de Entre Ríos, 5*, 24–40.

Bracco, R, del Puerto L, Inda H (2008) Prehistoria y Arqueología de la Cuenca de la Laguna Merín. En D. Loponte y A. Acosta (Eds.), Entre la tierra y el agua: arqueología de humedales de Sudamérica, pp. 1-59. Asociación Amigos del Instituto de Antropología y Pensamiento Latinoamericano (AINA), Buenos Aires.

Bracco, R., del Puerto, L., Inda, H., Castiñeira, C., Garcia-Rodriguez, F., & Panario, D. (2009). Cambios ambientales y arqueología en el actual territorio del Uruguay. *Cazadores-Recolectores del Cono Sur., 3*, 15–27.

Bracco, B. R., Inda, H., & del Puerto, L. (2015). Complejidad en montículos de la cuenca de la laguna Merín y análisis de redes sociales. *Intersecciones en antropología, 16*, 271–286.

Bracco, R., Mazarino, J., del Puerto, L., Inda, H., Panario, D., Bazzino, A. (2016). Cambios ambientales y estructuras monticulares en la cuenca de la Laguna Merín. *In: 15 Actas del XIX Congreso Nacional de Arqueología Argentina. Serie Monográfica y Didáctica* (pp. 1148–1154). Facultad de Ciencias Naturales e Instituto Miguel Lillo de la Universidad Nacional de Tucumán, San Miguel de Tucumán.

Bracco, R., C. Duarte, O. Gutiérrez, M. Clara, Panario D. (2021). Reflexiones sobre montículos, técnicas de procesamiento de alimentos y construcción de nichos. *Arqueología*, 27(2), mayo-agosto: 109–130.

Brown, A. (1997). *Alluvial geoarchaeology: Floodplain archaeology and environmental change.* Cambridge University Press.

Campos, S., del Puerto, L., & Inda, H. (2001). Opal phytolithsanalysis: Ilsapplication to the archaeobotanical record in the east of Uruguay. In F. Colin (Ed.), *Meunier J* (pp. 129–142). Phytoliths.

Capdepont, I., & Pintos, S. (2006). Manejo y aprovechamiento del medio por parte de los grupos constructores de montículos: cuenca de la Laguna de Castillos, Rocha-Uruguay. *Relaciones de la Sociedad Argentina de Antropología, XXXI*, 117–132.

del Puerto, L. (2009). *Silicofitolitos como indicadores paleoambientales: Bases comparativas y reconstrucción paleoclimática a partir del Pleistoceno tardío en el SE del Uruguay*. Editorial Académica Española.

del Puerto, L. (2011). Ponderación de recursos vegetales silvestres del este del Uruguay: rescatando el conocimiento indígena tradicional. *Trama Revista de Cultura y Patrimonio, 3*, 22–41.

del Puerto, L. (2015). *Interrelaciones humano-ambientales durante el Holoceno tardío en el este del Uruguay: Cambio Climático y Dinámica Cultural*. Tesis Doctoral. Programa de Desarrollo de las Ciencias Básicas. Universidad de la República.

del Puerto, L., & Inda, H. (2005). Paleoetnobotánica de los constructores de túmulos del Noreste del Uruguay: Análisis de silicofitolitos de la estructura monticular Yale 27 y su entorno. *Tapa, Traballos de Arqueoloxía e Patrimonio, 36*, 109–120.

del Puerto, L., & Inda, H. (2008). Estrategias de Subsistencia y Dinámica Ambiental: Análisis de silicofitolitos en sitios arqueológicos de la cuenca de Laguna de Castillos, Rocha, República Oriental del Uruguay. In A. Zucol, M. Osterrieth, & M. Brea (Eds.), *Fitolitos en Suelos, Sedimentos y Sitios Arqueológicos: Estado Actual de sus Conocimientos en América del Sur* (pp. 221–236). Universidad Nacional de Mar del Plata.

del Puerto, L., Bracco Boksar, R., Inda, H., Gutiérrez, O., Panario, D., & García-Rodríguez, F. (2013). Assessing links between late Holocene climate change and paleolimnological development of Peña lagoon using opal phytoliths, physical, and geochemical proxies. *Quaternary International, 287*, 89–100.

del Puerto, L., Capdepont, I., & Inda, H. (2016). Paleoetnobotánica y subsistencia de los constructores de Cerritos del Holoceno Tardío en el Este del Uruguay: análisis fitolítico en sedimentos y artefactos arqueológicos. *Tessituras, Pelotas, 4*, 117–160.

del Puerto, L., Gianotti, C., Bortolotto, N., Gazzán, N., Cancela, C., Orrego, B., & Inda, H. (2021). Geoarchaeological signatures of anthropogenic soils in southeastern Uruguay: Approaches to formation processes and spatial–temporal variability. *Geoarchaeology, 37*(1), 1–18.

Duarte, C., Bracco Boksar, R., Panario, D., Tassano, M., Cabrera, M., & Bazzino, A. (2017). Datación de estructuras monticulares por OSL / TL. *Revista de Antropología del Museo de Entre Ríos, 3*, 14–26.

Erickson, C. (2003). Historical ecology and future explorations. In J. Lehmann, D. Kern, B. Glaser, & W. Woods (Eds.), *Amazonian dark earths. Origins, properties and management* (pp. 455–500). Kluwer.

Erickson, C. (2006). The domesticated landscapes of the Bolivian Amazon. In W. Balée & C. Erickson (Eds.), *Time and complexity in historical ecology*. Columbia University Press.

Ferres, C. (1927). Los terremotos de los indios. *Revista de la Sociedad de Amigos de la Arqueología, Montevideo, 1*, 139–151.

Finlayson, M., Horwitz, P., & Weinstein, P. (Eds.). (2015). *Wetlands and human health*. Springer.

Fraser, J., Cardoso, T., Junqueira, A., Clement, F. N., & C. (2009). Historical ecology and dark earths in Whitewater and Blackwater landscapes: Comparing the middle Madeira and lower negro Rivers. In W. Woods, W. Texiera, J. Lehman, C. Steiner, & L. Rebellato (Eds.), (pp. 229–264). Amazonian Dark Earths: Wim Sombroek's Vision.

García-Rodríguez, F., Stutz, S., Inda, H., del Puerto, L., Bracco Boksar, R., & Panario, D. (2010). A multiproxy approach to inferring Holocene paleobotanical changes linked to sea-level variation, paleosalinity levels, and shallow lake alternative states in Negra lagoon, SE Uruguay. *Hydrobiologia, 646*, 5–20.

Gianotti, C. Y. Bonomo, M. De. (2013). *montículos a paisajes: procesos de transformación y construcción de paisajes en el sur de la Cuenca del Plata*. Número 17, segundo semestre 2013, pp. 129–163, Córdoba.

Inda, H. (2009). *Paleolimnología de cuerpos de agua someros del sudeste del Uruguay: evolución holocénica e impacto humano*. Unpublished doctoral thesis, PEDECIBA-Biología, Universidad de la República, Montevideo.

Inda, H., del Puerto, L. (2007). Antracología y Subsistencia: Paleoetnobotánica del Fuego en la Prehistoria de la Región Este del Uruguay. Puntas del San Luis, Paso Barrancas, Rocha, Uruguay. En Marconetto, M. B., Babot, M. del P. Y Oliszewsk, N. eds. – Paleoetnobotánica del Cono Sur: Estudios de casos y propuestas metodológicas Córdoba, : Museo de Antropologia FFyH-UNC. Ferreyra Editor, pp. 137–152.

Inda, H., García-Rodríguez, F., del Puerto, L., Acevedo, V., Metzeltin, D., Castiñeira, C., & Bracco, R. (2006). Relationships between trophic state, paleosalinity and climatic changes during the first Holocene marine transgression in Rocha lagoon, southern Uruguay. *Journal of Paleolimnology, 35*, 699–712.

Inda, H., del Puerto, L., Bracco Boksar, R., Castiñeira, C., Capdepont, I., Gascue, A., & Baeza, J. (2011). Relación Hombre-Ambiente para la costa estuarina y oceánica del Uruguay durante el Holoceno. Reflexiones y Perspectivas. In F. García-Rodríguez (Ed.), *El Holoceno en la zona costera del Uruguay* (pp. 229–260). D – Universidad de la República.

Iriarte, J. 2006. Landscape transformation, mounded villages and adopted cultigens: The rise of early formative communities in south-eastern uruguay. *World Archaeology, 38*, 644–663.

Iriarte, J., Hoist, I., López Mazz, J., & Cabrera, L. (2001). Subtropical wetland adaptations in Uruguay during the mid-Holocene: An archaeobotanical perspective. In B. A. Purdy (Ed.), *Enduring records: TheEnvironmental and cultural heritage of wetlands* (pp. 61–70). Oxbow Books.

Krtshnamurthy, K. (1990). The apiary of the mangroves. In D. Whigham,, R. Good & J. Kvet (Eds.). *Wetland Ecology and Management: Case Studies*. Kluwer Academic Publisher.

López Mazz, J., Dabezies, J., & Capdepont, I. (2014). La gestión de recursos vegetales en las poblaciones prehistóricas de las tierras bajas del sureste del Uruguay: un abordaje multidisciplinar. *Latin American Antiquity, 25*, 256–277.

López-Mazz, J. (2001). Las estructuras tumulares (cerritos) del Litoral Atlántico uruguayo. *Latin American Antiquity, 12*(3), 1–25.

Mazarino, J., Duarte, C., & Bracco Boksar, R. (2015). Datación por OSL: un ejemplo de aplicación. In L. Cabrera & J. López Mazz (Eds.), *Anuario de Arqueología* (pp. 72–87). Universidad de la República.
Meggers, B. J. (1954). Environmental limitation on the development of culture. *American Anthropologist, 56*, 801–824.
Meggers, B. J. (1971). *Amazonia: Man and nature on a counterfeit paradise*. Aldine.
Milheira, R., & Gianotti, C. (2018). The earthen mounds (Cerritos) of southern Brazil and Uruguay. In C. Smith (Ed.), *Encyclopedia of global archaeology Cham*. Springer.
Miller, E. T. (1999). A limitação ambiental como barreira à transformação do período Formativo no Brasil. Tecnologia, produção de alimentos e formação de aldeias no sudoeste da Amazônia. In P. Ledergerber-Crespo (Ed.), *Formativo Sudamericano, una Revaluación* (1st ed., pp. 331–339). Abya-Yala.
Montaña, J., Bossi, J. (1995). *Geomorfología de los humedales de la cuenca de la Laguna Merín en el departamento de Rocha*. In PROBIDES Work documents.
Moraes, C. P. (2015). *Estudos Avançados 29*(83), 25–43.
Neves, E. G., & Petersen, J. (2006). The political economy of pre-Columbian Amerindians: Landscape transformations in Central Amazonia. In W. Balée & C. L. Erickson (Eds.), *Time and complexity in historical ecology: Studies in the Neotropical lowlands*. Columbia University Press.
Panario, D. (1988). *Geomorfología del Uruguay. Publicación de la Facultad de Humanidades y Ciencias*. Universidad de la República.
Panario, D., Gutiérrez, O., Sánchez-Bettucci, L., Peel, E., Oyhantçabal, P., & Rabassa, J. (2014). Ancient landscapes of Uruguay. In J. Rabassa & C. Ollier (Eds.), *Ancient Gondwana landscapes in southern South America: Argentina, Uruguay and southern Brazil* (pp. 161–199). Springer.
Peel, M., Finlayson, B., & Mcmahon, T. (2007). Updated world map of the Köppen-Geiger climate classification. *Hydrology and Earth System Sciences Discussions, 4*, 439–473.
Pintos, S. (2000). Economía "Húmeda" del este del Uruguay: el manejo de recursos faunísticos. En: A. Durán Y R. Bracco (Eds), Arqueología de las Tierras Bajas, pp. 249–266. Imprenta Americana.
Ramsar. (2008). Ramsar Convention on Wetlands; FAO; International Water Management Institute (IWMI). (2014). Wetlands and agriculture: partners for growth. Gland, Switzerland.
Reed, K. (1997). Early hominid evolution and ecological change through the African Plio-Pleistocene. *Journal of Human Evolution, 32*, 289–322.
Schmitz, P. I. (1976). Sitios de Pesca Lacustre em Rio Grande, RS, Brasil. In *Tese de livre docencia, Pontificia Universidad Católica Río Grande do Sul (PUCRGS)*. UNISINOS, São Leopoldo.
Simoes, M. (1972). *Índice das fases arqueológicas brasileiras 1950–1971*. Conselho Nacional de Pesquisas. Instituto Nacional de Pesquisa da Amazonia. Museu Paraense Emilio Goeldi. Publicações Avulsas Nro. 18.
Sombroek, W. (1969). *Soil studies in the Merim Lagoon Basin*. Organización de las Naciones Unidas para la Alimentación y la Agricultura.
Soriano, A. (1991). Río de la Plata grasslands. In R. Coupland & D. Goodall (Eds.), *Natural grasslands* (pp. 367–407). Introduction and western hemisphere.
Tedeschi, A. (2011). Acts, practices, and the creation of place: Geoarchaeology of a Terra Preta de Índio site in the Central Amazon. In *Dissertation submitted in partial satisfaction of the requirements for the degree of doctor of philosophy in anthropology*. University of California.
Wilkinson, S. (1958). 5-Metkoxy-N-methyltryphmine: A new indole alkaloid from phalaris arundinacea L. *Journal of the Chemical Society, 2*, 2079–2081.

Chapter 12
Charcoal Production, Social Invisibility and the Genesis of a Landscape in the Pedra Branca Massif (Rio de Janeiro, Brazil)

Rúbia Graciele Patzlaff, Rogério Ribeiro de Oliveira, and Rita Scheel-Ybert

Abstract The association of different approaches, such as Environmental Archaeology, Anthracology, Historical Ecology, Plant Ecology, and Landscape Archaeology, offers new perspectives into recovering the histories of invisible people and of the forests in which they lived. The more than 2000 historical kiln sites and several other remains located in the urban forests of Rio de Janeiro city testify of the fuel economy during the colonial and imperial periods of the Brazilian society. Charcoal, allowing for easier handling and wider uses than firewood, was used in large amounts. Anthracological analysis associated to phytosociological surveys carried out in these historical charcoal kilns aims to bring a new understanding of charcoal burners who exploited these forests. Floristic similarity between the current and past forest formations is discussed, as well as the charcoal burners practices, such as location and species selection (or absence of). The main actors in this

R. G. Patzlaff (✉)
Archeobotany and Landscape Lab, Anthropology departament, Museu Nacional, Universidade Federal do Rio de Janeiro, Rio de Janeiro, Brazil
e-mail: patzlaff@mn.ufrj.br

R. R. de Oliveira
Geography Posgraduated Program, Pontifícia Universidade Católica do Rio de Janeiro, Rio de Janeiro, Brazil
e-mail: rro@puc-rio.br

R. Scheel-Ybert
Archeobotany and Landscape Lab, Anthropology departament, Museu Nacional, Universidade Federal do Rio de Janeiro, Rio de Janeiro, Brazil

Archeology Postgraduated Program, Museu Nacional, Universidade Federal do Rio de Janeiro, Rio de Janeiro, Brazil
e-mail: scheelybert@mn.ufrj.br

A. C. Colonese, R. G. Milheira (eds.), *Historical Ecology and Landscape Archaeology in Lowland South America*, Interdisciplinary Contributions to Archaeology, https://doi.org/10.1007/978-3-031-32284-6_12

process – the charcoal workers – experienced intense social invisibility; their work, despite essential to society, showed signs of marginality and illegality. The studied landscape keeps the marks of this cycle of energy provision in several of its attributes

Keywords Atlantic forest · Landscape · Charcoal kilns · Historical ecology · Anthracology · Brazil

12.1 Introduction

The written history of the Atlantic Forest, focused on the recent impact of the great economic cycles, does not consider the results of pre-colonial population activities, indigenous groups, and enslaved Africans. The "history of the vanquished", of those who left no written documentation, is left with no more than the material culture for its rescue.

In many cases, however, the material culture itself is minimal or non-existent. But the concept of material culture goes beyond the more common notion of artifacts found at the archaeological sites and can encompass the landscape as a whole (Bender, 2006). Modern archaeology interprets landscape as an expression of human agency, a manifestation of the way people think and act upon the world (Renfrew & Bahn, 2005). In environmental history, the landscape has long been interpreted as a historical document (Nash, 1970).

There are many landscape concepts, which vary according to the discipline or theoretical framework considered, but most people accept that landscape results from the interaction between natural processes and human activities. By considering the culture-environment pair as inseparable, the landscape concept rejects any kind of binary relationship between humans and nature (Ingold, 1993). Landscapes are dynamic, changing according to socioeconomic and biophysical systems, and evolving at various temporal and spatial scales. They reflect cultural, social, and economic histories. Thus, they carry strong historicity, which adds a condition of multidisciplinarity to their study.

Elements of forest structure and composition make it possible to retrieve the history of a landscape and identify traces of human agency. This approach has been widely applied in the Amazon. Recent studies have attributed the existence of forests with hyperdominance of useful species to management and cultivation practices by populations before the arrival of Europeans in the Americas (Junqueira et al., 2011; Ter Steege et al., 2013; Levis et al., 2017). Studies associating anthracology and historical ecology have also allowed identifying patterns of landscape modification. Secondary vegetation areas and possible concentration of useful species were identified around the settlements from late Holocene populations in the Amazon, Southeastern, and Southern Brazil (Scheel-Ybert et al., 2016). In the Atlantic Forest, the creation of such areas has been demonstrated in association with Proto-Tupinambá groups from at least about 3200 calibrated years BP. These groups lived in settlements surrounded by secondary vegetation, where they practiced cultivation and retrieved firewood from, but also exploited areas of mature forest. Despite the long history of environmental transformation in the Atlantic Forest, this did not result in biodiversity loss, as the forest tends to regenerate after the abandonment of clearings and settlements, even though it retains marks of its past uses (Scheel-Ybert et al., 2013, 2016).

Forests, as part of the landscape and geographic space, are also perceived as territories. Over time, successional land uses leave marks, become specialized, and are overlapped, constituting paleo-territories. Those, in constant social and ecological transition, form a repository of profound stories about relations between human communities and nature (Oliveira, 2007, 2015). This repository is particularly important for the study of populations or social segments that left few material traces of their passage and little or no written documentation related to their existence.

This is the case of charcoal burners who exploited the Pedra Branca Massif in the western part of Rio de Janeiro city (Brazil) since at least the eighteenth century, in an activity that continued at least until the 1940s. Documentary records about these people and their activities are virtually inexistent. The intense process of social invisibility to which they were subjected is also manifested in the scarce material culture found in the archaeological sites – which, except for a large number of ecofacts[1] (charcoal), is restricted to a few tools and bottles (Fraga & Oliveira, 2012; Oliveira et al., 2020).

This chapter presents data and discusses the relevance of studying the marks left in forests by the people who lived in and from these lands. Our work is based on the analysis of ecofacts and considers the charcoal burners' paleo-territory, the landscape built by them, and their material culture.

12.2 The Historical Context of Charcoal Production and Landscape Transformations in Rio de Janeiro

The history of charcoal production in Brazil is mainly related to the development of large urban centers, and still little known. As are the people involved in this process. Historical charcoal kiln sites are quite common in Southeastern Brazil. Their occurrence has been recorded in several locations in the states of Rio de Janeiro,

[1] Ecofacts are archaeological remains of biological origin that have a cultural significance, but that have not been directly modified by human action, including plant, animal, and landscape remains. For Emery (2007), ecofacts are both environmental and cultural indicators, being the most diverse and comprehensive archaeological remains. Ecofacts reflect the interrelationships of man with nature, with the climate, with different forms of life and with the landscape (Renfrew and Bahn, 2005). However, they also witness the choices of each group in the universe of possibilities that the environment presents to them. Therefore, the study of ecofacts, as well as artifacts of biological origin, goes beyond the interest in the biological trace itself, but seeks to identify in these remains the actions and motivations that may have led to their presence in an archaeological context.

São Paulo, Minas Gerais, and Espírito Santo, especially on forested slopes, where these vestiges are better preserved. They evidence how the charcoal production is part of the history of several Brazilian forests remnants (Oliveira et al., 2020).

The earliest record of charcoal production in Rio de Janeiro is a report by the Marquis of Lavradio from 1779 (Lavradio, 1913); in 1830, Taunay recorded this activity in a painting (Fig. 12.1). In the Itatiaia Massif, in the southern Rio de Janeiro State, there are reports of charcoal production in the 1930s (Deffontaines, 1937).

Researches such as those by Solórzano et al. (2005), Santos et al. (2006), and Beauclair et al. (2011) suggest that charcoal played an important role as an energy source for the Rio de Janeiro city until the beginning of the twentieth century. Charcoal was the main energy matrix of that time. It was consumed domestically, for transportation, in the production of tools and utensils, in smelting, in civil construction, and later in the textile and glass industries (Oliveira, 2005; Fraga & Oliveira, 2012).

The economic activity that most consumed charcoal was probably civil construction, more specifically stonemasonry – the practice of carving stones to produce construction pieces such as cobblestones, portals, façades, sidewalks, etc. (Fraga & Oliveira, 2012). This work required that the tools used, such as chisels, be sharpened frequently – but to keep their edge, the chisels needed to be taken to the forge, which was powered by charcoal. At the turning of the nineteenth and the twentieth centuries, the exponential growth of civil construction generated an increase in the

Fig. 12.1 The painting "View of a forest being turned into charcoal", by Félix-Émile Taunay (1830). (Source: Museum of Fine Arts of Rio de Janeiro; public domain image)

number of forges in Rio de Janeiro (Abreu, 2006). The city maintained in this period a complex socio-economic system, in which the main source of energy came from the forest, turned into charcoal by the charcoal burners. The intensification of charcoal production at that time was due to the conjunction of several factors, among which: (a) energy demand; (b) availability of nearby forest resource; (c) production facility; (d) low cost required, and (e) large available human contingent after the abolition of slavery in 1888 (Fraga & Oliveira, 2012).

The forest from the Pedra Branca Massif, near to the city of Rio de Janeiro, became an important charcoal production center. Historical records on this activity are rare, but Magalhães Corrêa (1936) reports that a considerable extension of the eastern part of this massif has been used for charcoal production. This author gave detailed descriptions of various workers and trades of the time. Two categories of work in the forests were performed by the same people: the woodcutters and the charcoal burners. The woodcutters, mostly small land invaders and squatters, were responsible for cutting down and selling the wood. They became charcoal burners by necessity, on their own, as hired laborers or as salaried employees. Charcoal burners and woodcutters are described as locals (born in the state of Rio de Janeiro) and mentioned as black, white, and mixed-race people.

The scarcity of historical references about this work and its workers shows that they were socially invisible people, who probably lived on society margins. In this society, with a strong slavery ideology, manual labor was considered degrading. It was therefore relegated only to the poorest individuals. The rare reports found in the literature confirm that these people were poor and reveal a strong prejudice concerning their skin color and occupation. The work performed inside the forests was even more prejudiced; forests were far from the city and considered as remote, dangerous, and dirty places, as well as a source of diseases. Besides, the forested mountains surrounding Rio de Janeiro were refuges for enslaved people who tried to escape from the system, forming *quilombos*. In these refuges, they became small producers but also conducted business with white society (Fraga & Oliveira, 2012).

Thus, it is not known for a fact who these people were, but they certainly belonged to the less privileged spheres of society. From the end of the nineteenth century, these people may have comprised the enslaved that were freed and their descendants, who would have found in this work a means of subsistence.

12.3 Landscape Archaeology, Anthracology, and Historical Ecology at Pedra Branca Massif

Currently, the entire area of this Massif that was exploited by charcoal burners has been recovered by the forest. Only small farms of residents still remain. The current, as well as the past vegetation, belong to the Atlantic Forest domain, specifically a slope forest (Submontane Dense Ombrophylous Forest). The area has a hilly topography, with deep valleys and eventual exposure of the crystalline massif with areas of rocky outcroppings and shallow soils (Firme et al., 2001; Oliveira, 2005).

Results from charcoal production can still be observed in this paleo-territory. Archaeological surveys have already located more than 1200 sites of ancient charcoal kilns in the region (Oliveira et al., 2020). Different analyses have been applied in this area, especially floristic and phytosociological surveys and, when possible, interviews with local inhabitants (Santos et al., 2006; Solórzano et al., 2007; Sales et al., 2014, 2016). The direct remains of this activity are still poorly explored, but anthracological analyses have been conducted by Beauclair et al. (2011) and by this work.

Our researches have shown that most of the historical charcoal kilns are now in areas of dense forest (Fig. 12.2). The time of forest regeneration varies spatially, reaching up to 150 years in some areas (Oliveira et al., 2013).

There was, therefore, intense use of the forested space. However, this intense use is accompanied by a relative archaeological invisibility of the activities performed in this paleo-territory. Despite intensive surface prospecting and pit tests carried out for site delimitation, only a few items of mobile material culture were retrieved. Among them, the most conspicuous remains were Port wine bottles, fragmented or intact, always found in the charcoal kilns vicinity. We believe that they were reused to store water, coffee, etc. A few tools, also found near the charcoal kilns, showed

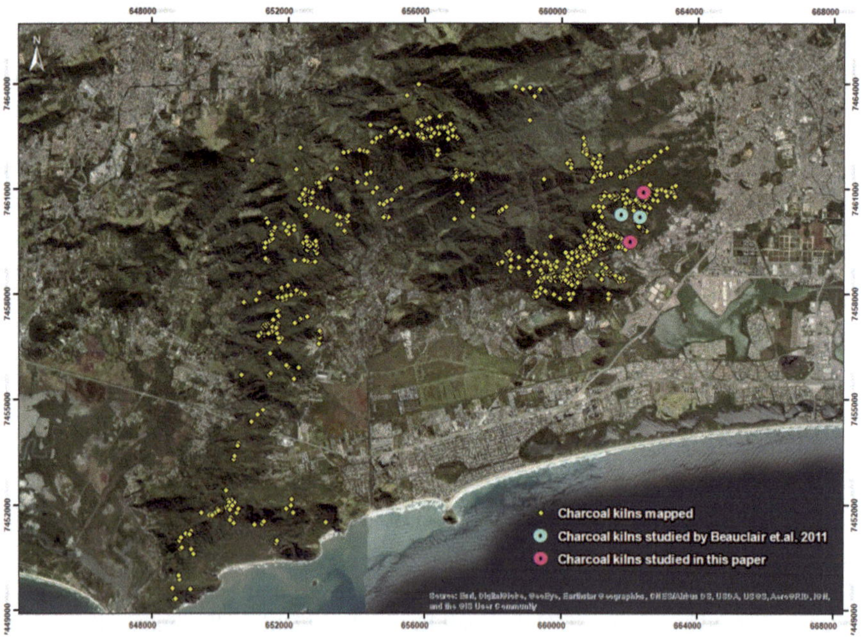

Fig. 12.2 Historical charcoal kilns located in the Pedra Branca Massif, Rio de Janeiro, RJ. Each *yellow dot* corresponds to a charcoal kiln site; *blue dots* correspond to charcoal kilns studied by Beauclair et al., 2011 and *pink dots* correspond to charcoal kilns studied in this paper. (Source: Modified from Oliveira, 2015 (by Joana Fraga, PUC-RJ, Laboratory of Biogeography and Historical Ecology))

evidence of severe deterioration and were unusable, possibly having been abandoned (Oliveira et al., 2020).

By contrast, this paleo-territory is filled with other categories of remains that offer different sets of information. Not only can the current landscape be read as a record, but the immense abundance of archaeological charcoal provides alternative interpretation possibilities for reconstructing the history of the forest and of the charcoal burners.

Anthracology is the study and interpretation of charred wood remains (charcoal) recovered at archaeological sites. The taxonomic identification is possible because the wood anatomical structure is perfectly preserved after carbonization (Fig. 12.3). Besides that, dendroanthracological analyses allow inferring wood features (such as diameter, phenology, and conditions such as rotting, fungal, or insect attack, among others) and environmental conditions (temperature, water availability) (Scheel-Ybert, 2020).

Analyses of charcoal from charcoal kilns allow reliable paleoecological reconstructions because charcoal burners did not select trees for production: they simply chose and logged an area of forest (Marziani & Citterio, 1999; Ludemann, 2002, 2008). Historical charcoal kilns are archaeological units marking the forest exploitation, thus bearing the memory of a forest that has now disappeared (Bonhote & Vernet, 1988). Therefore, anthracological analysis provide a picture of the past

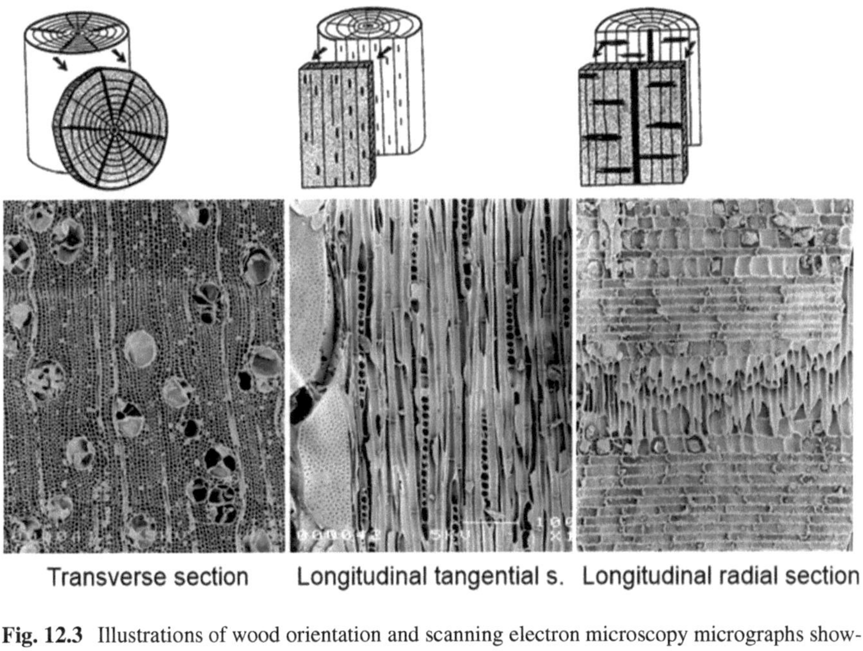

Fig. 12.3 Illustrations of wood orientation and scanning electron microscopy micrographs showing the wood anatomical structure in the three fundamental sections of a Leguminosae Caesalpinioideae (*Senna macranthera* (DC. ex Collad.) H.S. Irwin & Barneby). (Adapted from Scheel-Ybert, 2004)

landscape and, depending on the chronology, can provide an evolutionary view of the forest that was exploited (Marziani & Citterio, 1999).

The present work presents an unprecedented approach, in which anthracological and phytosociological analyses were performed in association at the same sites (Patzlaff et al., 2018).

Delimitation of Charcoal Kilns, Sampling, and Anthracological Analysis The remnants of historical charcoal kilns are recognized in the forest by the presence of platforms in the terrain – formed by plateaus excavated by the charcoal burners on slopes for *in situ* charcoal manufacture –, as well as by the abundance of charcoal fragments in the darkened soil. For anthracological sampling, charcoal kilns at different altitudes and forest contexts were selected (Fig. 12.2). After vegetation removal, the site surface was divided into 1 m × 1 m squares. Site delimitation and sediment sampling were carried out in 20-cm-side pits in alternated squares over the entire area of the charcoal kilns. The sampling was performed at 10 cm artificial levels to the end of the archaeological sediment; the entirety of the sediment in each pit was transported to the laboratory for flotation (Vernet, 1997; Scheel-Ybert et al., 2006). In the laboratory, the samples were floated and dried. For analysis, a minimum of 200 fragments per charcoal kiln was defined (Scheel-Ybert, 2002). The charcoal fragments were manually broken into three wood sections (Fig. 12.3); charcoal anatomy features were analyzed under a reflected light microscope with darkfield and brightfield. The taxonomical identification was performed by comparing the observed anatomical structures with the specialized literature and with anthracology and wood anatomy databases (InsideWood, 2004; Scheel-Ybert et al., 2014a, b; Scheel-Ybert & Gonçalves, 2017), along with the charcoal comparative collection from the Laboratory of Archaeobotany and Landscape of the Museu Nacional/UFRJ (Scheel-Ybert, 2016). Identifications were made at the genus level, and sometimes only at the family level. The high biodiversity of tropical regions usually prevents more precise determinations. An exception was made only in the case of monospecific genera (or with the occurrence of a single species for Rio de Janeiro State), which could be identified down to species level.

Phytosociological Analysis A phytosociological survey was conducted in areas of direct influence from the charcoal kilns. In each charcoal kiln, a total of 0.15 h (1500 m^2) was sampled in 15 parcels (20 m × 5 m) installed in a semicircle with a 60-m radius upstream from where the plateau was built, with the site placed as the center (Fig. 12.4).

This methodology was developed specifically for this study (Patzlaff et al., 2018). For sample collection, data recording, and taxonomic identification the standard phytosociological methods were used. Phytosociological parameters were calculated using the FITOPAC 2.1 software (Shepherd, 1995). The similarity between areas was calculated by Sørensen's coefficient, which compares species similarity qualitatively along an environmental gradient (Wolda, 1981). All plants were identified, whenever possible, to the species level. However, for comparison purposes

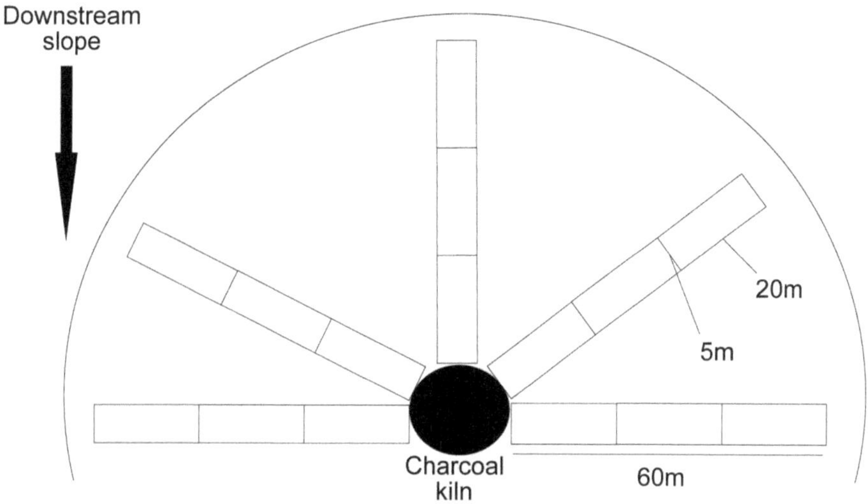

Fig. 12.4 Scheme showing the distribution of parcels for phytosociological survey in relation to the charcoal kiln location. (Adapted from Patzlaff et al., 2018)

with anthracological analysis, phytosociological analysis was reduced whenever necessary to the taxonomic level of genus or family.

12.4 Comparative Analysis: Past and Extant Forests in the Pedra Branca Massif

This study was based on anthracological and phytosociological sampling and analyses at two charcoal kilns located at Pedra Branca Massif, in the Camorim neighborhood, Rio de Janeiro:

1. Carvoaria Ruína do Cipó-2 (CRC2) (22°057′02″S, 43°024′38″W): located in Serra do Nogueira, in a forest whose regeneration time was estimated between 120–180 years, in a valley bottom, at 200 m altitude, with a slope of c. 30°. The soil, very dark, presented many charcoal fragments from the surface to c. 35 cm deep.
2. Carvoaria Soeiro-1 (Soeiro) (22°057′51″S, 43°025′13″W): located at the Soeiro farm, in a forest whose regeneration time was estimated between 60–100 years, at an altitude of 40 m, with a slope of c. 30°. The soil, very dark, presented large charcoal fragments from 2 cm from the surface to c. 30 cm deep.

In both sites, as well as in previous results obtained in the same area (Beauclair et al., 2011), the archaeological strata presented single layers, without any sedimentary hiatus or other sign of abandonment. Statistical analysis comparing the species composition in anthracological results also suggest that they correspond to single events (Beauclair et al., 2011). This indicates that each charcoal kiln represents a unique episode of forest logging and charcoal production.

In CRC2, the anthracological analysis identified 31 families, 37 genera, and 102 taxa, with 29 indeterminate. The phytosociological analysis identified 24 families, 49 genera, and 61 species, with 6 indeterminate specimens (Fig. 12.5).

The most abundant families in the anthracological analysis (ancient forest) were Celastraceae and Sapotaceae; the families with the highest taxonomic diversity were Euphorbiaceae, Leguminosae, and Sapotaceae. The most frequent taxa were *Maytenus*, Chrysobalanaceae, and *Pouteria*. The diversity of this sample was estimated at H′ = 4.06.[2]

In the phytosociological analysis (extant forest), families with the highest species richness were Leguminosae and Lauraceae; and with the highest Importance Value (IV) and Cover Value (CV)[3] were Moraceae and Euphorbiaceae. The species with the highest density was *Amphirrhox longifolia* (Violaceae, 38.74% of the individuals); and *Joannesia princeps* was the most dominant one. The diversity of this sample was estimated at H′ = 3.54.

At Soeiro, the anthracological analysis identified 27 families, 45 genera, and 69 taxa, with 5 indeterminate. The phytosociological analysis identified 21 families, 37 genera, and 38 species, with 7 indeterminate (Fig. 12.5).

The most well-represented families in the anthracological analysis (ancient forest), both in frequency and in taxonomic diversity, were Leguminosae and Euphorbiaceae. The most frequent taxon was *Pterocarpus rohrii*, followed by *Senegalia*, *Trichilia*, and *Maprounea*. The diversity of this sample was estimated at H′ = 3.73.

In the phytosociology analysis (extant forest), the Leguminosae family presented the highest species richness, while Meliaceae presented the highest IV and CV. Species with the highest densities were *Trichilia casaretti* and *Guarea guidonia* (Meliaceae), while *Gallesia integrifolia* and *Spondias mombin* showed the highest dominance. The diversity of this sample was estimated at H′ = 2.83.

In both sites, Leguminosae and Euphorbiaceae stand out to be the most prominent families for their importance in both ancient and extant forests. The genera

[2] The Shannon diversity index (H′) is one of the most used diversity indices in plant phytosociology. In Ecology, the diversity of a sample, or of a vegetation type, depends on the specific richness (number of species) and the frequency of each taxon in the community. A plant association has low diversity if the number of species is low and if the frequency distribution of each taxon is uneven. On the contrary, the diversity is high if the number of species is high and if its abundance is equivalent. At identical number of species, diversity will be higher if taxa frequency distribution is more uniform (Scheel-Ybert, 2005).

[3] The Importance Value (IV) and Coverage Value (CV) indices are frequently used in phytosociology to synthetically express the relative order between species (Curtis & McIntosh, 1950). Importance Value Index considers density, dominance, and relative frequencies; it provides an idea of the density, spatial dispersion and the size reached by a species, reflecting its ecological importance (Matteucci & Colma, 1982). The Coverage Value Index is obtained by adding the relative values of density and dominance, providing information related to the number of individuals and the biomass of each species (Moro & Martins, 2011).

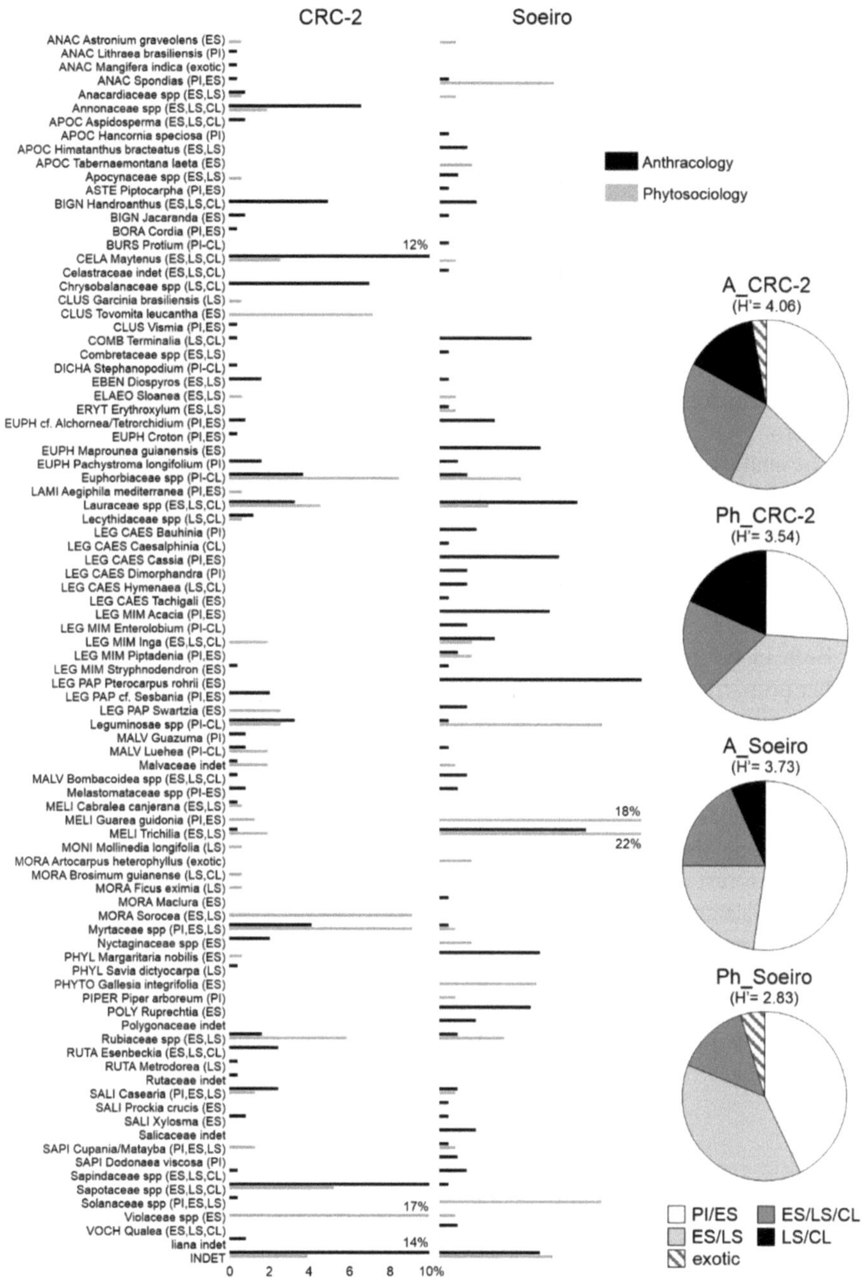

Fig. 12.5 Results of the analyses performed in two charcoal kiln sites of the Pedra Branca Massif, Rio de Janeiro, allowing to compare the ancient (anthracological analysis – *A*) and the extant (phytosociological analysis – *Ph*) forests. *Left*: histograms presenting the frequency of taxa identified in the anthracological (*black*) and phytosociological (*gray*) analyses in each studied site [*CRC2* = Carvoaria Ruína do Cipó-2; *Soeiro* = Carvoaria Soeiro-1]. Codes after the taxonomic names stand for the plants classification in ecological groups [*PI* = pioneer; *ES* = early secondary; *LS* = late secondary; *CL* = climax]. *Right:* pie charts presenting the relative proportion of taxa in different successional stages in the ancient (anthracology) and the extant (phytosociology) forests in each site

Maytenus (CRC2) and *Trichilia* (Soeiro) are the most prevalent in both forests, along with other important taxa such as Annonaceae, *Handroanthus*, Chrysobalanaceae, Lauraceae, *Casearia*, Myrtaceae, Rubiaceae, and Sapotaceae in CRC2 and *Spondias*, Combretaceae, Lauraceae, Rubiaceae, and *Casearia* in Soeiro.

In both charcoal kilns, the anthracological and phytosociological results showed higher diversity (cf. H′ index) in the ancient forest than in the present one, as well as higher richness, both in number of families and species. Additionally, a large diversity of taxa unique to the ancient forest as well as others unique to the extant forest attest that the composition of the extant vegetation is significantly different from the forest that was exploited by the charcoal burners. Only ten taxa are shared by the anthracological and phytosociological analyses in CRC-2, and only eight in Soeiro.

The analysis of successional stages suggests that in CRC-2 both the ancient and the extant forests represent intermediate to mature stages of ecological succession. In Soeiro, the ancient forest was in an intermediate to mature stage, while the extant forest is in an early to intermediate stage of ecological succession. In both sites, the proportion of taxa typical of more mature forests is higher in the ancient forest (Fig. 12.5).

Both in the past and in the present vegetation, the area of CRC-2 presents a higher proportion of taxa in more advanced successional stages than in Soeiro. The latter is situated in a lower elevation, therefore closer to the farmsteads (past) and urban areas (present), and as such more prone to anthropogenic perturbation. This explains the higher proportion of plants characteristic of early to intermediate successional stages in this site.

The strong dissimilarity between the forest exploited by the charcoal burners and the extant forest is demonstrated through a comparison of taxonomic types found in each analysis and their relative frequencies (Fig. 12.6). A statistical analysis using Sørensen's coefficient shows only 36% similarity between anthracological and phytosociological data. This indicates that anthracological results were more similar to each other than to phytosociology for the corresponding extant forests in each area. These results indicate that although the charcoal kilns studied are at different altitudes and have relatively different vegetation cover at present, the forest that once

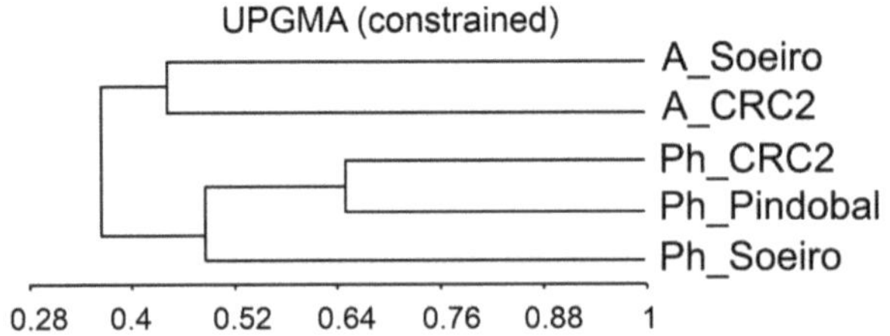

Fig. 12.6 Dendrogram of floristic similarity between anthracological (*A*) and phytosociological (*Ph*) analyses, using Sørensen's coefficient [*CRC2* = Carvoaria Ruína do Cipó-2; *Soeiro* = Carvoaria Soeiro-1; *Pindobal* = Carvoaria Pindobal. Pindobal charcoal kiln is located in a more elevated and well preserved area: only phytosociology analysis was performed in this area]

existed and was exploited by the charcoal burners (reflected in the charcoal fragments) had a greater intrinsic homogeneity, as a whole, than similarities with the forest currently existing.

It is important to highlight the identification of two particular species in the anthracological analysis, which means, in the ancient forest. One is *Pterocarpus rohrii*, a noble wood species with high commercial value, which could reach much higher prices than charcoal if it was sold as timber. The presence of this species in the sample (more than that, its dominance in one of the charcoal kilns) demonstrates that the charcoal burners made no species selection, cutting all trees found in the plot defined for exploitation indiscriminately. The other is *Mangifera indica*, the mango tree, an exotic fruit species that produces tasty and highly nutritious fruit. Its presence in the sample not only corroborates the absence of species selection but also confirms that the forest exploited by the charcoal burners was already a secondary formation, that is, it had already suffered anthropogenic modifications previously.

The previous land uses recorded for the Pedra Branca Massif are all associated with intense deforestation. Farms and sugar cane mills are recorded in the region since the seventeenth century. Orange farms were installed in the eighteenth century; in the nineteenth century, began the production of coffee and banana – all activities were accompanied by subsistence cultivation by enslaved people. From this century dates the first official records of charcoal production. The production of coffee and banana remained until the 1950s, associated with crop fields from small farmers (Solórzano et al., 2005). All of these economic activities, as well as other types of cultivation, may have occurred in parallel. Each time the forest was cut down to make room for crops, the trees were availed for firewood and charcoal production. It was a way to take advantage of all the effort generated, turning it into income.

12.5 The Histories of Forests and Charcoal Burners: An Interdisciplinary Approach

The study of historical charcoal kilns allows us to recover the histories of the charcoal burners and the forests in which they lived by associating different approaches: Environmental Archaeology, Anthracology, Historical Ecology, Plant Ecology, and Landscape Archaeology. Charcoal is one of the main evidences of landscape alteration by human populations. Unlike other plant remains that need special conditions for preservation, carbonized wood has great resistance in the soil, being present in most archaeological sediments (Scheel-Ybert, 2020).

The history of intentional charcoal production (i.e., not as a residue of other activities, such as hearths and slash-and-burn) has been very little explored. The little information available suggests that this production is related to metallurgy development and industrial activity, mainly in Europe (e.g. Vernet, 1997; Ludemann & Nelle, 2002), but also to urban centers, as in Brazil (Oliveira et al., 2011; Fraga

& Oliveira, 2012). The transformation of wood into charcoal via muffled combustion enables an increase in calorific power with a reduction in mass, producing an energy source that can be easily transported over longer distances. Charcoal can be produced locally and – a consideration of great relevance to the present work – is an energy source whose production cost is composed almost entirely of the labor invested in it (Olson, 1991).

Authors who have addressed the issue disagree about the impact of charcoal production on the forest. Kammen and Lew (2005) argue that it causes major impacts on forest formations for two reasons: the inefficiency of the artisanal carbonization process, where four to six times more firewood is needed to produce a volume of charcoal with the same amount of energy; and the fact that charcoal production, in general, is directed to commercialization, where the production of large quantities requires the cutting of large forest areas in short periods. In the contrast, Molina-Colón and Lugo (2006) argue that charcoal production by itself, at least when carried out on a restricted temporal and spatial scale, has minimal lasting effects on forest structure and composition. This would occur because this activity does not alter the seed bank in a significant way and because trees can regrow from cut trunks. It is noteworthy that multiple trunks, which indicate regrowth, are rare in undisturbed forests (Oliveira, 2002; Guariguata & Ostertag, 2001), but rather common in the Pedra Branca Massif (Santos et al., 2006).

Our results demonstrate that although the forest has recovered due to the ecological succession process that occurred after the discontinuation of charcoal production in the Pedra Branca Massif, its composition was significantly changed. This suggests that the charcoal production there was not an intensive practice, meaning that only small areas of forest were cut at a time and that sufficient time was given to these areas to recover.

All findings and evidence from the present work are corroborated by a previous study in the same area, in which anthracological analyses in two charcoal kilns, also at different altitudes, were compared with results from phytosociological surveys of the region (Fig. 12.2). This study had already shown that the forest exploited by the charcoal burners was already a secondary formation, but more diverse, and with a different structure than the extant forest. Also that there was no selection of species neither of wood diameter while cutting trees for charcoal production (Beauclair et al., 2011).

In addition to anthracological and phytosociological results, other vestiges offer key information about the paleo-territory of the charcoal burners. By using the present landscape as a document, various elements found in the extant forest can be interpreted to reconstruct its past uses. Besides, several plant species are indicators of human activity. Certain attributes of these species, such as density and dominance, are directly related to the historical uses to which the forest has been submitted. Differences in the composition of secondary forests offer an example.

Extensive areas covered with secondary forests, with a myriad of typical species, are common in many areas of Southeastern Brazil. These forests are characterized by being dominated by few species with large numbers of individuals. In these forest as a whole, the most frequent ecological groups are pioneer species, early secondaries, and, to a lesser extent, late secondaries (Budowski, 1965; Swaine & Whitimore, 1988). In contrast, mature (or climax) forest systems are characterized by a high number of species; a reduced number of pioneer and early-secondary species; a high contingent of late-secondary and climax species; a low number of individuals per species; and low density or dominance values per species (Oliveira, 2002).

In general, pioneer or early secondary species only germinate under maximum light conditions. They have rapid growth, early regeneration, continuous seed production, and a shorter life span. Typically, pioneer species do not exceed 10 years of age and early secondary species, 10–25 years (Budowski, 1965). However, some pioneer species can exhibit marked longevity. Long-lived pioneer species are of great interest to landscape archeology, particularly in areas of Atlantic Forest. By remaining in the system for a long time, they spatially indicate the occurrence of past human activities, where today the forest has completely recovered (Oliveira, 2008). When abandoned, natural or anthropogenic clearings are covered by pioneer plant communities. They act as facilitators for the entry of more advanced species, as they create more shaded spaces allowing these species to germinate. With time, the pioneer species disappear, and after a few decades, only the long-lived pioneers survive, coexisting with late secondary or climax species.

Some long-lived pioneers occur throughout the Pedra Branca Massif, such as *Guarea guidonia* (L.) Sleumer, which can reach more than 150 years. This species does not germinate in natural clearings, only under full sun or with the opening of large clearings – such as the ones made for charcoal production. Phytosociological analysis of different parcels demonstrated the existence of a density gradient and relative dominance in *Guarea guidonia* populations in the study area. In forest formations ranging from 20 to 60 years old, this species presents a relative density of up to 53.5% and a relative dominance of up to 45%. In more advanced successional formations, these values are, respectively, 2% and 5.2%, which indicates the permanence of few but large specimens (Oliveira et al., 2013). The anthracological analysis of a charcoal kiln neighboring one of the parcels studied where there was a dominance of *Guarea guidonia*, demonstrated the existence of this species in the forest exploited by charcoal burners, although in a smaller proportion (Beauclair et al., 2011).

Another important pioneer species in the massif is *Miconia cinnamomifolia* (DC.) Naudin, which can also reach more than 150 years of age (Delamonica et al., 2002). This species rarely colonizes natural clearings (such as those opened by falling trees or windstorms) and has been almost always associated with the existence of old cultivated fields. A phytosociological analysis conducted in a very well-preserved forest section of this study area, at an altitude of 600 m, revealed a forest in an advanced stage of ecological succession, with 95 tree species, a high proportion of climax and late secondary species, and a large basal area (43 m^2/ha). In this area, *Miconia cinnamomifolia* is the species with the highest importance value. However, all specimens found in the area showed very similar stem diameters (n = 22; mean diameter = 0.59 cm; standard deviation = 0.10 cm; coefficient of variation = 16.1%), which suggests that they are from the same generation, probably linked to the same past event, such as an abandoned field (Sales et al., 2020).

Schizolobium parahyba (Vell.) S.F. Blake is another long-cycle pioneer species that currently occurs in secondary forests in the Massif. This species is one of the fastest-growing forest trees in the Southeastern and Southern regions of Brazil (Callado & Guimarães, 2010). Measurements of wood growth rings in specimens of *S. parahyba* allowed to identify individuals with up to 69 and 73 years, respectively in Ilha Grande and Tinguá Biological Reserve, both in Southeastern Brazil (Callado & Guimarães, 2010; Latorraca et al., 2015). Although available data for this species does not reach the same time range as the previous ones, it is still a long-lived pioneer species with characteristics that enable the detection of previous uses of the landscape.

12.6 Final Considerations

The history of the people is the history of the forests. The Atlantic Forest, which is the second largest neotropical forest in the world (after the Amazon Forest), once stretched almost continuously from the Northeast to Southern Brazil. Today, completely devastated by urbanization and agriculture, it survives only in small patches, estimated at less than 12% of its original extent (SOS Mata Atlântica and INPE, 2019).

People have lived in the Atlantic Forest for thousands of years. During this time, people have maintained different relationships with the landscape, using and understanding the forest in different ways. They also changed the forest in many ways: opening clearings, making their homes, performing subsistence activities, exploiting economically, and others. Whenever it was given the chance, the forest regenerated itself but kept its history, which today can be told. The cycle of the charcoal burners may have been one of the last of these cycles that we can still tell. The history of people who lived in and from the forests and about whom there are practically no written references, although it happened during the colonial period. A history of invisible people, a history of the vanquished.

The spectrum of vestiges and information about the human past that can be retrieved from the landscape and the archaeological sites themselves when conducting an interdisciplinary study involving Landscape Archaeology, Anthracology, and Historical Ecology is extremely broad. In the present study, the comparison of anthracological and phytosociological analyses results in areas of charcoal kilns in the Pedra Branca Massif, Rio de Janeiro city, demonstrated that: (1) The extant forests are different in species composition from those exploited by the charcoal burners in both study areas; (2) The current dominance of certain species today may have been influenced by the charcoal production activity; (3) The areas exploited by charcoal burners had greater diversity than the same forested areas today; (4) This ancient forests also seem to be in more advanced successional stages; (5) The forest cutting for charcoal production was done without any selection, of species or diameters; (6) The forests exploited by the charcoal burners in the two study areas were already secondary forests, meaning that they had already been cut before, probably because of one of the several previous soil uses reported for the area (mostly monocultural and subsistence crops). All these results are corroborated by the study of other charcoal kilns in the same massif, which reached the same conclusions (Beauclair et al., 2011).

It has been shown that long-lived pioneer species are indicators of human intervention history on the forested environment and that their occurrence contributes to understanding, mapping, and dating the areas used. From this study, we were able to demonstrate that some charcoal kiln areas were abandoned in the mid-nineteenth century, and some areas may have been abandoned even before this time.

Furthermore, this study contributes to increasing our understanding about a population whose written records are too scarce, and oral traditions are being lost, or have already become extinct. Moreover, our results support the re-reading of the idea of a pristine forest and its worth representation, since a large number of forest remnants in Southeastern Brazil were intensely used in the past by several populations including charcoal burners. Studies from our team, not only in the Pedra Branca Massif but in other areas in Rio de Janeiro and other regions of Southeastern Brazil have highlighted the omnipresence of evidence of charcoal production in the forest landscape, at the same time that the agent (the charcoal burner) is situated as an element excluded from the society who required him to produce energy.

Acknowledgments We would like to thank the teams of Laboratório de Arqueobotânica e Paisagem (MN/UFRJ) and Laboratório de Biogeografia e Ecologia Histórica (PUC-Rio). This work benefited from funding from the Coordination for the Improvement of Higher Education Personnel (CAPES), National Counsel of Technological and Scientific Development (CNPq), and Foundation Carlos Chagas Filho for Research Support of the State of Rio de Janeiro (FAPERJ). The third author is a CNPq fellowship holder and a Senior Scientist from FAPERJ.

References

Abreu, M. A. (2006). *Evolução Urbana do Rio de Janeiro* (4th ed.). Instituto Pereira Passos.

Beauclair, M., Rangel, A., Oliveira, R. R., & Scheel-Ybert, R. (2011). Wood charcoal production and forest dynamics in the Pedra Branca massif, Rio de Janeiro, RJ, Brazil. *Saguntum, 11*, 121–122.

Bender, B. (2006). Place and landscape. In C. Tilley, W. Keane, S. Küchler, M. Rowlands, & P. Spyer (Eds.), *Handbook of material culture* (pp. 303–314). Sage.

Bonhôte, J., & Vernet, J. L. (1988). La mémoire des charbonnières. *Revue Forestière Française, 43*(1), 197–212.

Budowski, G. (1965). Distribution of tropical American rain forest species in the light of successional processes. *Turrialba, 15*(1), 40–42.

Callado, C. H., & Guimarães, R. C. (2010). Estudo dos anéis de crescimento de *Schizolobium parahyba* (Leguminosae: Caesalpinioideae) após episódio de mortalidade em Ilha Grande, Rio de Janeiro. *Brazilian Journal of Botany, 33*, 84–91.

Curtis, J. T., & McIntosh, R. P. (1950). The interrelations of certain analytic and synthetic phytosociological characters. *Ecology, 31*, 434–455.

Deffontaines, P. (1937). La vie montagnarde dans la montagne de l'Itatiaya (Serra de la Mantiqueira, Brésil). *Revue de Géographie Alpine, 25*(3), 496–508.

Delamonica, P., Lima, D. F., Oliveira, R. R., & Mantovani, W. (2002). Estrutura e funcionalidade de populações de *Miconia cinnamomifolia* (DC.) Naud. em florestas secundárias estabelecidas sobre antigas roças caiçaras. *Pesquisas. Botânica, 52*, 125–142.

Emery, K. F. (2007). Ecofacts, overview. In D. M. Pearsall (Ed.), *Encyclopedia of archaeology* (pp. 1111–1114). Academic Press.

Firme, R. P., Vicenz, R. S., Macedo, G. V., Silva, I. M., & Oliveira, R. R. (2001). Estrutura da vegetação de um trecho de mata atlântica sobre solos rasos (maciço da Pedra Branca, RJ). *Eugeniana, 24*(1), 3–10.

Fraga, R. R., & Oliveira, J. S. (2012). Social metabolism, cultural landscape, and social invisibility. In M. Canevacci (Ed.), *Polyphonic anthropology – Theoretical and empirical cross-cultural fieldwork* (pp. 139–156). INTECH.

Guariguata, M. R., & Ostertag, R. (2001). Neotropical secondary forest succession: Changes in structural and functional characteristics. *Forest Ecology and Management, 148*, 185–206.

Ingold, T. (1993). The temporality of the landscape. *World Archeology, 25*(2), 152–173.

InsideWood. (2004-onwards). *Published on the internet.* Retrieved July 13, 2019, from http://insidewood.lib.ncsu.edu/search

Junqueira, A., Shepard, G., & Clement, C. (2011). Secondary forests on anthropogenic soils of the middle Madeira River: Valuation, local knowledge, and landscape domestication in Brazilian Amazonia. *Economic Botany, 65*(1), 85–99.

Kammen, D. M., & Lew, D. J. (2005). *Review of technologies for the production and use of charcoal.* Renewable and Appropriate Energy Laboratory Report.

Latorraca, J. V. D. F., Souza, M. T. D., Silva, L. D. S. A. B. D., & Ramos, L. M. A. (2015). Dendrocronologia de árvores de *Schizolobium parahyba* (Vell.) SF Blake de ocorrência na REBIO de Tinguá-RJ. *Revista Árvore, 39*(2), 385–394.

Lavradio, M. (1913). Relação do Marquêz de Lavradio (Parte II): Relações parciais apresentadas ao Marquêz de Lavradio [1779]. *Revista do Instituto Histórico e Geográfico Brasileiro, 127*, 284–360.

Levis, C., Costa, F. R. C., Bongers, F., et al. (2017). Persistent effects of pre-Columbian plant domestication on Amazonian forest composition. *Science, 355*(6328), 925–931.

Ludemann, T. (2002). Anthracology and forest sites: The contribution of charcoal analysis to our knowledge of natural forest vegetation in south-West Germany. In S. Thiébault (Ed.), *Charcoal analysis: Methodological approaches, palaeoecological results and wood uses* (British Archaeological Reports International Series) (Vol. 1063, pp. 209–217). Archeopress.

Ludemann, T. (2008). Experimental charcoal-burning with special regard to anthrocological wood diameter analysis. In G. Fiorentino & D. Magri (Eds.), *Charcoals from the past: Cultural and palaeoenvironmental applications* (British Archaeological Reports International Series) (Vol. 1807, pp. 147–157). Archaeopress.

Ludemann, T., & Nelle, O. (Eds.). (2002). *Die Wälder am Schauinsland und ihre Nutzung durch Bergbau und Köhlerei.* Forstwissenschatliche Fakultät der Universität Freiburg und Forstlische Versuchs und Forschungsanstalt.

Magalhães Corrêa, A. (1936). *O sertão carioca.* Imprensa Nacional.

Marziani, G., & Citterio, S. (1999). The effects of human impact on the arboreal vegetation near ancient iron smelting sites in Val Gabbia, northern Italy. *Vegetation History and Archaeobotany, 8*(3), 225–229.

SOS Mata Atlantica, & INPE. (2019). *Atlas dos Remanescentes Florestais da Mata Atlântica.* Relatório Técnico, Período 2017–2018. Retrieved January 7, 2021, from https://www.sosma.org.br/wp-content/uploads/2019/10/Atlas-mata-atlanticaDIGITAL.pdf

Matteucci, S. D., & Colma, A. (1982). *Metodologia para el estudio de la vegetacion* (p. 22). Série Biologia – Monografia.

Molina-Colón, S., & Lugo, A. E. (2006). Recovery of a subtropical dry forest after abandonment of different land uses. *Biotropica, 38*(3), 354–364.

Moro, M. F., & Martins, F. R. (2011). Métodos de levantamento do componente arbustivo-arbóreo. In J. M. Felfili, P. V. Elsenlohr, M. M. d. F. Melo, L. A. Andrade, & J. A. A. M. Neto (Eds.), *Fitossociologia no Brasil: métodos e estudos de caso* (pp. 174–212). Editora da Universidade Federal de Viçosa.

Nash, R. (1970). Environmental history. In H. J. Bass (Ed.), *The state of American history* (pp. 249–260). Quadrangle Press.

Oliveira, R. R. (2002). Ação antrópica e resultantes sobre a estrutura e composição da mata atlântica na Ilha Grande, RJ. *Rodriguésia, 53*(82), 33–58.

Oliveira, R. R. (2005). Os cenários da paisagem. In R. R. Oliveira (Ed.), *As marcas do homem na floresta. História ambiental de um trecho da Mata Atlântica* (pp. 23–33). PUCRio.

Oliveira, R. R. (2007). Mata Atlântica, paleoterritórios e história ambiental. *Ambiente & Sociedade, 10*(2), 11–23.

Oliveira, R. R. (2008). Environmental history, traditional populations, and paleo-territories in the Brazilian Atlantic coastal Forest. *Global Environment, 1*, 176–191.

Oliveira, R. R. (2015). Fruto da terra e do trabalho humano: paleoterritórios e diversidade da Mata Atlântica no Sudeste brasileiro. *Revista de História Regional, 20*, 277–299.

Oliveira, R. R., Fraga, J. S., & Berck, D. E. (2011). Uma floresta de vestígios: metabolismo social e a atividade de carvoeiros nos séculos XIX e XX no Rio de Janeiro, RJ. *INTERthesis, 8*, 286–315.

Oliveira, R. R., Solórzano, A., Sales, G. P. S., Oliveira, M. B. D., & Scheel-Ybert, R. (2013). Ecologia histórica de populações da carrapeta (*Guarea guidonia* (L.) Sleumer) em florestas de encosta do Rio de Janeiro. *Pesquisas Botânica, 64*, 323–339.

Oliveira, R. R., Patzlaff, R. G., & Scheel-Ybert, R. (2020). A Floresta como Esconderijo: Arqueologia da Paisagem na Mata Atlântica do rio de Janeiro. *Revista Mosaico, 13*, 61–82.

Olson, S. D. (1991). Firewood and charcoal in classical Athens. *Hesperia, 60*(3), 411–420.

Patzlaff, R. G., Araújo, D. S. D., Oliveira, R. R., & Scheel-Ybert, R. (2018). Historical charcoal kilns: A method to compare the surrounding vegetation with the anthracological data in forested slopes of Rio de Janeiro estate, RJ, Brasil. In S. Paradis-Grenouillet, S. Burri, & R. Rouaud (Eds.), *Charbonnage, charbonniers, charbonnières: Confluence de regards autour d'un artisanat méconnu* (pp. 173–178). Aix Marseille University Press.

Renfrew, C., & Bahn, P. (Eds.). (2005). *Archaeology: The key concepts*. Routledge.

Sales, G. P. S., Solórzano, A., Patzlaff, R. G., & Oliveira, R. R. (2014). Resultantes ecológicas, práticas culturais e provisão de lenha para a fabricação de carvão nos séculos XIX e XX no Rio de Janeiro. *Pesquisas. Botânica, 65*, 389–402.

Sales, G. P. S., Patzlaff, R. G., Solórzano, A., & Oliveira, R. R. (2020). De cinzas para floresta: atividade carvoeira e trajetórias sucessionais no maciço da Pedra Branca. *GEOPUC., 13*, 364–385.

Salles, G. P. S., Oliveira, R. R., Solórzano, A., & Patzlaff, R. G. (2016). No caminho dos carvoeiros: composição florística e estrutura da floresta em um paleoterritório de exploração de carvão no Maciço da Pedra Branca, RJ. In *Anais do Simpósio Internacional de História Ambiental e Migrações* (pp. 951–966). UFSC.

Santos, V. S., Solórzano, A., Guedes-Bruni, R. R., & Oliveira, R. R. (2006). Composição do estrato arbóreo de um paleo-território de carvoeiros no Maciço da Pedra Branca, RJ. *Pesquisas. Botânica, 57*, 181–192.

Scheel-Ybert, R. (2002). Evaluation of sample reliability in extant and fossil assemblages. *British Archaeological Reports, 1063*, 9–16.

Scheel-Ybert, R. (2004). Teoria e métodos em antracologia. 2. Técnicas de campo e de laboratório. *Arquivos do Museu Nacional, 62*(4), 343–356.

Scheel-Ybert, R. (2005). Teoria e métodos em antracologia. 3. Validade amostral. *Arquivos do Museu Nacional, 63*(2), 207–232.

Scheel-Ybert, R. (2016). Charcoal collections of the world. *IAWA Journal, 37*, 489–505.

Scheel-Ybert, R. (2020). Anthracology (charcoal analysis). In C. Smith (Ed.), *Encyclopedia of global archaeology* (2nd ed., pp. 408–418). Springer.

Scheel-Ybert, R., & Gonçalves, T. A. P. (2017). *First anthracological atlas of Brazilian species* (p. 10). Museu Nacional, Série Livros Digital.

Scheel-Ybert, R., Carvalho, M. A., Moura, R. P. O., Gonçalves, T. A. P., Scheel, M., & Ybert, J. P. (2006). Coleções de referência e bancos de dados de estruturas vegetais: subsídios para estudos paleoecológicos e paleoetnobotânicos. *Arquivos do Museu Nacional, 64*(3), 255–266.

Scheel-Ybert, R., Buarque, A., & Oliveira, R. R. (2013). "Cariocas da Gema": evidências de presença humana na Ilha Redonda no período pré-colonial. In F. Moraes, A. Bertoncini, & A. Aguiar (Eds.), *História, Pesquisa e Biodiversidade do Monumento Natural das Ilhas Cagarras* (pp. 38–47). Série Livros Museu Nacional 48.

Scheel-Ybert, R., Beauclair, M., & Buarque, A. (2014a). The forest people: Landscape and firewood use in the Araruama region (southeastern Brazil) during the late Holocene. *Vegetation History and Archaeobotany, 23*(2), 97–111.

Scheel-Ybert, R., Boyadjian, C. H. C., Mateus, J., & Selle, Y. P. (2014b). Los sistemas Anthrakos y Phyton: Propues¬ta de creacion de bancos de datos en linea. In *Taller Micro paleoetnobotánica: Relevancia de una red interdisciplinaria de investigaciones en fitolitos y almidones* (pp. 98–104). Resumens.

Scheel-Ybert, R., Caromano, C. F., & Azevedo, L. W. (2016). Of forests and gardens: Landscape, environment, and cultural choices in Amazonia, Southeastern and southern Brazil from c. 3000 to 300 cal yrs BP. *Cadernos do LEPAARQ (UFPEL), 13*, 425–458.

Shepherd, G. J. (1995). *Fitopac: manual do usuário Departamento de Botânica*. Universidade Estadual de Campinas.

Solórzano, A., Oliveira, R. R., & Guedes-Bruni, R. R. (2005). História ambiental e estrutura de uma floresta urbana. In R. R. Oliveira (Ed.), *As marcas do homem na floresta. História ambiental de um trecho de mata Atlântica* (pp. 817–106). PUC-Rio.

Solórzano, A., Guedes-Bruni, R. R., & Oliveira, R. R. (2007). Composição florística e estrutura de dois trechos de Floresta Ombrófila Densa submontana, no Parque Estadual da Pedra Branca, RJ. *Revista Brasileira de Biociências, 5*, 609–611.

Swaine, M. D., & Whitmore, T. C. (1988). On the definition of ecological species groups in tropical rain forests. *Vegetatio, 75*, 81–86.

Ter Steege, H., Pitman, N. C., Sabatier, D., et al. (2013). Hyperdominance in the Amazonian Tree Flora. *Science, 342*(6156), 1243092. https://doi.org/10.1126/science.1243092

Vernet, J. L. (1997). *L'homme et la forêt méditerranéenne de la Préhistoire à nos jours*. Errance.

Wolda, H. (1981). Similarity indices, sample size and diversity. *Oecologia, 50*, 296–302.

Chapter 13
Mounds of Eastern Uruguay, Beyond the Causes of their Accretion

Roberto Bracco Boksar, Ofelia Gutiérrez, Christopher Duarte, and Daniel Panario

Abstract The articulation of the original inhabitants of the wetlands in the eastern lowlands of Uruguay with the environment was based on socio-ecological systems that have not yet been fully deciphered. This work aims to explain some interactions and synergies with the environment and food processing. We have proposed that the mounds in the India Muerta-Paso Barranca region of eastern Uruguay raised as a consequence of the secular accumulation of waste produced by pit or earth oven where sedimentary heat retainers were used. We examine the consistency of the hypothesis, but we no longer limit ourselves to explaining the cause of the accretion of the mounds, instead we focus on the use of a technology integrated into a subsistence system that would have been based on the processing and consumption of resources, very abundant, whose cooking increases its nutritional value. We propose that the jump from growth mechanisms to subsistence systems is valid and consistent if it allows us to explain further dimensions of the archaeological record. From this perspective, we try to explain the location of the mounds in the landscape, their shape, their aggregation habits, the textural characteristics of their matrix, their absolute chronologies and, finally, the dietary profile of the populations associated

R. Bracco Boksar (✉)
Facultad de Humanidades y Ciencias de la Educación, Universidad de la República, Montevideo, Uruguay

Laboratorio 14C, Ministerio de Educación y Cultura – Laboratorio de Luminiscencia, UNCIEP, Instituto de Ecología y Ciencias Ambientales (IECA), Facultad de Ciencias, Universidad de la República, Montevideo, Uruguay
e-mail: braccoboksar@montevideo.com.uy

O. Gutiérrez · D. Panario
UNCIEP, Instituto de Ecología y Ciencias Ambientales (IECA), Facultad de Ciencias, Universidad de la República, Montevideo, Uruguay
e-mail: oguti@fcien.edu.uy; panari@fcien.edu.uy

C. Duarte
Laboratorio de Luminiscencia, UNCIEP, Instituto de Ecología y Ciencias Ambientales (IECA), Facultad de Ciencias, Universidad de la República, Montevideo, Uruguay
e-mail: christopherduarte1@gmail.com

A. C. Colonese, R. G. Milheira (eds.), *Historical Ecology and Landscape Archaeology in Lowland South America*, Interdisciplinary Contributions to Archaeology, https://doi.org/10.1007/978-3-031-32284-6_13

with them. This allows us to approach the logic that governed the construction of the cultural landscape of eastern Uruguay from the middle Holocene.

Keywords Eastern Uruguay · Late Holocene · Pit ovens · Mounds · Stable isotopes palaeodiet

13.1 Introduction

In southern Brazil and eastern Uruguay, from the Patos lagoon to the Castillos lagoon, including the middle and upper basin of the Negro river, there is a type of archaeological site that is distinctive of wetland environments (Fig. 13.1). These are isolated or grouped mounds, which can reach up to 7,2 meters in height and circular or elliptical plants whose diameters are around 35 meters (Fig. 13.2). Similar archaeological sites are found in the nearby regions of the Brazilian Pantanal, the

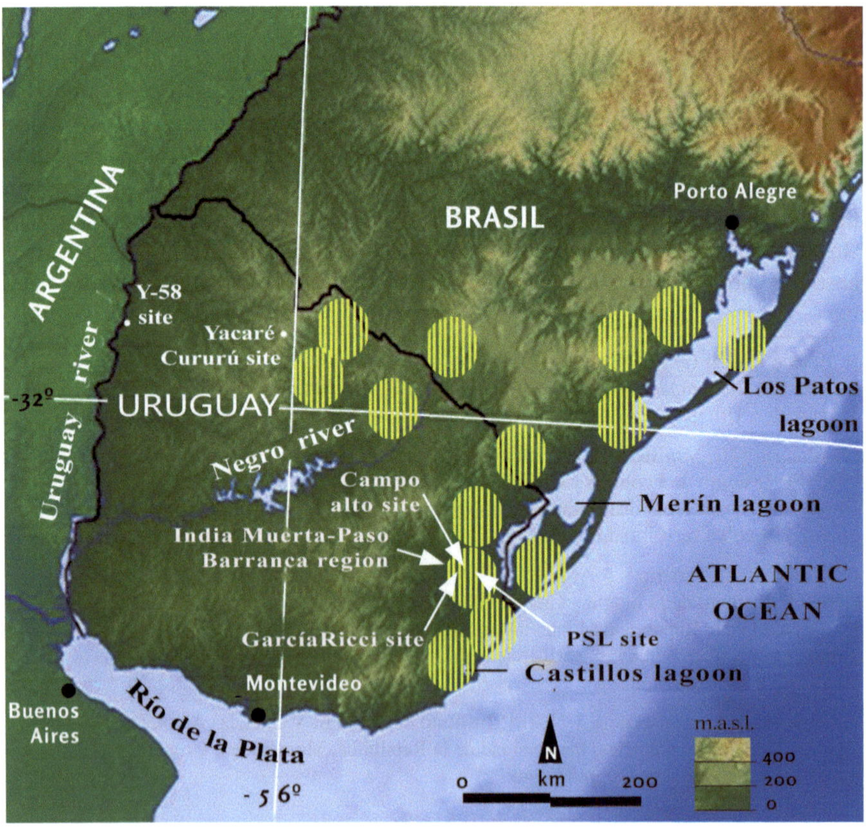

Fig. 13.1 Regions where the mounds are located in southern Brazil and eastern Uruguay

Fig. 13.2 Mound of the PSL site, located in the Paso Barranca-India Muerta region. (See Fig. 13.1)

delta of the Paraná river and the lower Uruguay river (Eremites de Oliveira, 1995; Bonomo et al., 2011; Loponte et al., 2016; Gascue et al., 2019). The early investigations of this last region, at the end of the nineteenth century and the beginning of the 20th, provided the main interpretative axes that dominate the archaeology of the mounds in these latitudes until the present time (see, among others, Zeballos & Pico, 1878; Torres, 1911; Outes, 1918; Frenguelli & Aparicio, 1923; Serrano, 1931).

For southern Brazil and eastern Uruguay, these landforms are recognized as anthropogenic earth structures. However, the controversy over their natural or artificial origin is still current in the Paraná delta region (Loponte et al., 2016; Politis & Bonomo, 2018). They have been assigned to different functions: platforms to occupy flooded areas, burial sites, ritual-monumental sites, territorial markers, discard sites or multifunctional systems that would even have included horticultural purposes (Schmitz, 1981; Baeza & Panario, 1999; Bracco et al., 2000c; Gianotti, 2000; López Mazz, 2001; Iriarte et al., 2004; Villagran, 2006; Loureiro, 2008; Milheira et al., 2016; Milheira & Gianotti, 2018). Their ^{14}C and luminescent dates cover the period 6100–200 years BP (see below). According to archaeofauna, macro and microbotanical and isotopic records, its builders would have developed an economy with a low level of food production (sensu Smith, 2001) which, according to the lapses shown by the dating, would be closer to a long-term strategy than to a transition to a productive economy (sensu Greaves & Kramer, 2014).

Currently, two interpretative-explanatory lines dominate the investigations on the mounds in the south of the Merín Lagoon basin. One of these lines perceives them as architectural or engineering products: "monuments" (Pintos Blanco, 1999;

López Mazz et al., 2016; Milheira et al., 2016; Milheira & Gianotti, 2018, among others). The second line, based mainly on the behaviour of numerical chronologies, proposes that they are the unintentional consequence of a secular recursive behaviour (Bracco & Ures, 1999; Bracco, 2006; Duarte & Bracco, 2020; Bracco et al., 2020). From this last position, and observing the multiple evidence that indicate that fire was present during its elevation, we have put forward, taking the Australian oven mounds as analogous, that the recursive behaviour which raised the mounds of the India Muerta-Paso Barranca region (Fig. 13.1) was the use of pit or earth ovens (Bracco et al., 2019b, c, 2020, 2021a, cf. 2022).

In this chapter we examine the explanatory power of the hypothesis by analyzing its ability to account not only for its accretion mechanism, but also for other different dimensions of its archaeological record, some of them directly related to environmental adaptation. Following the definition of Crumley's Historical Ecology, the proposal offers a new perspective "to understand the dialectical relationships that exist between human acts and the acts of nature that are manifested in the landscape" (Crumley, 1994). The dialectical relationships that would have occurred between a way of processing food and the environment, leading to one of the greatest anthropogenic modifications that occurred in the Pampa biome from the middle Holocene.

13.1.1 The Validity of the Analogy Used

As well as in other historical sciences, analogical reasoning is one of the pillars of archaeology. Since this type of reasoning has the same limitations as induction; great efforts have been made in order to strengthen inferences. Along these lines, Wylie (1985), in a classic research paper, distinguished different types of analogy, pointing out that it is the relational analogy that offers the greatest inferential certainty, because it is based on the fact that the properties shared by source and subject are the consequence of a "principle of causal or quasi-causal connection" (Wylie, 1985, p. 99). In the immediate term, we have proposed that the connection principle that validates the relational analogy between Australian oven mounds and mounds from the India Muerta-Paso Barrancas region is a consequence of its genesis, mainly linked to the processing of food resources present in wetland environments. Oven mounds are a type of mound, similar in shape and dimension to those in the southern sector of the Merín Lagoon basin, associated with groups of hunter-gatherers-fishermen, found in the Australian wetland areas of the West, in the South Alligator and Murray rivers. Ethnohistorical sources and archaeological evidence indicate that they originated as the consequence of the accumulation of waste produced by oven pits (Beveridge, 1869; Brockwell, 2006; Martin, 2011; Jones et al., 2017).

13.1.2 The Evidence

Different lines of evidence indicate that the fire was present in the accretion processes of the mounds of India Muerta-Paso Barrancas. In the immediate term, it is pointed out by the finding of combustion areas and charcoals inside it, as well as the geochemical properties of the sediment that forms them (Bracco et al., 2019b). The action of fire is also manifested by the abundance of burnt earth present in its matrix; fragments of sedimentary nodules whose external rubefaction and hardness prove that they were cooked (Duarte et al., 2017; Duarte & Bracco, 2020). For the PSL (Puntas de San Luis) and García Ricci sites (Fig. 13.1), 25–30% of the matrix of the mounds is formed by burned earth fragments (Bracco et al., 2000c; Duarte et al., 2017). These fragments, which reach up to 7 cm in their long axis, frequently exhibit galleries (Fig. 13.3) which, together with their microstructural characteristics, indicate that fragments of epigeal nests of *Camponotus punctulatus* (ants) were used to

Fig. 13.3 Fragments of burner earth from mound II at the PSL site. (Galleries are indicated)

make them. These amyhill are very abundant in the environments where the mounds are found and easy to break down to obtain heat resistant and dry sediment fragments (Duarte & Bracco, 2021). Experimental tests have allowed to infer that, during cooking, the burned earth reached temperatures between 400 and 600 °C (Bracco et al., 2020). But the strongest evidence that fire was widely present in the mound elevation processes comes from an unexpected sector of the record: the OSL-TL-^{14}C dates. The fact that the different fractions of the matrix (burned earth, gravel, or finer fractions) for same levels produce similar OSL and TL ages, which are consistent with ^{14}C ages, leads to the conclusion that the matrix of the mounds was exposed to fire, directly or indirectly. If this were not the case, the TL ages would not be consistent with the ^{14}C ages and would be older than the OSL ages. In other words, what the agreement between the TL/OSL and ^{14}C ages tells us is that all (or almost all) the matrix that makes up the analysed mounds was heated, at least, to temperatures in the range of 350–380 °C, the temperature of luminescent record bleaching (Duarte et al., 2017; Bracco et al., 2020).

The question is the following: what human behaviour produces a record with these characteristics, where on the one hand the dates indicate extensive exposure to heat, and on the other that it was formed gradually over extensive periods?

13.1.3 Pit Ovens

Cooking food is a specific practice that has had important biological and social implications for the humans evolution (Wrangham & Conklin-Brittain, 2003; Wrangham, 2009). Throughout prehistory, the ways of cooking evolved (Thoms, 2008, Fig. 3) in parallel with the resources that necessarily made up the diet. Pit ovens are a very extended way of processing food in time and space; its earliest records, circa 32,000 years ago, are found in the European Upper Palaeolithic (Thoms, 2017). For North and South America, these features appear around 11,000 years BC (Rivas & Ocampo, 2005; Thoms, 2017). Although there were only two antecedents for the current territory of Uruguay (Bracco et al., 2020), "culinary pits" at site Y-58, excavated during the Archaeological Rescue of the Salto Grande Dam (Guidón, 1989, p. 446) and "combustion structures formed by wells filled with stones", at the Yacaré Cururú site, a mound in the department of Artigas (Consens, 2001, p. 84). For neighbouring regions of Argentina its presence is documented in the provinces of La Pampa, Entre Ríos, Córdoba, La Rioja, Santiago del Estero, Tucumán, Chaco and Formosa, and the Cuyo region (provinces of Mendoza, San Juan and San Luis), (see, among others, Ceruti & González, 2007; Cornero et al., 2013). For their part, Milheira and collaborators (2016, p. 49) identify in the PSG-02 mound from Pontal da Barra, Brazil, a structure that they interpret as a possible culinary pits. Also the hearths indicated for the base of the mounds can be similar structures (Milheira et al., 2019, Fig. 2). Analogous features have been recognized in mounds from eastern Uruguay (Bracco et al., 2021b, p. 163 and Fig. 8).

Pit ovens operate through the accumulation of the heat produced by fire in heat retainers. These can be rock fragments or sediment fragments. This is how Curr (1886, p. 108–109, cited in Frankel & Major, 2017, p. 69), a European settler, describes how the indigenous Australians made and used pit ovens in the nineteenth century:

> The way in which these ovens were used was as follows: – When there was food to be baked, the women, with their hands and yamsticks, scooped a hole in the mound; if in doing so they came upon any lumps of clay (for there was no stone in those parts), they roughly lined the bottom of the hole with them. If none were met with, they quickly dug up a quantity for the purpose with their yamsticks, from somewhere near at hand. These lumps were about the size of a man's fist. The bottom of the hole being lined with them, a fire was made on top of them, and on the fire was thrown more lumps of clay. When the fire had burnt down, these last lumps were removed to one side, and the hot embers to the other. The hole being thus cleared of every-thing except its flooring of hot lumps of clay, the latter were strewn thinly with grass, or with the leaves of a herb called pennyroyal, green if possible, and well damped with water. On this were laid, neatly packed, the animals or roots to be cooked; then came another coating of wet grass, next the remaining lumps of heated clay, and then the burning members. These were often covered with a sheet of bark, and on top of all these was a quantity of earth. In an hour or two the food was taken out well-cooked and clean.

The prolonged cooking offered by pit ovens at a moderate temperature and in a humid environment, causes beneficial changes in food, facilitates its ingestion, improves its taste, its sanitary quality and in some cases increases its nutritional value (Wandsnider, 1997). This is the case for long-chain carbohydrates, they are broken down, increasing their bioavailability for humans (Leach, 2007; Thoms, 2009). Plants that store carbohydrates in rhizomes, bulbs, tubers or succulent roots (geophytes) have significantly vegetative periods and many of them dominate plant communities, constituting resources of high supply and high temporal-spatial predictability. These properties, which are inherent to multiple wild plant species, allow a true harvest of non-cultivated foods (e.g. Metraux, 1944; Sušnik, 1978), reducing the costs of obtaining them and, consequently, increasing the energy return compared to agriculture, which is an option with a lower energy yield (Bowles, 2011).

Pit ovens are often made in the same locations for centuries or millennia, transforming the site into an oven station, characterized by large accumulations of waste, mainly discarded heat retainers (see Thoms, 2008, Fig. 10). From the perspective of the formation processes of the archaeological record, the repetition of the practice, particularly digging pits, even if they are not very deep, produces a continuous removal of the previous records, forming true palimpsests (Black & Thoms, 2014).

13.2 Beyond Explaining How Mounds Were Raised

If we look beyond the processes that led to accretion and recognize the mounds as oven stations focused on the processing of food, mainly vegetables, we face a hypothesis (close to a theory) of a different order and of greater scope. In order to evaluate its explanatory force, we will analyse six dimensions of the registry of the mounds of India Muerta-Paso Barranca. Namely: location, shape, aggregation habits, characteristics of the matrix, behaviour of the chronologies and process of intensification of the use of space and finally, dietary profile of the inhabitants who were buried in them. All of this brings us closer to understanding the process that led to the formation of the cultural landscape of eastern Uruguay.

13.2.1 Location

From the beginning of mounds investigation, it was recognized that most of the mounds in eastern Uruguay are located in temporarily or permanently flooded areas (Arechavaleta, 1892; Ferrés, 1927). The others are found on hills that project or are surrounded by wetlands (Bracco et al., 2000a). In the plains of India Muerta-Paso Barranca, the mounds are located close to the courses, exhibiting a greater concentration where the water network is denser, in the central part of the wetland (Bracco et al., 2015). For their part, the mounds located in the hills, particularly the most numerous groups such as Los Ajos, Martínez Damonte and Campo Alto, are located where there is a stream with its widened channel or an oxbow lake (a lagoon according to local terminology) between the hills and floodplain areas (Bracco et al., 2021a).

From very early moments, the location of the mounds was interpreted in economic terms, proposing that the groups linked to them had developed an economy based on the high supply of resources (mainly faunal) of the wetlands (see e.g. López Mazz & Bracco, 1992). However, and assuming that wildlife resources are important conditioning factors, the location of the mounds, both in the plain and in the hills, show a certain degree of contradiction (Bracco et al., 2015). In the first case, they would be competing for the spaces where the animal species that would have been most exploited are found, according to the archaeofaunal record (Pintos Blanco, 2000; Moreno, 2014). In the second, they would be located far from these resources, while replicating a ground uplift behaviour that does not offer greater benefits in places that are not flooded.

Now, if the mounds are oven stations where plants with high spatio-temporal predictability are mainly processed, they are expected to be found close to where the plants grow. Three of the main plants with high predictability and spatio-temporal supply present in the India Muerta-Paso Barranca region are *Canna glauca* (achira), *Typha dominguensis* and *Typha latifolia* (totora). Their silicophytolites and starches have been recovered from artefacts and from the matrix of the mounds (Capdepont et al., 2004, 2016; Capdepont & Pintos Blanco, 2006; del Puerto, 2011a, 2015; Gianotti & Bonomo, 2013). The management and/or cultivation of *Canna glauca* is widely referred to by American ethnographic, ethnohistorical and archaeological sources (Ugent et al., 1984; del Puerto, 2011a; Piperno, 2011; Watling et al., 2015). Together with the totora, they are recognized as resources of high nutritional value

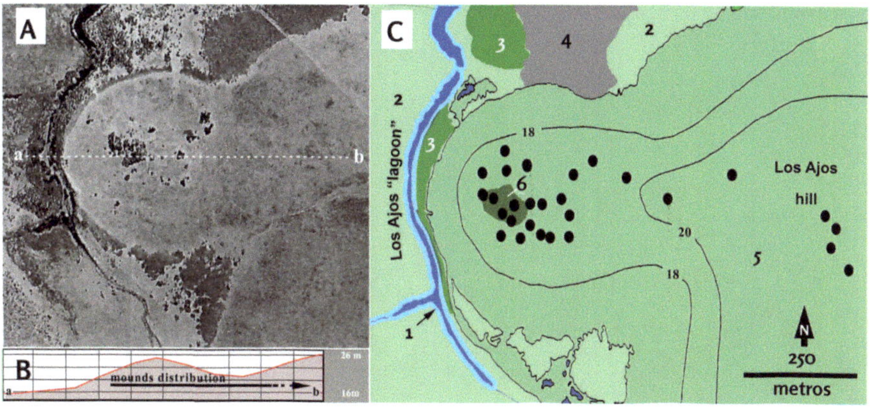

Fig. 13.4 Los Ajos site located at the end of the Sierra de Los Ajos. (**a**) SGM (Military Geographical Service) aerial photograph, flight 1966–68; original scale 1:20,000. (**b**) Topographic profile indicating mound distribution (on Google Earth image, 08/22/2020). (**c**) Diagram of the distribution of the dominant vegetation: (1) Rooted aquatic plants with floating leave (permanent wetland); (2) *Tiririca*l and *Totoral* (permanent and semi-permanent flooded area); (3) Riparian mount; (4) Grassland (middle plains); (5) Summer grasslands; 6. Native tree vegetation. The black circles indicate the mounds location

(Rapoport et al., 2009; del Puerto, 2011b). These species develop in extensive areas that occupy, permanently or temporarily floodable soils and in the alluvial plains of lagoons and streams (Barreneche & Zarucki, 2017).[1] They grow very close to where the mounds are both on the plain and on the hills (Figs. 13.4 and 13.5).

13.2.2 *Shape*

Despite what has been proposed for eastern Uruguay, "earthen buildings" exhibit different categories according to their height, and the flatness of the top (Gianotti et al., 2009; Villagran & Gianotti, 2013) their basic shape (consequence of the actions of cultural and natural processes) has little variations and conforms to the definition of mound according to the RAE (2020): "heap, set of things put out of order" (see Ingold, 2013, p. 75 and following).[2] This agrees with the observations made, for example, by López Mazz (2001, p. 238) during his excavation: "It has a

[1] These spaces would have had a wide extension even in the conditions of less precipitation that occurred at the end of the Middle Holocene (Bracco et al., 2012).

[2] "*Yet most prehistorians, convinced that any kind of making or building involves the projection of form onto the material world, are predisposed to think of the mound as a sculptured earthwork, answering to the specifications of a preconceived design. It is as though the mound builders, as they piled up earth and stones, had an idea of how the work would finally look, and kept on piling until the height, diameter and contours of the accumulated material matched their expectations.*" (Ingold 2013, p. 77).

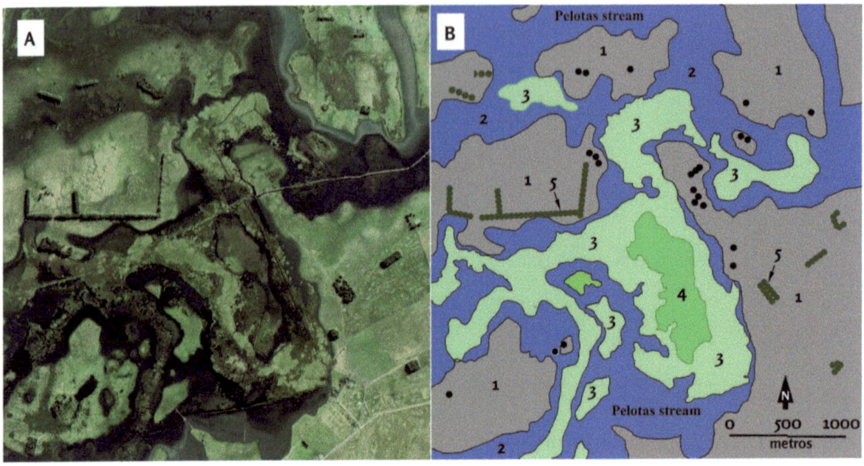

Fig. 13.5 Pelotas site located in the plains zone. A. Satellite image (Google Earth, 08/22/2020). B. Diagram of the distribution of the dominant vegetation: (1) Grassland (middle plains); (2) Rooted aquatic plants with floating leaves (permanent dipping); (3) *Tirirical* (semi-permanent flooded area); (4) *Totoral* (semi-permanent wetland); (5) Recent afforestation. The black circles indicate the mounds location

'nougat' physiognomy, with a heterogeneous composition (gravel, burned earth, archaeological material, sediments, etc.)...". Without denying the possibility that other contributions or conditions related to other functions have occurred, this is expected -in form and content- as a consequence of the secular accumulation of waste from pit oven where heat-retaining sedimentary thermophores are used.

13.2.3 Aggregation Habits

In the India Muerta-Paso Barranca region 636 mounds form 174 groups (Bracco et al., 2021b, Table 1). The most numerous groups are found on hills. The Campo Alto site, the one with the largest population, has 57 mounds (Bracco et al., 2015). In the plains, the largest groups reach populations of 20 mounds. Together with their clear grouping habit, it is observed that the distances between mounds are repeated very frequently, as well as certain spatial arrangements (Fig. 13.6). We have proposed that the distances and physical arrangements would correspond to the distances and social arrangements maintained and reproduced by the social actors who produced and occupied them (Bracco et al., 2015). Although the interpretation has difficulties to be corroborated, it increases the plausibility if we link it to socially ordered functional units, committed to the exploitation of fixed resources subject to collective property rights. This is observed in Australian groups that exploit vegetables that grow in floodplain areas and are processed in oven mounds (Brockwell, 2006; Westell & Wood, 2014; Jones et al., 2017).

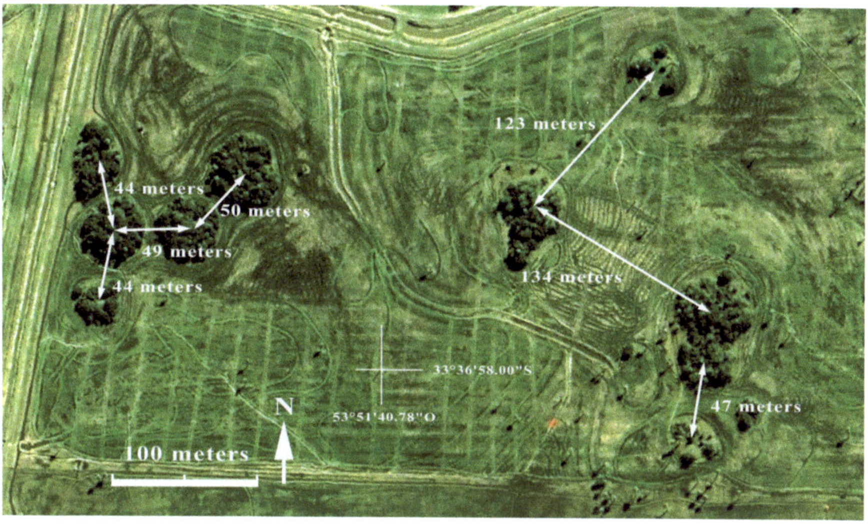

Fig. 13.6 Satellite image of a set of mounds located on the left bank of the San Luis river (Google Earth, 10/22/2020). As an example the distances between mounds are indicated

13.2.4 Matrix Characteristics

The difference of texture between the matrix of the mounds and the nearby soils or sedimentary deposits, mainly a greater amount of silt, sand and gravel fractions in the former, led to infer that the sedimentary material that constitutes them was transported, totally or partially, from sources that present similar grain size composition (Durán, 1990; López Mazz, 1992; Castiñeira & Piñeiro, 2000; Castiñeira et al., 2015). However, this approach does not pay much attention to the fact that the differences could have been the consequence, not of transport and mixing, but of human behaviour capable of modifying the size of the particles that make up the matrix. The burned earth shows that this behaviour was present in the accretion processes. From sediments with fine to very fine textures (those used by the ant *Camponotus punctulatus* to build their ant hills, for example) gravel-size aggregates were produced by exposure to fire (Bracco et al., 2000c, 2019c; Duarte et al., 2017). At first, the burned earth was interpreted as an artificial fraction added to the matrix in order to increase the resistance of the mounds as non-cemented structures and / or as a way to improve the quality of the floors (Bracco et al., 2000c, 2017).[3,4]

[3] A similar idea was proposed by Politis et al. (2011) for the mounds of the Paraná delta. For these authors, both the ceramic fragments and the burned earth would have fulfilled this function.

[4] From a fragment of burned earth that included a small piece of vegetation recovered in the excavation carried out at the La Viuda site, López et al. (2022) infer that the burnt earth fragments could be related to a domestic construction technique. This technique would be based on the use of adobes. This deserves two comments. First of all, anthills are colonized by different types of vegetation, including shrubs. Therefore, it is possible that plant remains or their traces appear in frag-

However, if we admit that the burned earth corresponds to the heat retainers (cf del Puerto et al., 2022; López Mazz et al., 2022 among others) we observe that during the operation of an pit ovens the heat retainers are heated as well as the walls of the oven and the sediments that are used as a seal on the lid. Therefore, we are faced with massive exposure of sediments to heat. If a sufficient temperature is reached, its clays will form larger aggregates, from silt to gravel, producing a modification of the texture of all matrix (Bracco et al., 2019a).

Recently del Puerto and collaborators (2022) have presented a soil characterization of the mounds of the La Tapera site, located in the India Muerta-Paso Barranca region, just 5 km from the García Ricci site. In their paper they show the importance of fire in the growth process, indicated by the presence in the matrix of burnt earth fragments, charcoals, thermally altered fauna remains, high levels of available phosphorus and the presence of pyrogenic SOM of herbaceous vegetation in every anthropogenic mound layers analyzed by pyrolysis (PY) – gas chromatography (GC) – mass spectrometry (MS). Although the authors speak of waste management through fire, only this practice can hardly explain how the burned earth was formed (for the highest mounds of the La Tapera site it constitutes 30% of the matrix (del Puerto et al., 2022, p. 195) equal to what was observed at the Garcia Ricci and PSL sites). We must also bear in mind that herbaceous vegetation in many cases can be used as fuel. Black and Thoms (2014) have pointed out that low caloric capacity fuels can be used in ovens pit. Anthracological studies show that the occupants of the mounds would have followed this behaviour (Inda & del Puerto, 2007).

13.2.5 Chronologies and the Process of Intensifying the Use of Space

The behaviour of the available chronologies led us to rethink three central questions in the discussion about the mounds of eastern Uruguay: What were the processes involved whilst built to them?, what was their function or functions? and finally, what were the lapses and rhythms of the phenomenon?, both at the scale of a singular mound, groups of mounds or at a regional scale (Bracco & Ures, 1999; Bracco, 2006; Duarte et al., 2017). On the one hand, the series of the ages of the individual mounds indicate that the elevation occurred as a consequence of a slow, gradual and secular growth mechanism (Duarte & Bracco, 2020); on the other, the concordance between ages ^{14}C, OSL and TL for the same levels indicates that during the accretion process, coetaneous charcoals were deposited at the time of luminescent bleaching of the minerals (Bracco et al., 2020). New OSL, TL and ^{14}C ages of mound I of the archaeological locality La Viuda also showed the same

ments of anthills. Second, the adobes are not cooked. As much as we admit that in some circumstance these could have been exposed to fire, it would be difficult for fragments to be produced with their entire external surface showing rubefaction, oxidation produced by having been exposed to temperatures above 300 °C.

Table 13.1 Luminescence dates from La Viuda I, stratigraphic unit 12. For the stratigraphic unit 12 and 13, López Mazz et al. (2022, Table 1) report an age of ^{14}C: 3026 ± 32 (cal BP (2 sigma) 3726–3748 (p. 0.026), 3768–3794 (p. 0.045), 3823–3988 (p. 0.903), 4044–4069 (p. 0.026) (Hogg et al., 2020)

Lab. #	Method	Paleodose (Gy)	σ	Age	σ
UY 182	OSL	4.18	0.29	3130	250
UY 183	OSL	3.94	0.29	2950	250
UY 184	Tl	7.47	0.60	3340	300
UY 185	Tl	7.00	0.60	3130	300
UY 186	OSL	4.41	0.36	3310	300
UY 187	OSL	4.01	0.36	3010	300
UY 188	OSL	4.12	0.25	3090	230
UY 189	Tl	7.94	0.60	3550	300
UY 190	Tl	7.22	0.60	3230	300

concordances. This mound is the oldest and tallest (7.27 m) in the India Muerta-Paso Barranca region (Bracco & Ures, 1999). It has been recently excavated by López Mazz and collaborators (2022). The news luminescent ages come from excavation I, stratigraphic unit 13 (López Mazz et al., 2022, Fig. 4), from two sampler points separated horizontally by approximately 0.80 m. The OSL ages were made from the sand fraction of the matrix. TL ages from silt fraction of burned earth fragments. The OSL and TL ages of each of the sampler point and the two points are very similar (Table 13.1). The ^{14}C age reported for stratigraphic units 12 and 13 (López Mazz et al., 2022, Table 1) appears to be slightly older.

The concordance between ages ^{14}C, OSL and TL for the same levels a slow, gradual and secular accretion are fully consistent with the expected record produced by pit ovens and with the accretion rates and lapses of the waste deposits that are formed in the oven stations.[5] This is how Beveridge (1869, p. 187) described it for the oven mounds of Australia:

> When the cooking has been completed, the covering is scraped off, and this debris (calcined clay, ashes, and earth) becomes the nucleus of a black fellow's oven. This process being continually repeated for many years, perhaps centuries, results in the heaps which are actually black fellows' ovens, although often improperly designated tumuli.

On the other hand, the chronologies on a regional scale show two characteristics of the phenomenon: its extension and its intensification.[6] It has been pointed out (Iriarte et al., 2000; Bracco et al., 2011a; del Puerto, 2015) that the appearance of the mounds in the India Muerta-Paso Barranca region coincided with a climatic deterioration characterized by the advent of colder and drier conditions that occurred at the end of the Middle Holocene (Bracco et al., 2011b). This allowed proposing

[5] *Oven mounds* have grown in periods ranging from 300 to 2500 years (Brockwell, 2006, Table 2).

[6] The chronological contexts that provided information leading to the historical period are linked to human burials, which do not necessarily have to be related to the contribution of sediment that caused the elevation of the mounds (see Cabrera, 2013; Capdepont et al., 2016).

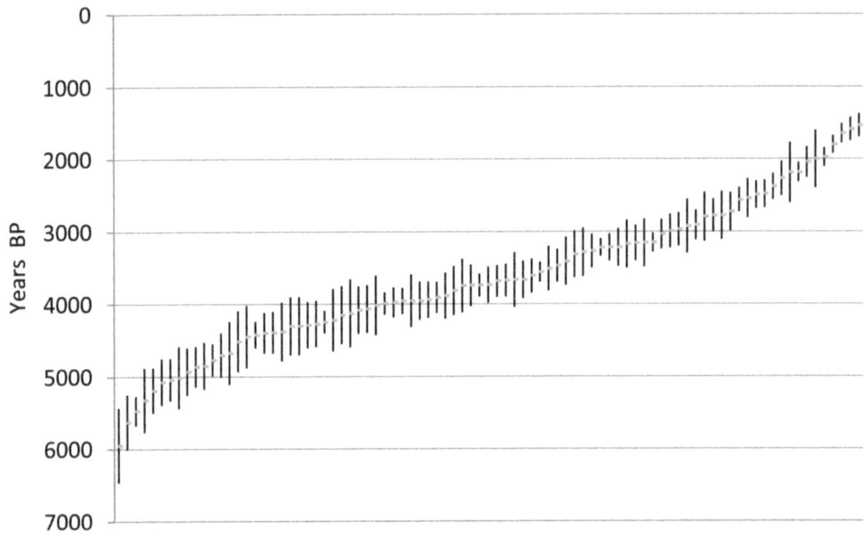

Fig. 13.7 Luminescent ages (TL and OSL) from the India Muerta-Paso Barranca and Pelotas region mounds, eastern Uruguay (taken from Duarte et al., 2017; Duarte & Bracco, 2020; Bracco et al., 2020)

that the intensification in the occupation of wetlands was generated as a response to environmental changes (Bracco et al., 2011a). However, the luminescent dates and the revision of ^{14}C ages in the light of the first ones, which appeared as anomalous because they are very old, could indicate that the mounds would have begun to rise at least 1000 years earlier, around 6000 solar years BP (Figs. 13.7 and 13.8);[7] therefore, climate change could even have occurred after the adoption of pit ovens. Although precedence invalidates causality, the possibility that climatic deterioration led to an increase in pit ovens use cannot be ruled out. In any case, the dates show that the behaviour that produced the mounds intensifies markedly from 4500 years ^{14}C BP, and continues throughout the cold and dry period (Bracco et al., 2011a).

In the recent end of the chronological series, it is observed that very few ages of the ^{14}C approach the historical period. Luminescent ages show a similar behaviour. This could be due to a decrease in the frequency of the practice around 1000 years BP. If that were true, we again find a correlation with a climatic period, the medieval optimal climatic period, which has been identified in the regional palaeoclimatic record (Bracco et al., 2011a). However, what is stated in the previous paragraph forces us to be extremely cautious in interpreting this last correlation not being able to rule out that we are again facing a sampling bias.

Likewise, the chronologies discriminated by areas show how the mounds were gradually expanded, increasing in number and gradually beginning to appear in new

[7] It is very unlikely that the earliest available dates correspond to the first mounds, so we must admit that the behaviour that gave rise to them most likely originated earlier.

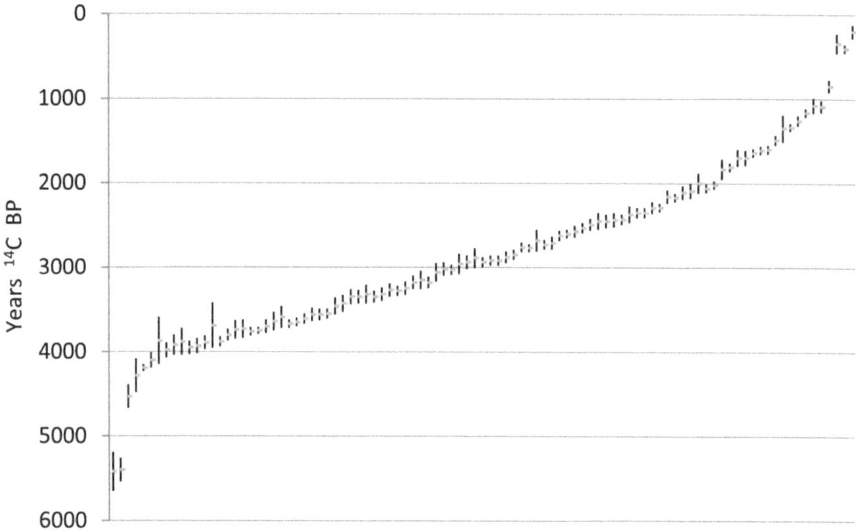

Fig. 13.8 ^{14}C ages from mounds in eastern Uruguay and southern Brazil (taken from Bracco et al., 2015; Milheira & Gianotti, 2018)

areas. This would reveal an increase in redundancy in the occupation of the mounds (a decrease in mobility) along with a process of intensification of the use of the territory manifested by an enlargement in their number and a greater dispersion (Fig. 13.9). Following Thoms (2009, p. 122) "land use intensification" refers to a trend over the millennia towards spending more energy per unit area to obtain more food from the same landscape with the goal of feeding more people. In our case, it would have focused on an increase in investment in processing. For the U.S.A., the author called the intensification process that would have started around 4000 years BP as the "carbohydrate revolution". It was based on the adoption of a subsistence system that included mainly wild plant resources of high temporal-spatial supply (mainly geophytes); whose return increases when they are processed, cooked in pit ovens, for example (Thoms, 2008, 2009, 2017).

13.2.6 Pit ovenPit ovens and Dietary Profile

If it is correct that the process of implantation and proliferation of the mounds is part of a process of intensification of the use of space that was based on an increase in the consumption of plant resources, then it should be expressed in the stable isotope dietary profiles of the populations linked to the mounds. The first stable carbon isotope values of collagen (hereafter $\delta^{13}C_{collagen,}$ a proxy for the isotope composition of dietary protein, and to a lesser extent carbohydrates and lipids) from human bone remains from east Uruguay, southern sector of the Merín lagoon basin (sCLM),

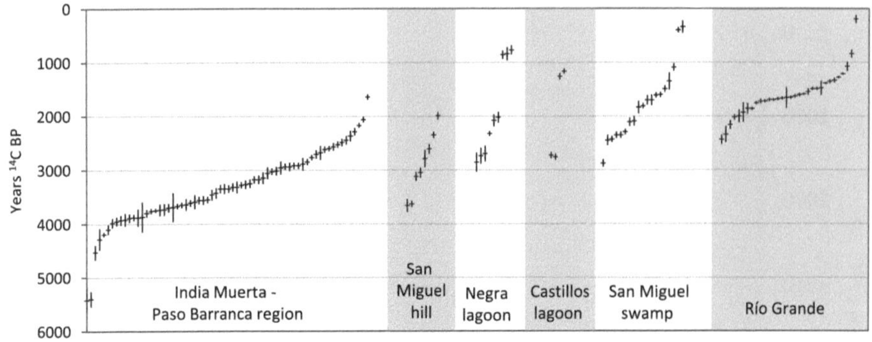

Fig. 13.9 ^{14}C ages from mounds in eastern Uruguay and southern Brazil, distributed by region (taken from Bracco et al., 2015; Milheira & Gianotti, 2018)

indicated a continental diet based on the C3 photosynthetic pathway (Bracco et al., 2000b). New isotopic data, combining other parameters, including the δ^{13}C of bone apatite (hereafter $\delta^{13}C_{apatite,}$ a proxy of the isotope composition of the whole diet) and the stable nitrogen isotope composition (hereafter δ^{15}N, a proxy of the isotope composition of dietary protein only) have allowed a more precise approach to the subject, while resolving the apparent contradictions with palaeobotanical records, presence of cultigens and particularly corn (del Puerto, 2015; Mut, 2015a). The average $\delta^{13}C_{collagen}$ value available today is −20.48 ± 1.42‰ (n = 30, VC (variation coefficient) = 6.94%), including an individual from a mound located near the Castillos Lagoon ($\delta^{13}C_{collagen}$ – 14.2 ± 0.2‰). If this anomalous value is eliminated, the average $\delta^{13}C_{collagen}$ value becomes −20.70 ± 0.79‰ (n = 29, VC = 0.80%). The average $\delta^{13}C_{apatite}$ value is −10.57 ± 1.55‰ (n = 9, VC = 14.64%). Excluding the individual from Castillos lagoon, the average $\delta^{13}C_{apatite}$ is −9.17 ± 0.51‰ (n = 8, VC = 5.51%). The average δ^{15}N value is 10.45 ± 2.19 (n = 9, VC = 20.99). Again, if we do not consider the individual from Castillos lagoon, we have a δ^{15}N value of 9.94 ± 0.65 (n = 8, VC = 6.53%). These values that cover the period 2530 ± 60–290 ± 75 BP present low dispersion, which would reflect diets with few spatial-temporal variations and different from that of other close and contemporary groups that occupied similar environments (Fig. 13.10) (del Puerto, 2015; Mut, 2015b; Mut & Bracco, 2022, Table 1).

Mut (2015b, p. 38 and following), applying the model of Froehle et al. (2010) and Kellner and Schoeninger (2007) also inferred that both the protein and energy sources (carbohydrates and lipids) would have been from the C3 line. However, four individuals would show a mixed C3/C4 consumption of energy sources. Likewise, these observations are consistent with the results that she obtained when applying the multivariate model of Froehle et al. (2012), which allowed her to conclude that we would be facing a diet based on the C3 photosynthetic line, but with a small contribution from C4 foods (Mut, 2015b, p. 52). In turn, she warned that the Δ^{13}C values are very high (average Δ^{13}C 10.4 ± 1.2‰, n = 8, VC = 11.5%) (Mut, 2015b, Fig. 10), noting that large Δ^{13}C have been reported for diets where the plant

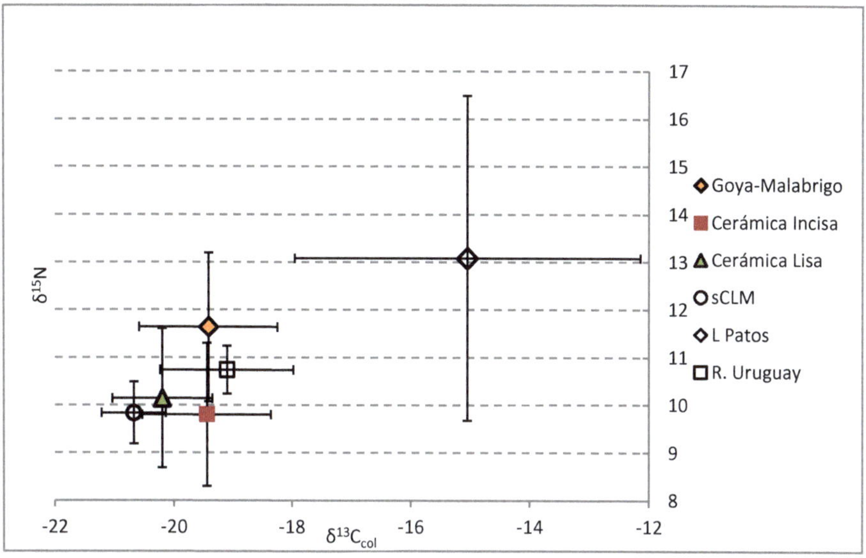

Fig. 13.10 Comparison of dietary profiles of individuals from three archaeological units of hunter-gatherers in the Paraná Delta, sites on the left bank of the lower Uruguay River, south of Laguna de los Patos and sCLM. Based Chanca and collaborators (2021), Loponte (2020) and Loponte et al. (2016). Those of the sCLM differ mainly by their more depleted values of $\delta^{13}C$ and lower values of $\delta^{15}N$. Taken from Mut and Bracco (2022, Fig. 6)

component is important (see Ambrose, 1993; Schwartz, 2000; Harrison & Katzenberg, 2003). The similarity of the Sr/Ca relationship between humans and local herbivores was interpreted in the same sense (Bracco et al., 2000b, Table VI). The foregoing is consistent with the information provided by other sectors of the registry. The archaeofaunal record indicates that the proteins would come mainly from terrestrial mammals and from freshwater fish. The importance of deer (*Blastocerus dichotomus,* commonly called "*ciervo de los pantanos*" and *Ozotoceros bezoarticus uruguayensis, Mazama gouazoubira* commonly called "*venado de campo*") in the subsistence system was early pointed out by Pintos Blanco (2000). Moreno (2014) indicated that the *apereá* (*Cavia aperea*), the otter (*Myocastor coypus*) as well as fish would also have played an important role in their diet. Bica (2020), based on the analysis of ichthyofauna from the CH2D01 site (San Miguel swamp, Rocha), indicates that the fish would have contributed 40% of the proteins, being the most represented taxa the black catfish (*Rhamdia aff. quelen*), tararira (*Hoplias* spp.) and eel (*Synbranchus aff. marmoratus*). Beside silicophytoliths and starches of *phaseolus*, *cucurbitacea* and *Zea maiz*, recovered from the matrix of the mounds and on the surface of artifacts (Iriarte, 2003; Gianotti et al., 2013; del Puerto, 2015) showed the consumption of cultigens. In the case of the first two since before 3500 years BP. Corn would have appeared later, according to Iriarte (2007) and Iriarte et al. (2004), around 3000 years BP., and according to del Puerto (2015) around 2000 years BP. Likewise, in practically all the records analysed, silicophytoliths of plant genera that integrate edible plants have been identified, such as *Syagrus romanzoffianum, Butia odorata, Bromelia antiacantha, Canna glauca, Cyperus esculentus* and *Typha domingensis* and/or *Typha latifolia* (del Puerto, 2011a, 2015). Based on the above, it has been suggested that we would be facing an economy in which hunting, fishing and collecting was complemented with small-scale horticulture, and where the C3 plant component made up a significant proportion of the diet (Bracco, 2006; del Puerto, 2015; Mut, 2015a, b).

The application of statistical models that are based on the transfer of the isotopic signal from food to the consumer (food reconstruction using isotopic transferred signals) not only corroborate the conclusions but can also allow us to understand furthermore the reason for the intense use of pit ovens. Seeking to contrast what has been deduced, we have applied the *Fruit* v 3.0 Beta program (Food Reconstruction Using Isotopic Transferred Signals [https://sourceforge.net/projects/fruits/files/], Fernandes et al., 2014) a Bayesian mixing model that reconstructs individual dietary profiles from the human isotopic record ("target") and likely food sources. We chose the simplest option of the model (unweighted, independent of compensations and concentrations model) since the objective is to obtain an independent test that allows estimating the weight of vegetables in the diet.

Based on the archaeobotanical and archaeofaunal records, we have defined four main food sources: (1) C3 plants, (2) herbivores in the C3 photosynthetic line, (3) freshwater fish, and finally 4) C4 plants (Mut & Bracco, 2022). Not having enough isotopic data to calculate the isotopic signal from each source, data from nearby and ecologically similar regions were used (see Mut & Bracco, 2022, Table 4). Two targets were chosen, those with the highest and lowest $\Delta^{13}C$, noting that such spacing would indicate the incidence of plants in the diet. It has been pointed out that high $\Delta^{13}C$ values would indicate a greater importance of plants in the diet (Ambrose, 1993; Harrison & Katzenberg, 2003; Schwartz, 2000). For the first individual we have a direct ^{14}C dating of 1350 ± 160 years BP (Mut & Bracco, 2022). The results are presented graphically in Figs. 13.11, 13.12 and Table 13.2.

On the one hand, a diet with such an important plant component C3 would explain why, following the classical approaches of isotopic analysis, the corn signal, present according to the microbotanical record, was not visualized. On the other, it directs attention to the local existence of wild C3 plants of high supply, which due to physiological characteristics, predictability and time-spatial abundance could provide an even greater or equivalent return to that of cultigens (see Smith, 2001). Along with the two species of palm tree (*Syagrus romanzoffiana* and *Butia odorata*), as we already mentioned, three are the geophytes that stand out: *Typha dominguensis*, *Typha latifolia* (totoras) and *Canna glauca* (achira). Canna has rhizomes and Typha thick roots rich in carbohydrates, whose bioavailability would increase when cooked (Wandsnider, 1997; Leach, 2007; Thoms, 2009).

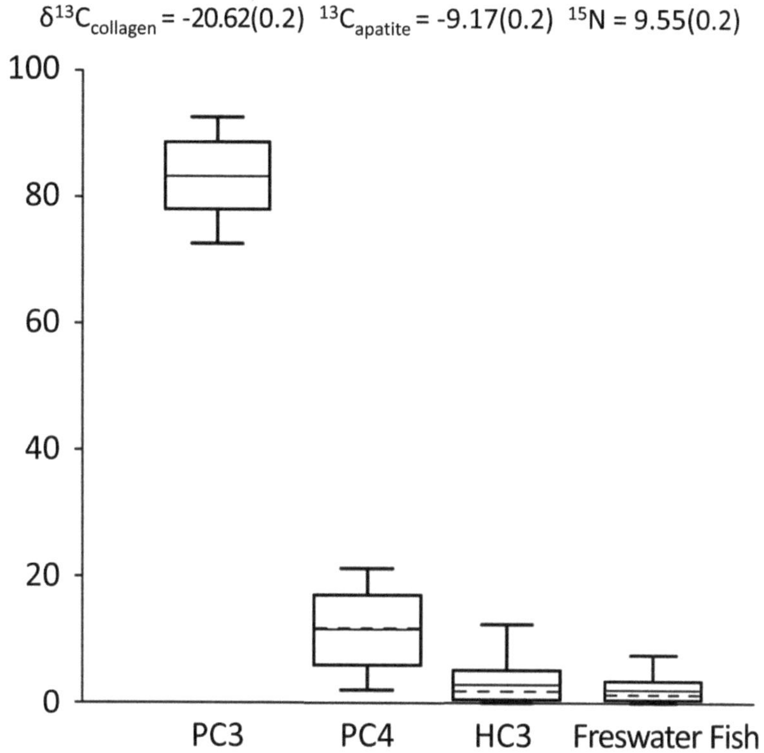

Fig. 13.11 Graphic representation of estimates of the source/food contribution for the individual of the series with the highest $\Delta^{13}C$ (Mut & Bracco, 2022, Table 1)

13.3 Final Remarks

Different lines of evidence show that the fire was present during the formation of the mounds of India Muerta-Paso Barranca. The most robust comes from the luminescent registers of the matrix. The verification that its different fractions (burned earth, gravel or finer fractions) for the same level, produce similar OSL and TL ages, and consistent with ages of ^{14}C, allows to conclude that the matrix was heated to at least bleaching temperatures of its luminescent record, circa 350 °C (Duarte et al., 2017; Bracco et al., 2020). Seeking to answer the question: what human behaviour could produce a record with these characteristics, thermo-altered sediment mounds? We come to the oven mounds of Australia. From the analogy we have proposed that the recursive behaviour (Bracco, 2006) that was behind the elevation of the mounds of the India Muerta-Paso Barranca region was the use of earth or pit ovens. The proposal does not start from a general and therefore weak analogy, but rather from a relational analogy that is based on a "principle of causal or quasi-causal connection" (Wylie, 1985, p. 101). This would be the use of a technology integrated into a

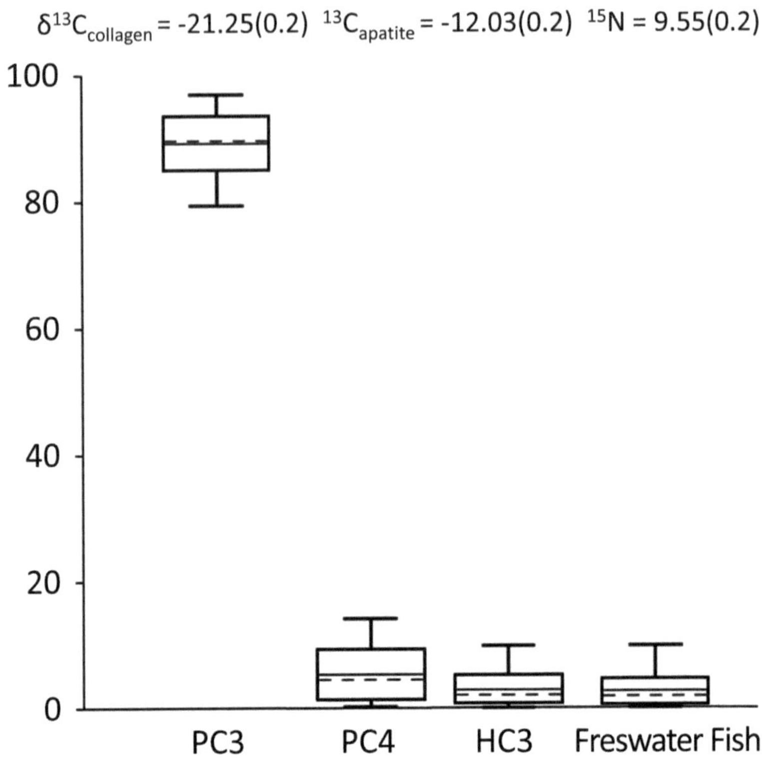

Fig. 13.12 Graphic representation of estimates of the source/food contribution for the individual of the series with the lowest $\Delta^{13}C$ (Mut & Bracco, 2022, Table 1)

Table 13.2 Estimates on the contribution of sources/foods

Estimates on source/food contribution							
Source/Food	Average	Sd	2.5pc	Median	97.5pc		
Individual of the series with the highest Δ13C							
PC3	0.8325	0.05291	0.7272	0.8332	0.9265	0.7805	0.8880
PC4	0.116	0.05218	0.02065	0.1176	0.2123	0.0586	0.1706
HC3	0.02972	0.03427	6.22E-01	0.01862	0.1252	0.0045	0.0532
Freshwater fish	0.02171	0.02297	5.04E-01	0.01461	0.08422	0.0037	0.0391
Individual of the series with the lowest Δ13C							
PC3	0.8925	0.0455	0.7942	0.8967	0.97	0.8488	0.9365
PC4	0.05239	0.03845	0.002084	0.04476	0.1408	0.0131	0.0950
HC3	0.02885	0.02639	8.61E-01	0.02105	0.09792	0.0058	0.0526
Freshwater fish	0.0263	0.02719	6.52E-01	0.01806	0.09846	0.0044	0.0479

subsistence system that focuses on the processing and consumption of the vegetable resources in the wetland.

Consequently, the proposal not only gives an account of how the mounds were erected, furthermore an economic option that combines technological with environmental factors. As far as it is possible to explain other diverse dimensions of the archaeological record, the jump shown from accretion mechanisms to subsistence systems is valid and coherent.

In this sense, we have tried to explain the location of the mounds, their shape, their aggregation habits, textural characteristics of their matrix, their chronologies and, finally, the dietary profile of the populations linked to them. We conclude that these dissimilar dimensions are reasonably explained if we admit that the mounds are the product of the secular accumulation of waste produced by pit ovens. This practice would have been articulated with an intensification process that focused on the exploitation of high supply C3 plants, the return of which increases when cooked. This leads us to observe the importance that wild plants would have had in these economic systems, to the point that resources such as corn remained in the background for long periods. But the explanation does not necessarily have to be given by a simple cost-benefit equation.

By focusing on the causes that led to the elevation of the mounds and on the subsistence systems that would have been at their base, only some aspects of their reality is prioritized, without denying the probability of their diversity and multifunctionality. Their probable character of "total social fact" (Mauss, 2009) which is evidenced by sectors of the registry that we have only mentioned laterally, such as burials.

So far, our analogy-based hypothesis has provided a credible explanation for various and diverse known aspects of the archaeological record of the India Muerta-Paso Barranca region. The next step is to exploit its potential as a heuristic tool to generate new hypotheses to augment our knowledge about the peculiar dialectical relationship that existed between the human groups that occupied the region and nature. The final objective is to understand how, over five millennia, the unique prehistoric landscape of this region was formed, where in approximately 250 km^2 more than 600 mounds are present.

Acknowledgments We thoroughly thank Dr. Jorge Rabassa for the critical reading of the draft.

References

Ambrose, S. H. (1993). Isotopic analysis of paleodiets: Methodological and interpretative considerations. In M. K. Sandford (Ed.), *Investigations of ancient human tissue* (pp. 59–130). Gordon & Breach Science Publishers.

Arechavaleta, J. (1892). Viaje a San Luis. In *El Uruguay en la exposición histórico-americana de Madrid. Memoria de los trabajos realizados por la comisión Nacional encargada de organizar los elementos de concurrencia* (pp. 65–106). Imprenta Artística de Dornaleche y Reyes.

Baeza, J., & Panario, D. (1999). La horticultura indígena en las estructuras monticulares. In C. Goso, C. Castiñeira, & S. Martínez (Eds.), *Actas de las Primeras Jornadas sobre Cenozoico en Uruguay (20–21 de diciembre de 1999)* (pp. 1–2). SUG, INGEPA y UNCIEP, Facultad de Ciencias.

Barreneche, J. M, & Zarucki, M. (2017). *Mapeo y clasificación de formaciones vegetales del sitio Ramsar Bañados del Este y Franja Costera. Documentos de Trabajo N° 52*, Programa de Conservación de la Biodiversidad y Desarrollo Sustentable en los Humedales del Este (PROBIDES).

Beveridge, P. (1869). On aboriginal ovens. *Journal of the Anthropological Society of London, 7*, clxxxvii–clxxxix. http://sci-hub.tw/10.2307/3025359

Bica, C. (2020). *Peces y pesca en las tierras bajas de la Laguna Merín. Análisis de la ictiofauna recuperada en el sitio arqueológico CH2D01 (Rocha, Uruguay)*. Master's thesis, Instituto de Ciências Humanas, Universidade Federal de Pelotas (Pelotas).

Black, S. L., & Thoms, A. V. (2014). Hunter-gatherer earth ovens in the archaeological record: Fundamental concepts. *American Antiquity, 79*, 204–226. http://sci-hub.tw/10.7183/0002-7316.79.2.204

Bonomo, M., Politis, G. G., & Gianotti, C. (2011). Montículos, jerarquía social y horticultura en las sociedades indígenas del Delta Del Río Paraná (Argentina). *Latin American Antiquity, 22*, 297–333. http://sci-hub.tw/10.7183/1045-6635.22.3.297

Bowles, S. (2011). Cultivation of cereals by the first farmers was not more productive than foraging. *Proceedings of the National Academy of Sciences of the United States of America, 108*, 4760–4765. http://sci-hub.tw/10.1073/pnas.1010733108

Bracco, R. (2006). Montículos de la cuenca de la laguna Merín: Tiempo, espacio y sociedad. *Latin American Antiquity, 17*, 511–540. http://sci-hub.tw/10.2307/25063070

Bracco, R., & Ures, C. (1999). Ritmos y dinámica constructiva de las estructuras monticulares. Sector sur de la cuenca de la laguna Merín. Uruguay. In J. M. López & M. Sanz (Eds.), *Arqueología y Bioantropología de las Tierras Bajas* (pp. 13–33). Facultad de Humanidades y Ciencias de la Educación, Universidad de la República, Montevideo.

Bracco, R., Cabrera, L., & López Mazz, J. M. (2000a). La prehistoria de las tierras bajas de la cuenca de la laguna Merín. In A. Duran & R. Bracco (Eds.), *Arqueología de las Tierras Bajas* (pp. 13–38). Ministerio de Educación y Cultura, Comisión Nacional de Arqueología.

Bracco, R., Fregeiro, M. I., Panarello, H., et al. (2000b). Dieta, modos de producción de alimentos y complejidad. In A. Duran & R. Bracco (Eds.), *Arqueología de las Tierras Bajas* (pp. 227–248). Ministerio de Educación y Cultura.

Bracco, R., Montaña, J. R., Nadal, O., & Gancio, F. (2000c). Técnicas de construcción y estructuras monticulares, termiteros y cerritos: de lo analógico a lo estructural. In A. Duran & R. Bracco (Eds.), *Arqueología de las Tierras Bajas* (pp. 285–300). Ministerio de Educación y Cultura.

Bracco, R., del Puerto, L., Inda, H., et al. (2011a). The relationship between emergence of mound builders in SE Uruguay and climate change inferred from opal phytolith records. *Quaternary International, 245*, 62–73. http://sci-hub.tw/10.1016/j.quaint.2010.05.025

Bracco, R., García-Rodríguez, F., Inda, H., et al. (2011b). Niveles relativos del mar durante el Pleistoceno final-Holoceno en la costa de Uruguay. In F. García-Rodríguez (Ed.), *El Holoceno en la zona costera de Uruguay* (pp. 65–92). Departamento de Publicaciones, Unidad de Comunicación de la Universidad de la República (UCUR).

Bracco, R., del Puerto, L., Inda, H., et al. (2012). Evolución ambiental y constructores de cerritos en la región de India Muerta. Un replanteo. In *III Jornadas del Cenozoico (14 y 15 de junio de 2012), CD-Rom* (pp. 14–15). Sociedad Uruguaya de Geología/DINAMIGE.

Bracco, R., Inda, H., & del Puerto, L. (2015). Complejidad en montículos de la cuenca de la laguna Merín y análisis de redes sociales. *Intersecciones en Antropología, 16*, 271–286.

Bracco, R., Bazzino, A., Duarte, C., et al. (2017). Estructuras monticulares, hormigueros y formigueros. La seducción de la analogía. In C. Scabuzzo, A. F. Zucol, R. C. Angrizani, & Colobig M de los M (Eds.), *Scientia Interfluvius Suplemento Resúmenes VII Encuentro de Discusión*

Arqueológica del Nordeste Argentino, 7 EDAN (20 al 23 de septiembre de 2017) (p. 37). Universidad Autónoma de Entre Ríos.

Bracco, R., Duarte, C., Gutiérrez, O., et al. (2019a). El fuego en la génesis de los montículos de la cuenca de la Laguna Merín. Su visualización a través de las técnicas de datación por luminiscencia. In M. Bonnin, A. Laguens, & M. B. Marconetto (Eds.), *XX Congreso Nacional de Arqueología Argentina, Córdoba (15 al 19 de Julio de 2019)* (pp. 1536–1541). Universidad Nacional de Córdoba.

Bracco, R., Panario, D., Gutiérrez, O., et al. (2019b). Mounds and landscape in the Merín lagoon basin, Uruguay. In H. Inda & F. García-Rodríguez (Eds.), *Advances in coastal Geoarchaeology in Latin America* (pp. 103–129). Springer.

Bracco, R., Panario, D., Gutiérrez, O., et al. (2019c). Estructuras monticulares y hormigueros en el sur de la Cuenca de la Laguna Merín: ¿ingenieros ambientales y/o la estrategia del bricoleur? *Revista de Antropología del Museo de Entre Ríos, 5*, 24–40.

Bracco, R., Duarte, C., Gutiérrez, O., et al. (2020). El fuego en los procesos constructivos de los montículos del sur de la cuenca de la Laguna Merín (Uruguay): Un aporte de la datación por luminiscencia (OSL/TL). *Latin American Antiquity, 31*, 498–516. http://sci-hub.tw/10.1017/laq.2019.98

Bracco, R., Duarte, C., Gutiérrez, O., et al. (2021a). Reflexiones sobre montículos, técnicas de procesamientos de alimentos y construcción de nichos. *Arqueología, 27*, 109–130. http://sci-hub.tw/10.34096/arqueologia.t27.n2.7635

Bracco, R., Duarte, C., Gutiérrez, O., & Panario, D. (2021b). El fuego, los hornos de tierra y la elevación de los montículos de la cuenca de la Laguna Merín. *Revista de Arqueologia, 34*, 152–176. http://sci-hub.tw/10.24885/sab.v34i1.734

Bracco, R., Duarte, C., Gascue, A., et al. (2022). Comparación de los procesos de acreción de los montículos de Cañada Saldaña y cuenca de la Laguna Merín a través de dataciones. *Cadernos do CEOM, 35*(57), 29–51.

Brockwell, S. (2006). Earth mounds in northern Australia: A review. *Australian Archaeology, 63*, 47–56. http://sci-hub.tw/10.1080/03122417.2006.11681837

Cabrera, L. (2013). Construcciones en tierra y estructura social en el Sur del Brasil y Este de Uruguay (Ca. 4.000 a 300 a. A.P.). *Techne, 1*, 25–33.

Capdepont, I., & Pintos Blanco, S. (2006). Manejo y aprovechamiento del medio por parte de los grupos constructores de montículos: Cuenca de la Laguna de Castillos, Rocha– Uruguay. *Relaciones de la Sociedad Argentina de Antropología, 31*, 117–132.

Capdepont, I., Castiñeira, C., Del Puerto, L., & Inda, H. (2004). 40X: arqueología de lo micro. In L. Beovide, I. Barreto, & C. Curbelo (Eds.), *La arqueología uruguaya ante los desafíos del nuevo siglo* (pp. 1–15). X Congreso de Arqueología Uruguaya. CD-ROM interactivo.

Capdepont, I., Castiñeira, C., del Puerto, L., & Fernández, G. (2016). Desarrollo de las ocupaciones humanas durante el Holoceno en la cuenca de La Laguna de Castillos (Uruguay): síntesis y actualización de las investigaciones arqueológicas. *Tessituras: Revista de Antropologia e Arqueologia, 4*, 53–93. http://sci-hub.tw/10.15210/tes.v4i1.6156

Castiñeira, C., & Piñeiro, G. (2000). Análisis estadístico textural para el estudio de las columnas estratigráficas de la excavación I y II del Bañado de los Indios. In A. Duran & R. Bracco (Eds.), *Arqueología de las Tierras Bajas* (pp. 467–478). Ministerio de Educación y Cultura, Comisión Nacional de Arqueología.

Castiñeira, C., Capdepont, I., del Puerto, L., & Blasi, A. M. (2015). Aportes de la geoarqueología para el estudio de la variabilidad constructiva prehispánica de cerritos del este uruguayo y delta del rio Paraná. In C. M. Favier Dubois & R. T. da Silva (Eds.), *Rubin De Rubin JC* (pp. 55–90). Geoarqueologia na América do Sul. Editora da PUC Goiás.

Ceruti, C., & González, M. I. (2007). Modos de vida vinculados con ambientes acuáticos del Nordeste y Pampa bonaerense de Argentina. *Relaciones de la Sociedad Argentina de Antropología, 32*, 101–140.

Chanca, I., Borges, C., Colonese, A. C., et al. (2021). Food and diet of the pre-Columbian mound builders of the Patos lagoon region in southern Brazil with stable isotope analysis. *Journal of Archaeological Science, 133*, 105439. http://sci-hub.tw/10.1016/j.jas.2021.105439

Consens, M. (2001). Yacaré-Cururú: 18 años después. In *Arqueología Uruguaya Hacia el Fin del Milenio. Anales del IX Congreso Nacional de Arqueología (Colonia del Sacramento, 16 a 19 de junio de 1997)* (pp. 115–123). Ministerio de Educación y Cultura.

Cornero, S., del Rio, P., & Ceruti, C. (2013). Sitios con 'hornitos' del Holoceno tardío en el Chaco austral: Colonia Dolores, Dpto. San Justo, Pcia. de Santa Fe. *Anuario de Arqueología, Rosario, 5*, 103–115.

Crumley, C. L. (1994). *Historical ecology: Cultural knowledge and changing landscapes*. School of American Research Press.

del Puerto, L. (2011a). *Silicofitolitos como indicadores paleoambientales. Bases comparativas y reconstrucción paleoclimática a partir del Pleistoceno tardío en el SE del Uruguay*. Editorial Académica Española.

del Puerto, L. (2011b). Ponderación de recursos vegetales silvestres del este del Uruguay: rescatando el conocimiento indígena tradicional. *Trama Revista de Cultura y Patrimonio, 3*, 22–41.

del Puerto, L. (2015). *Interrelaciones humano-ambientales durante el Holoceno tardío en el este del Uruguay: Cambio Climático y Dinámica Cultural*. Unpublished doctoral dissertation, PEDECIBA, Universidad de la República (Montevideo).

del Puerto, L., Gianotti, C., Bortolotto, N., et al. (2022). Geoarchaeological signatures of anthropogenic soils in southeastern Uruguay: Approaches to formation processes and spatial–temporal variability. *Geoarchaeology, 37*, 180–197. http://sci-hub.tw/10.1002/gea.21854

Duarte, C., Bracco, R. (2020). Procesos de elevación de los montículos del este de Uruguay y sur de Brasil. In *Modelo de crecimiento y dataciones luminiscentes*. Anuario de Arqueología 2020 (In press).

Duarte, C., & Bracco, R. (2021). La tierra quemada como artefacto en los montículos de India Muerta-Paso Barranca (Departamento de Rocha-Uruguay). *Cuadernos del Instituto Nacional de Antropología y Pensamiento Latinoamericano, 30*, 42–55.

Duarte, C., Bracco, R., Panario, D., et al. (2017). Datación de estructuras monticulares por OSL/TL. *Revista de Antropología del Museo de Entre Ríos, 3*, 14–26.

Durán, A. (1990). *Observaciones sobre los suelos del sitio arqueológico de San Miguel*. Informe de trabajo.

Eremites de Oliveira, J. (1995). *Os argonautas guató: aportes para o conhecimento dos assentamentos e da subsistência dos grupos que se estabeleceram nas áreas inundáveis do pantanal matogrossense*. Unpublished master's thesis, Pontificia Universidade Catolica do Rio Grande do Sul (Porto Alegre, RS).

Fernandes, R., Millard, A. R., Brabec, M., et al. (2014). Food reconstruction using isotopic transferred signals (FRUITS): A Bayesian model for diet reconstruction. *PLoS One, 9*, e87436. http://sci-hub.tw/10.1371/journal.pone.0087436

Ferrés, C. (1927). Los 'terremotos de los indios'. *Revista de la Sociedad Amigos de la Arqueología, 1*, 139–149.

Frankel, D., & Major, J. (Eds.). (2017). *Victorian aboriginal life and customs: Through early European eyes*. La Trobe University.

Frenguelli, J., & de Aparicio, F. (1923). Los paraderos de la margen derecha del Río Malabrigo (Departamento de Reconquista, Prov. de Santa Fe). In *Anales de la Facultad de Ciencias de la Educación* (pp. 7–112). Paraná.

Froehle, A. W., Kellner, C. M., & Schoeninger, M. J. (2010). FOCUS: Effect of diet and protein source on carbon stable isotope ratios in collagen: Follow up to Warinner and Tuross (2009). *Journal of Archaeological Science, 37*, 2662–2670. http://sci-hub.tw/10.1016/j.jas.2010.06.003

Froehle, A. W., Kellner, C. M., & Schoeninger, M. J. (2012). Multivariate carbon and nitrogen stable isotope model for the reconstruction of prehistoric human diet. *American Journal of Physical Anthropology, 147*, 352–369. http://sci-hub.tw/10.1002/ajpa.21651

Gascue, A., Bortolotto, N., Loponte, D., et al. (2019). Contextos geomorfológicos y tecno-económicos del registro arqueológico del bajo río Uruguay (margen izquierda). *Resultados preliminares de nuevas prospecciones. Arqueología, 25*, 87–117. http://sci-hub.tw/10.34096/arqueologia.t25.n3.7325

Gianotti, C. (2000). Paisajes monumentales en la región meridional sudamericana. *Gallaecia, 19*, 43–72.

Gianotti, C., & Bonomo, M. (2013). De montículos a paisajes: procesos de transformación y construcción de paisajes en el sur de la cuenca del Plata. *Comechingonia Virtual: Revista Electrónica de Arqueología, 17*, 129–163.

Gianotti, C., Criado Boado, F., Seoana, Y., et al. (2009). Dinámica constructiva y formación de un asentamiento monumental en el valle de Caraguatá (Tacuarembó). *Informes y Trabajos, 3*, 265–274.

Gianotti, C., del Puerto, L., Inda, H., & Capdepont, I. (2013). Construir para producir. Pequeñas elevaciones en tierra para el cultivo de maíz en el sitio Cañada de los Caponcitos, Tacuarembó (Uruguay). *Cuadernos del Instituto Nacional de Antropología y Pensamiento Latinoamericano – Series Especiales, 1*, 12–25.

Greaves, R. D., & Kramer, K. L. (2014). Hunter–gatherer use of wild plants and domesticates: Archaeological implications for mixed economies before agricultural intensification. *Journal of Archaeological Science, 41*, 263–271. http://sci-hub.tw/10.1016/J.JAS.2013.08.014

Guidón, N. (1989). *Misión de Rescate Arqueológico de Salto Grande, República Oriental del Uruguay* (Vol. II, 1st. part). Ministerio de Educación y Cultura.

Harrison, R. G., & Katzenberg, M. A. (2003). Paleodiet studies using stable carbon isotopes from bone apatite and collagen: Examples from southern Ontario and san Nicolas Island, California. *Journal of Anthropological Archaeology, 22*, 227–244. http://sci-hub.tw/10.1016/S0278-4165(03)00037-0

Hogg, A. G., Heaton, T. J., Hua, Q., et al. (2020). SHCal20 southern hemisphere calibration, 0–55,000 years cal BP. *Radiocarbon, 62*, 759–778. http://sci-hub.tw/10.1017/RDC.2020.59

Inda, H., & del Puerto, L. (2007). Antracología y Subsistencia: Paleoetnobotánica del Fuego en la Prehistoria de la Región Este del Uruguay. Puntas del San Luis, Paso Barrancas, Rocha, Uruguay. In M. B. Marconetto, M. P. Babot, & N. Oliszewsk (Eds.), *Paleoetnobotánica del Cono Sur: Estudios de casos y propuestas metodológicas* (pp. 137–152). Museo de Antropología FFyH-UNC. Ferreyra Editor.

Ingold, T. (2013). *Making anthropology, archaeology, art and architecture*. Routledge.

Iriarte, J. (2003). *Mid-Holocene emergent complexity and landscape transformation: The social construction of early formative communities in Uruguay, La Plata basin*. Doctoral dissertation, University of Kentucky (Lexington).

Iriarte, J. (2007). La construcción social y transformación de las comunidades del Periodo Formativo Temprano del sureste de Uruguay. *Boletín de Arqueología PUCP, 11*, 143–166.

Iriarte, J., Holst, I., López Mazz, J. M., & Cabrera, L. (2000). Subtropical wetland adaptation in Uruguay during the mid-Holocene: An archaeobotanical perspective. In B. A. Purdy (Ed.), *Enduring records: The environmental and cultural heritage of wetlands* (pp. 61–70). Oxbow Books.

Iriarte, J., Holst, I., Marozzi, O., et al. (2004). Evidence for cultivar adoption and emerging complexity during the mid-Holocene in the La Plata basin. *Nature, 432*, 614–617. http://sci-hub.tw/10.1038/nature02983

Jones, R., Morrison, M., Roberts, A., & The River Murray and Mallee Aboriginal Corporation. (2017). An analysis of indigenous earth mounds on the Calperum floodplain, Riverland, South Australia. *Journal of the Anthropological Society of South Australia, 41*, 18–61.

Kellner, C. M., & Schoeninger, M. J. (2007). A simple carbon isotope model for reconstructing prehistoric human diet. *American Journal of Physical Anthropology, 133*, 1112–1127. http://sci-hub.tw/10.1002/ajpa.20618

Leach, J. D. (2007). Prebiotics in ancient diets. *Food Science and Technology Bulletin: Functional Foods, 4*, 1–8.

López Mazz, J. M. (1992). Aproximaciones a la génesis y desarrollo de los cerritos de la zona de San Miguel (Dpto de Rocha). In R. Pi Hugarte (Ed.), *Ediciones del Quinto Centenario* (pp. 76–96). Universidad de la República.

López Mazz, J. M. (2001). Las estructuras tumulares (cerritos) del litoral atlántico uruguayo. *Latin American Antiquity, 12*, 231–255. http://sci-hub.tw/10.2307/971631

López Mazz, J. M., & Bracco, R. (1992). Relación Hombre-Medio Ambiente en las poblaciones prehistóricas de la Zona Este del Uruguay. In O. Ortiz-Troncoso & T. v. d. Hammen (Eds.), *Archaeology and environment in Latin America* (pp. 259–282). Instituut voor Pre- en Protohistorische Archeologie Albert Egges van Giffen, Universiteit van Amsterdam.

López Mazz, J. M., Rostain, S., & McKey, D. (2016). Cerritos, tolas, tesos, camellones y otros montículos de las Tierras Bajas de Sudamérica. *Revista de Arqueología, 29*, 86. 10.24885/sab.v29i1.444.

López Mazz, J. M., Moreno, F., Machado, A., et al. (2022). Cambio ambiental, respuesta humana y emergencia de complejidad cultural: primeros resultados de la investigación en la localidad arqueológica de La Viuda (Bañado de India Muerta, Rocha, Uruguay). *Arqueología, 28*, 9912. http://sci-hub.tw/10.34096/arqueologia.t28.n2.9912

Loponte, D. (2020). La agricultura prehispánica en el nordeste argentino analizada a través de los isótopos estables. *Anuario de Arqueología, Rosario, 12*, 25–64. http://sci-hub.tw/10.35305/aa.v12i12.42

Loponte, D. M., Acosta, A., & Tchilinguirián, P. (2016). Estructuras "monticulares", unidades arqueológicas y falsas premisas en la arqueología del noreste argentino. *Anuario de Arqueología, Rosario, 8*, 45–78.

Loureiro, A. G. (2008). *Sítio PT-02-Sotéia: análise dos processos formativos de um cerrito na região sudoeste da Laguna dos Patos/RS*. Unpublished Master,s thesis, Museu de Arqueologia e Etnologia, Universidade de São Paulo (San Pablo).

Martin, S. (2011). Palaeoecological evidence associated with earth mounds of the Murray riverine plain, South-Eastern Australia. *Environmental Archaeology, 16*, 162–172.

Mauss, M. (2009). *Ensayo sobre el don: forma y función del intercambio en las sociedades arcaicas*. (J. Bucci, Trans.). Katz editores.

Metraux, A. (1944). Estudios de etnografia chaquense. *Anales del Instituto de Etnologia Americana Universidad Nacional de Cuyo, 5*, 263–311.

Milheira, R. G., & Gianotti, C. (2018). The earthen mounds (Cerritos) of southern Brazil and Uruguay. In C. Smith (Ed.), *Encyclopedia of global archaeology* (pp. 1–9). Springer.

Milheira, R. G., Garcia, A. M., Ribeiro, B. L. R., et al. (2016). Arqueologia dos Cerritos na Laguna dos Patos, Sul do Brasil: uma síntese da ocupação regional. *Revista Cadernos do CEOM, 29*, 33–63. http://sci-hub.tw/10.22562/2016.45.02

Milheira, R. G., Attorre, T., & Borges, C. (2019). Construtores de cerritos na Laguna Dos Patos, Pontal da Barra, sul do Brasil: Lugar persistente, território e ambiente construído no Holoceno recente. *Latin American Antiquity, 30*, 35–54. http://sci-hub.tw/10.1017/laq.2018.66

Moreno, F. (2014). *La gestión de los recursos animales en la prehistoria del Este de Uruguay (4000 años AP-Siglo XVI)*. Doctoral dissertation, Universitat Autònoma de Barcelona (Barcelona).

Mut, P. (2015a). Paleodieta de los pobladores prehistóricos del este del Uruguay: un retrato isotópico. *Anuario de Arqueología, 2015*, 147–178.

Mut, P. (2015b). *Paleodieta de los pobladores prehistóricos del este del Uruguay: un retrato isotópico*. Unpublished degree thesis, Facultad de Humanidades y Ciencias de la Educación, Universidad de la República (Montevideo).

Mut, P., & Bracco, R. (2022). Dieta de los constructores de montículos del este de Uruguay, una aproximación biogeoquímica. *Revista de Antropología del Museo de Entre Ríos, 7*, 11–29. http://sci-hub.tw/10.5281/zenodo.7234183

Outes, F. F. (1918). Nuevo jalón septentrional en la dispersión de representaciones plásticas de la cuenca paranaense y su valor indicador. In *Anales de la Sociedad Científica Argentina* (pp. 53–66). Imprenta y Casa Editora Coni.

Pintos Blanco, S. (1999). Túmulos, caciques y otras historias. Cazadores recolectores complejos en la cuenca de la Laguna de Castillos, Uruguay. *Complutum, 10*, 213–226.
Pintos Blanco, S. (2000). Economía "húmeda" del este del Uruguay: el manejo de los recursos faunísticos. In A. Duran & R. Bracco (Eds.), *Arqueología de las Tierras Bajas* (pp. 249–270). Ministerio de Educación y Cultura, Comisión Nacional de Arqueología.
Piperno, D. R. (2011). The origins of plant cultivation and domestication in the New World tropics. *Current Anthropology, 52*, S453–S470. http://sci-hub.tw/10.1086/659998
Politis, G. G., & Bonomo, M. (Eds.). (2018). *Goya-Malabrigo Arqueología de una sociedad indígena del noreste argentino*. Editorial UNICEN.
Politis, G. G., Bonomo, M., Castiñeira, C., & Blasi, A. (2011). Archaeology of the Upper Delta of the Paraná River (Argentina): Mound construction and anthropic landscapes in the Los Tres Cerros locality. *Quaternary International, 245*, 74–88. http://sci-hub.tw/10.1016/j.quaint.2011.02.007
Rapoport, E. H., Marzocca, A., & Drausal, B. S. (2009). *Malezas comestibles del Cono Sur y otras partes del planeta*. Instituto Nacional de Tecnología Agropecuaria.
Rivas, P., & Ocampo, C. (2005). El antiguo curanto chilote. In G. Badal, F. Balcella, M. Montanari, et al. (Eds.), *Chile, País Oceánico* (pp. 70–71). Editorial Ocho Libros.
Schmitz, P. I. (1981). Contribuciones a la prehistoria de Brasil. *Pesquisas, Série Antropologia, 32*, 1–243.
Schwartz, H. P. (2000). Some biochemical aspects of carbon isotopic paleodiet studies. In S. H. Ambrose & M. A. Katzenberg (Eds.), *Biogeochemical approaches to Paleodietary analysis* (pp. 189–209). Academic/Plenum.
Serrano, A. (1931). Arqueología del Litoral. In *Arqueología* (pp. 1–5). Memorias del Museo de Paraná. Paraná.
Smith, B. D. (2001). Low-level food production. *Journal of Archaeological Research, 9*, 1–43. http://sci-hub.tw/10.1023/A:1009436110049
Sušnik, B. (1978). *Los aborigenes del Paraguay. Tomo I. Etnología del Chaco boreal y su periferia (siglos XVI y XVII)*. Museo Etnográfico Andrés Barbero.
Thoms, A. V. (2008). Ancient savannah roots of the carbohydrate revolution in south-Central North America. *Plains Anthropologist, 53*, 121–136. http://sci-hub.tw/10.1179/pan.2008.008
Thoms, A. V. (2009). Rocks of ages: Propagation of hot-rock cookery in western North America. *Journal of Archaeological Science, 36*, 573–591. http://sci-hub.tw/10.1016/J.JAS.2008.11.016
Thoms, A. V. (2017). Burned-rock features. In A. S. Gilbert (Ed.), *Encyclopedia of earth sciences series* (pp. 89–94). Springer.
Torres, L. M. (1911). *Los primitivos habitantes del Delta del Paraná*. Imprenta De Coni Hermanos.
Ugent, D., Pozorski, S., & Pozorski, T. (1984). New evidence for ancient cultivation of Canna edulis in Peru. *Economic Botany, 38*, 417–432. http://sci-hub.tw/10.1007/BF02859081
Villagran, X. S. (2006). Existió la monumentalidad en tierra entre los cazadores-recolectores del Este de Uruguayo? Propuesta Metodológica para el estudio de construcciones Antrópicas en Tierra. *Arqueologia Suramericana/Arqueología Sul-Americana, 2*, 263–290.
Villagran, X. S., & Gianotti, C. (2013). Earthen mound formation in the Uruguayan lowlands (South America): Micromorphological analyses of the Pago Lindo archaeological complex. *Journal of Archaeological Science, 40*, 1093–1107. http://sci-hub.tw/10.1016/J.JAS.2012.10.006
Wandsnider, L. (1997). The roasted and the boiled: Food composition and heat treatment with special emphasis on pit-hearth cooking. *Journal of Anthropological Archaeology, 16*, 1–48. http://sci-hub.tw/10.1006/JAAR.1997.0303
Watling, J., Saunaluoma, S., Pärssinen, M., & Schaan, D. (2015). Subsistence practices among earthwork builders: Phytolith evidence from archaeological sites in the southwest Amazonian interfluves. *Journal of Archaeological Science: Reports, 4*, 541–551. http://sci-hub.tw/10.1016/J.JASREP.2015.10.014
Westell, C., & Wood, V. (2014). An introduction to earthen mound sites in South Australia. *Journal of the Anthropological Society of South Australia, 38*, 30–66.
Wrangham, R. (2009). *How cooking made us human*. Basic Books.

Wrangham, R., & Conklin-Brittain, N. (2003). Cooking as a biological trait. *Comparative Biochemistry and Physiology Part A: Molecular & Integrative Physiology, 136*, 35–46. http://sci-hub.tw/10.1016/S1095-6433(03)00020-5

Wylie, A. (1985). The reaction against analogy. In M. B. Schiffer (Ed.), *Advances in archaeological method and theory* (pp. 63–111). Academic.

Zeballos, E., & Pico, P. (1878). Informe sobre el túmulo de Campana. *Anales de la Sociedad Científica Argentina, 6*, 244–260.

Conclusion: Historical Ecology Travels South

Eduardo Góes Neves

Archaeologists love to disagree, and it is easy to understand why. Despite the remarkable advances of the recent years, it will always remain a discipline based on the interpretation of a normally scanty and fragmentary record, and such ambiguity, together with other factors, may render it prone to deep dissent. There is a place, however, where a firm theoretical consensus has emerged among practicing archaeologists, despite the usual conflicting perspectives on particular historical reconstitutions. This place is the Amazon, and the theoretical perspective that holds archaeology together there is Historical Ecology. In providing a robust conceptual framework that after 30 years is still going strong, Historical Ecology has deeply transformed Amazonian archaeology.

The history of Amazonian archaeology goes back to the nineteenth century, thanks to the pioneering contributions of Charles Hartt and Domingos Ferreira Penna, among others. Together with research on coastal shellmounds of the southern Atlantic coast and in the rockshelters of Lagoa Santa, in Minas Gerais, research in the Amazon is at the core of Brazilian archaeology; but, unlike these other regions, there has been 150 years of continuous research in the Amazon. This history has already been written and it is not the case to rephrase it here, but one important thing that has set Amazonian archaeology apart in Brazil is the fact that, since its beginnings, it has been associated with the testing of hypotheses generated in other realms of knowledge. This is the case, for instance, of Hartt, who was brought as a geologist by Louis Agassiz in his Thayer Expedition of 1865–1866. Agassiz, as it is known, was a leading voice defending catastrophist and anti-Darwinian perspectives in the mid-nineteenth century. His expedition to the Amazon was funded with the goal of identifying evidence of large glaciers whose expansion led to massive extinctions, and hence the need for a geologist. Hartt, by his turn, found no

E. G. Neves
Museum of Archaeology and Ethnology, University of São Paulo, São Paulo, Brazil
e-mail: edgneves@usp.br

A. C. Colonese, R. G. Milheira (eds.), *Historical Ecology and Landscape Archaeology in Lowland South America*, Interdisciplinary Contributions to Archaeology, https://doi.org/10.1007/978-3-031-32284-6

such evidence but rather had his attention drawn by the numerous archaeological sites he found along his travels, leading to the first scientific publications on the topic.

After World War II, there was an increase in the influence of US archaeology in South America. Such influence manifested itself with well-funded research projects, as well as with publications of syntheses, such as the Handbook of South American Indians, edited by Julian Steward. Together with scientific diplomacy, at the heart of this effort was an attempt to understand what seemed to be a major divide that marked the deep ancient history of the South American continent: the fact that apparently just a small core area, the Central Andes, was the center for the emergence of civilization in the continent. In this view, places such as the Amazon had a marginal role in the deep history of South America. Much of the archaeology done in the Amazon in the second half of the twentieth century was aimed at testing this hypothesis. Within the general framework of Brazilian Archaeology, Amazonian archeology has stood out, since the 1950s, for maintaining a consistent pattern of research aimed at testing anthropological hypotheses. Classic examples of these works can be found in the publications of the 1950s and 1960s by authors such as Betty Meggers, Donald Lathrap, Irving Rouse, Robert Carneiro and Eduardo Galvão. For instance, to explain such a supposed peripheral role, Amazonian archaeologists adopted a strong environmental deterministic paradigm. In this reasoning, factors such poor soil fertility, lack of animal protein or drastic climatic change would have presented a ceiling for population growth.

The environmental deterministic paradigm prevailed until the 1990s, when it was eventually replaced by Historical Ecology. Among the reasons for its dismissal was the fact that it could not account for the abundant evidence of large archaeological sites, as well as the mounting data showing the modification of nature by contemporary and ancient Indigenous populations. Initially, archaeological data were used as an empirical basis for testing general hypotheses about the influence of the environment on the structure and change of social formations in the past. Over the years, however, the accumulation of archaeological, ethnographic and historical data has contributed to the critique of the classic deterministic paradigm. The decline of this paradigm and the political changes that occurred in Brazil from the 1980s onwards opened new research possibilities for Amazonian archeology.

Historical Ecology was eagerly taken up by archaeologists because it focuses on the relationships between human populations and the environment from a diachronic perspective, without assuming *a priori* a deterministic perspective. I was privileged to be a young student living in the city of Belém in the late 1980s, while scholars such as William Balée and Darrel Posey were based at the illustrious Goeldi Museum and the ideas that converged into Historical Ecology were being brewed. Since the turn of the century, archaeological research in the Amazon has increased exponentially. The reasons for this are many: large scale projects demanding contract work, the creation of new undergraduate programs in the Amazon, the emergence of graduate programs in Brazil, and the establishment of a strong network connecting scholars working in different Amazonian countries. These changes, however, would not have been possible if it were not for the conceptual support provided by Historical Ecology. As a consequence, it is accepted today that the Amazon was a major independent center of plant domestication; that highly fertile soils – known as *terras pretas* in Portuguese – were created by Indigenous people thousands of years ago; that the oldest pottery in the Americas was produced in the Amazon, and so forth.

The Historical Ecology paradigm has been so strong and important to the Amazon, I have always wondered why it was not adopted by colleagues working in other biomes in lowland South America or even the Andes. This excellent book, *"Historical Ecology and Landscape Archaeology in Lowland South America"*, edited by Andre Colonese and Rafael Milheira, was conceived as an attempt to fill this gap, and in my view the authors have been successful in this attempt. Although there are chapters devoted to Amazonian topics, most of them bring case studies from places such as Central Brazil, the Southern Brazilian Plateau, and the Southern Atlantic shore. Many of the contributors are archaeologists, but there are also authors who come from fields such as Human Ecology, Human Geography, and Social Anthropology. The book is also welcome in its attempt to establish connections between the past and the present, as archaeologists working in the Amazon are trying to do.

"Historical Ecology and Landscape Archaeology in Lowland South America" will stand as an important contribution that demonstrates the strength of Historical Ecology as a guiding paradigm for archaeology beyond its original applications in the Amazon. Hopefully the chapters presented here will inspire other authors to pursue similar approaches for case studies in other tropical settings, beyond lowland South America.

Index

A. C. Colonese, R. G. Milheira (eds.), *Historical Ecology and Landscape Archaeology in Lowland South America*, Interdisciplinary Contributions to Archaeology, https://doi.org/10.1007/978-3-031-32284-6

MIX
Papier aus verantwortungsvollen Quellen
Paper from responsible sources
FSC® C105338

If you have any concerns about our products,
you can contact us on
ProductSafety@springernature.com

In case Publisher is established outside the EU,
the EU authorized representative is:
Springer Nature Customer Service Center GmbH
Europaplatz 3, 69115 Heidelberg, Germany

Printed by Libri Plureos GmbH
in Hamburg, Germany